Wings Over Madison

From The Journals of Clyde H. Beyer at the
Dixie Flying Service in Madison, Indiana
1931-1933

Clyde H. Beyer

Bloomington, IN  Milton Keynes, UK

authorHOUSE®

AuthorHouse™
1663 Liberty Drive, Suite 200
Bloomington, IN 47403
www.authorhouse.com
Phone: 1-800-839-8640

AuthorHouse™ UK Ltd.
500 Avebury Boulevard
Central Milton Keynes, MK9 2BE
www.authorhouse.co.uk
Phone: 08001974150

First published by AuthorHouse 7/18/2006

ISBN: 1-4208-7342-3 (e)
ISBN: 1-4208-3980-2 (sc)
ISBN: 1-4259-4139-7 (dj)

Printed in the United States of America
Bloomington, Indiana

This book is printed on acid-free paper.

Prologue

In August 1999, while on a visit from Richmond, Virginia to Madison for my Dad's 87th birthday, he told me that the Madison Historical Society was interested in his notes and pictures from his days at the Dixie Flying Service. He told me that he had been transcribing his diary notes (handwritten) to tell a story, not only about learning to be a pilot, but to describe what is was like for a young man who moved from the big city of Pittsburgh, PA, at the age of 18, to a small town of Madison on the Ohio River in Southern Indiana and the personal experiences during this time.

I casually said that I thought it was a great idea and that I would be happy to type his story for him. Little did I know at the time how serious he was. He pulled out the first 100 handwritten pages, pictures, newspaper articles, brochures that he had kept for seventy years, autograph book, the Dayton School of Aviation Correspondence Course, and a tremendous recollection of those days that he loved so much.

The past three years has been a labor of love for me and Dad as the project expanded from a mere typing of the notes, to a professional memoir (at least as much as I could make it) that would give him a little piece of history.

It was a labor of love for me because I have come to understand how much this period of his life meant to him. Why else would you keep old faded pictures, newspaper articles, brochures, if it they weren't significant.

I have learned more about my Dad's early years typing this memoir that I did in all the years growing up in Madison and as I moved away to start my own family. Reading and typing every page was like reading an adventure story, wanting the hero to win the girl and fly the plane.

Thanks to my sisters Sallie, for help in typing Dad's notes, and Diane for proofreading and making suggestions. Thanks to my Dad for his patience over the past three years while I assembled the book to his satisfaction.

Martin W. Beyer
August 11, 2002

Introduction

I first became interested in aviation when Charles Augustus Lindbergh flew alone, May 20, 1927, on a transatlantic flight from Roosevelt Field, L.I. to Paris, France non-stop, 3,600 miles in 33 ½ hours. Lindbergh was born in Detroit, Michigan, on February 4, 1902. His father, Charles Augustus Lindbergh, was a congressman from Minnesota.

Lindbergh graduated from high school in 1918 and studied mechanical engineering at the University of Wisconsin from 1920-1922. He then enrolled in the Flying School at Lincoln, Nebraska. On March 19, 1924, he enlisted in the U.S. Air Service, and later joined the Air National Guard, 1925-1926. He entered the U.S. Air Mail Service and piloted his first mail plane on April 15, 1926, from Chicago to St. Louis. While flying this route he decided to compete for the "Raymond B. Orteig Prize" of $25,000 offered to the first pilot to make a N.Y. to Paris non-stop flight. He secured the necessary financial backing in St. Louis and on February 1927 ordered a Ryan plane built in San Diego, California in a tuna cannery in sixty days for $10,580, which he named the "Spirit of St. Louis". Lindbergh was involved in the building of the plane everyday and made numerous modifications he thought necessary for a successful flight. The plane was a "Ryan NYP" parasol monoplane, powered by a radial "Wright Whirlwind" air-cooled engine, license No. N-X-211.

After establishing a record of 21 hours, 20 minutes for a trans-continental flight from San Diego to St. Louis to Long Island on May 12,1927, he took off alone on May 20, 1927, for Paris, France.

Lindbergh received, in addition to the $25,000, many honors, medals, and awards from all over the world, and the media began referring to him as "The Lone Eagle," although a few called him "Lucky Lindy." He received a tremendous "ticker tape" reception when he returned to New York. This flight created a renewed interest in aviation, and the "Lone Eagle" devoted the rest of his illustrious career to the advancement of aviation. He was and is my hero.

My Story...The Beginning...

At this time, I was in the ninth grade at Peabody High School in Pittsburgh, Pa. and didn't have the least idea what I wanted to do with my life. After listening to the news about Lindbergh and other famous aviators on the radio and reading about them in the paper I thought this would be an exciting way to live, and began to read everything I could find on flying.

I ordered model airplane kits through the mail and began building them. It was difficult and took a long time. These kits consisted of sheets of thin, light balsam wood, rice paper, cellulose nitrate ("dope"), glue, patterns, instructions,

a very large rubber band to power the plane (there were no miniature gas engines at that time), and a block of balsam wood for the propeller. Each wing, there were two because this one was a biplane, about 3 feet long and 4" or 5" wide, ailerons, fuselage, stabilizer, rudders, and elevators, consisted of many individual pieces that had to be cut exactly according to the pattern. Building these model planes was educational and taught me the fundamentals of building airplanes as they were designed to fly.

After ruining some pieces, I finally got the plane together, and before painting it I wound up the rubber band with the hand carved propeller and launched it from a standing position. As there was not enough flying space it flew only a short distance. I decide to paint the plane before taking it where there was plenty of safe flying space.

After painting the plane it looked pretty nice, but there were some spots that needed touching up. By the time I finished painting it I had put five or six coats of paint on it, and it did look great. I took the plane to a park where there was plenty of room and I wound the

rubber band as tight as it would go and launched the plane. It fell like a rock. I learned another valuable lesson; the additional coats of paint made the plane too heavy and changed the lifting surfaces for the mount of power the rubber band furnished. So I put the plane on top of the piano in the parlor where it sat for quite a few years.

I took a business course in high school and when I graduated in 1930, I didn't have any money, and my dad's bakery was losing money due to the depression and an A&P grocery that had just started up directly across the street. They sold everything cheaper than my dad could make it. Also, they started selling sliced bread; it was really poor quality and didn't taste like bread, but everyone wanted sliced bread. I will never forget when a neighbor came in the store and told my mother she liked our bread and asked if my mother would slice a loaf for her. After this happened several times my mother said "No More". The slicing machine was out of the question.

At that time there were no jobs. Some men were standing on the street corners selling apples and anything else that they thought might sell. A lot of them became paperboys, selling papers on streetcars and street corners. I tried that for a short time but quit, as it got too dangerous. I was always getting into someone's territory.

It was a bad time to be looking for a job. It was amazing things people did to make a dollar or two. There was no unemployment checks or welfare, as we now know it. I answered every ad in the paper that I thought I could handle. During this time I would go to several private airfields that were close to the city and hang around. Finally, I answered an ad for an office boy to work in a wholesale store selling U.S. Rubber Co. tires and accessories. I got the job – it paid $16.00 a week. Practically all the newspaper ads were for salesmen. After working there for a year I decided to look for a flying school. There were no flying schools in

2

Pittsburgh, only individual flying lessons at the private airports. I had an uncle living in Cleveland, Ohio who told me that there was an excellent school there sponsored by T.W.A.

I went to Cleveland to visit the school and after the management showed me the hangers, class rooms, and the training planes, a pilot fitted me with goggles, helmet, and ear phones and I got in the front seat of a Waco biplane with the pilot sitting behind me. It was a free sample flight to see if I would make a good pilot, as I had never been up in the air before. We took off and after reaching a safe altitude, the pilot told me to push the left pedal, move the stick slightly to the left, use the throttle a little; and we turned left. After several different maneuvers he told me to lightly rest my feet on the pedals, hold the stick and let him do the flying. He did several stunts then said it was time to go down. I was in heaven and didn't want to go down but we did. The only course they offered at the time was a Transport Pilots Course, which cost $3,000. They were primarily interested in training pilots to fly their passenger planes. I went back home and to work, trying to find a way to get enough money to go to that school.

A couple of weeks later I was reading a "Popular Mechanics" magazine, and I came across a 1" x 3" ad in the back stating that "The first student that signed up to take the L.C. Pilots Course, $398.00, and received a license was guaranteed a flying job." As I have saved up a little over $400 I thought this would get me started. The address was D.F.S., 216 E. Second Street, Madison, Indiana. I cut out the ad and wrote to them right away and they sent me a two-page pamphlet showing a hanger with a workshop and airplanes to be worked on, an office, and east-west, north-south runways. I wrote right back that I was on my way, and quit my job.

Grow Fast
with the World's
Fastest Growing Industry

THE Aviation industry, in spite of the depression, is developing almost beyond belief. Each day brings new developments, new airports, new manufacturing. Capitalists are backing Aviation as they have backed no other industry. Those who get into Aviation now are the ones who will become the leaders of tomorrow. In the development of every new industry the men who get in first are the ones who make the big money. Take the automobile industry, for example, as it developed its giant factories to produce enough cars to take care of people's needs, huge fortunes were made by those who had the foresight to become trained. Think, too, of the many thousands of others who have made fine big incomes directing sales agencies, operating garages, repair shops, accessory stores and service stations. Years ago these men were in much the same position you are in today—considering whether they dare take the chance to enter a new industry. Those that had the courage became highly successful. Yet their opportunities then in the development of the automobile did not begin to compare with yours as you look forward to your future in the development of the airplane. Aviation has scarcely scratched the surface as compared to what it will be three, five or ten years from today. And—there is no reason why you can't get started. We are willing to do our part—if you sincerely want to get into this giant industry of the future—We can help you.

Our—EARN as YOU LEARN—Plan is the Short-Cut to Success in Aviation

OURS is an Aviation company which specializes in airplane repairing, flying instruction, airplane taxi service and airplane distribution. That is why we are able to supply part time work to our students to help pay a part of their flying tuition. Students at our airport are allowed to help with work on airplanes, engines, etc.; under the direct supervision of our Government Licensed Mechanic. In this work actual instruction is also given so that this combination with the actual practice of doing these things makes you thoroughly qualified in every respect when you have completed your training with us. This practical method of ours trains you to have iniative. It will make a real leader out of you. You will have a good start toward your future because you will already have become experienced in everything connected with the airplane and in the operation and management of an airport. This is going to mean a lot to you—more than you perhaps realize right now.

JOE D. VAIL
Transport Pilot

To Become a Good Pilot You Need Training That is Practical—the Kind Employers Demand

ONLY the fully trained man succeeds. Only by preparing seriously for your work will you reach the top in Aviation during the next few years. When you go out on a job you want to be sure of yourself—sure of what you can do. That is why we give you actual, practical work while you are taking our training. Employers recognize this type of training and it gives you the advantage over others who have not received this thorough up-to-date method of preparation. To get and hold a job today you must be more than just a pilot—you must be capable of making repairs, in case of a forced landing, and you must have business judgment. That is why our method of training is the best. That is why we see to it that you learn everything an employer expects of you before he hires you.

TURN THE PAGE AND LEARN ABOUT OUR GREAT OFFER

1932

We'll Give You Part Tin
to Help You Pay for You

WE realize it might be hard for you to raise enough money to pay your flying tuition in cash. We have therefore arranged the work here at our airport in such a manner that we can use a certain amount of student labor. If you are ambitious and willing to help with work around the airport we in turn will credit what you do toward your tuition. As we have already stated we conduct a Flying Service, specializing in Airplane repairing, Flying instruction, Taxi Service and Airplane distribution. Instead of hiring a lot of help to take care of the work in connection with our airport activities we employ our students, giving them a chance to work out a part of their tuition in return for their help. You do this work under the direct supervision of our Government Licensed mechanic and when you have completed your training you are not only thoroughly qualified to be a pilot, but you have received training in the operation, maintenance and management of an airport.

HARVAY KATTELMANN
Our Government
Licensed Mechanic

The Actual Practical Work You Will Do
at Our Airport Will Quickly Qualify You for a Job

YOU will get a chance to do practically every kind of work in connection with the aircraft industry. For instance when we buy and rebuild planes you get the opportunity to help our Licensed Mechanic on all this type of work. Often times these jobs are completely overhauled, engine, wings, fuselage, covering, doping, etc. This is the kind of training you need most, for as a pilot you will undoubtedly want to become an airport manager, or possibly you will want to become test pilot for a manufacturer. Regardless of what kind of work you want to do this experience will be very valuable to you.

In the above photo you see a section of a wing covering department. In the photo below they are working on ribs.

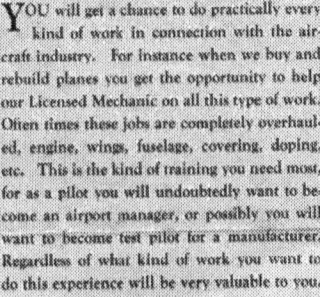

As a good pilot you should know every part of your plane, from the propeller to the tail surfaces. You should know every part and piece which goes into it.

Our Employment Method
Saves You Time—and
Saves You Expense

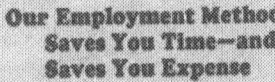

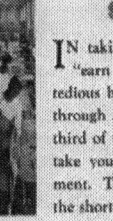

IN taking your training with us under our "earn as you learn" plan you save many tedious hours you would otherwise have to go through and in addition you save at least one-third of what it would cost you if you were to take your training without part-time employment. This saving to you in time and expense is the shortest, quickest road to sure success.

September 30, 1931 – Wednesday

My mother and I left home (4121 Main Street, Pittsburgh, Pa., Fisk 1136J) at 7:30am in my 1925, 4-cylinder Chevrolet Coupe that I had bought in April 1931 for $80.00. I was as proud of my Chevy as my dad was proud of his new 8-cylinder Marmon 4-door sedan that cost him $3,200.

We were going to Madison, Indiana to enroll in a "Limited Commercial Flying Course" at the Dixie Flying School. My mother took this opportunity to go with me as she rarely had the chance to get away from home due to her responsibility of taking care of the living quarters above the bakery, as well as being in charge of the store that my father owned and operated.

The weather was perfect, and the car cruised along at 40 mph without a lick of trouble; this was about as fast as the car could go. We drove approximately 300 miles and had six detours, using 20 gallons of gas and one quart of oil, and made all of the hills in high gear. We stopped for the night at 7:30pm at Grand View Inn, 15 miles east of Cincinnati, Ohio. It was a nice cabin; this was before motels came into being.

October 1, 1931 – Thursday

At 9:00am we left the Grand View Inn and stopped at several gas stations for information on how to get to Madison, Indiana. The advice was to "just follow the Ohio River." The roads were narrow, rough, and nothing but curves, they were putting tar on a section of the road, and as the road was narrow I had to follow behind the truck, splashing tar on my car as fast as they were putting it on the road. It was not a very pleasant trip, but we finally arrived at Madison at 2:30 PM. No detours today and the car used 5 gallons of gas. The car ran fine and the weather was pleasant.

We met Joe D. Vail, V.P. and flying instructor of Dixie Flying School, then rode around town sightseeing, which didn't take long as it is a very small town situated between the Ohio River on the south and high hills on the north, east, and west.

Joe D. Vail, VP and Instructor – Dixie Flying Service

Madison is 420 feet above sea level, and the surrounding hills are 910 feet above sea level. They say the population is around 6,000 and it seems like a nice, quiet, sleepy town, quite different than the heart of Pittsburgh, where I was born and raised. All cars park at an angle facing the sidewalk, on Jefferson and Broadway Streets cars also park in the middle of the street.

I was surprised to find that the two movie houses were open on Sundays. Pennsylvania has what is called "Blue Laws", nothing is open on Sunday; no stores, no movies, no work places, nothing of entertainment, no gas stations, etc.

All the streets going east and west beginning at and parallel to the Ohio River are Front Street, now Vaughn Drive; all the streets are 40 feet wide except Main Street. Next is First Street, then Second Street, Main Street, which is 80 feet wide, runs east to west the length

of the city, 1.8 miles, at the east end is S.R. 56 going to Vevay, and at the west end is S.R. 56 going to Hanover. Then Third Street, Presbyterian Ave., Fourth Street, Fifth Street, and finally Sixth Street, which ends at the entrance to Springdale Cemetery on the west, and at the bottom of Michigan Hill, which is 490 feet above Madison. At the top of Michigan Hill is farmland.

All the streets going North and South are 40 feet wide except Jefferson Street and Broadway. Beginning at the east city limits and going west is Ferry Street; next is Harrison Street, which ends at the Ohio River toll bridge to Milton, Kentucky. Then Clay Street, Baltimore Street, Church Street (now St. Michael's Avenue), East Street, Walnut Street, Jefferson Street (80 feet wide), Mulberry Street, West Street (it divides East and West Madison), and all the buildings numbers begin at 100 going west and 100 going east. It begins at the Ohio River and ends at the bottom of Michigan Hill road, S.R. 29 to Versailles, and is 0.6 miles long. Next is Central Avenue, Poplar Lane, Broadway Street (80 feet wide), Elm Street, Vine Street, Mill Street, Plum Street, Vernon Street, and Pearl Street. Cragmont Street joins S.R. 7, on the north and goes up Hanging Rock Hill to North Madison. Then: Wall Street, McIntire Street, Madison Street, and Marine Street. Madison Country Club is the west city limit.

We stayed at the Madison Hotel for the night. It was on the SW corner of Mulberry and Second Street.[1]

October 2, 1931 – Friday

We got up bright and early for a meeting with Mr. Richard Oglesby, Manager and Promoter of Dixie Flying School. The office is in a two-story brick home at 216 E. Second Street.[2]

I was the first student to enroll at Dixie Flying Service, and, incidentally, I enrolled one day before Dixie Flying Service was incorporated. I enrolled as a student in the "Limited Commercial Pilots Course," $398.00 (room and board not included), and work at the airport when not taking lessons. The course consisted of 50 solo hours and enough dual instruction to qualify for a L.C. license. As I was the first student to enroll in the L.C. course, I was guaranteed a flying job if I received a L.C. pilots license.

---

[1]   It was later demolished and replaced by a Kroger grocery store and parking lot, and is now the J.C. Food Store.

[2]   Later became "Doc" Peters Funeral Home, and is now an apartment house.

The Dixie Flying Service was incorporated October 3, 1931. This is the story about how it began:

During Joe Vail's high school years he spent all of his time down on the Ohio River racing powerful speedboats. I was told the story that when Joe graduated his mother was worried that he would get himself killed racing, and encouraged him to take up something safer. So he went to Park Air College and finished up at Purdue University where he bought his first airplane. If this story is true then it will turn out to be ironic.

The Vail family owned a furniture store (211 E. Main Street), and a funeral home located at the NW corner of Main and Broadway streets. When Joe came home with his plane and a transport pilot license he used the Sachleben airstrip to haul passengers and teach three students how to fly. The family then decided to set up a flying service.

Dixie Flying School Office 216 E. Second Street

DIXIE FLYING SERVICE

MADISON, INDIANA

FLYING AGREEMENT

This is to certify that _Clyde H. Beyer_ is enrolled as a Student in the _Limited Commercial_ course as offered by The DIXIE FLYING SERVICE.

For and in consideration of the amount of $ _398.00_ payable as follows: $ _398.00_ down, cash in hand, paid this date, receipt of which is hereby acknowledged.

I. The DIXIE FLYING SERVICE hereby agrees:

a. To make availabe to the student complete flying instruction as described below:

50 Solo Hours of Flying time

b. To grant to the student the use of the DIXIE FLYING SERVICE library in accordance with the library rules, with permission to borrow the latest aviation books so long as said rules are adhered to.

c. To lend such personal aid and cooperation as it may be able to give in assisting the student to place himself or herself in the aeronautical industry upon the completion of this course.

II. _Clyde H. Beyer_, the student hereby agrees:

a. To abide by the rules of the field as laid down by the instructors in all matters pertaining to the course.

b. That the DIXIE FLYING SERVICE be freed from all responsibility, personal or financial, in case of accident involving the student.

c. That the DIXIE FLYING SERVICE reserves the right to discharge the student upon a refund of the amount paid to date, less Fifteen dollars, ($15.00) per hour for the time actually spent in the air, providing such action is necessary, due to insubordination to instructors or other just cause.

B. The above items are hereby specifically agreed to this _2_ day of _Oct_, 19_31_.

_____
Student

_____   DIXIE FLYING SERVICE
Witness   By _____

12

These three students were:

J. P. Wilkerson, a contractor who rebuilt the Hanover Hill road. He bought a Warner-Cessna monoplane, but was never a student of Dixie Flying School.

Paul Wallace, co-owner of "Bucknell-Wallace Chevrolet dealership at 115-117 E. Second Street,[3]. He never received a pilot's license and was never a student of the Dixie Flying Service. He was never able to learn to land the plane.

David G. Gentrup, Box 153, N. Madison. He was a student at N. Madison High School. Although he was never a student of the D.F.S. he continued taking individual flying lessons after D.F.S. started operating.

The Vails, John A. Naill, owner of Naill Lumber Co., Dale Wilson, bookkeeper at Vail's Furniture Store, Curtis Marshall, a judge, Richard Oglesby, manager and promoter of the flying school at Milan, Indiana, all contributed financially to creating Dixie Flying Service. I do not know what percentage each contributed.

Coincidently, at about this time the flying school at Milan was closing, as was the small Cloud Coupe Corp. (airplane company). Richard Oglesby,

---

[3]   Later became a roller skating rink, and is now a Chinese fitness center for the martial arts

manager and promoter at Milan and the Madison group got together and formed Dixie Flying Service, and he brought everything from the Milan Flying School, office equipment and supplies, and plane parts, tools, junk, and one training plane, a "Monoprep", to Madison.

This is how and why Dixie Flying Service was born.

The officers of Dixie Flying Service were:

- Harry L. Vail: President. He was the owner of Vail's Furniture Store, 209-211 E. Main Street.[4]
- Joe D. Vail: VP and flying instructor. He was the son of Harry L. Vail. He graduated from Madison High School in 1928, and attended Parks Air College at E. St. Louis, Mo. for his early training. He finished his training at Shambaugh airport at Purdue University, West Lafayette, Indiana. He bought his first plane there, a 3 place "Waco 9" biplane and named it "Peg," and he had a Transport Pilots License.
- John A. Naill, Treasurer, owner of Naill's Lumber Company on the SW corner of Second and West Street.[5]
- Dale Wilson, Secretary, bookkeeper at Vail's Furniture Store.

Other officials of the Dixie Flying Service were:

- Curtis Marshall, Judge.
- Richard Oglesby, Manager & Promoter of D.F.S. He came to D.F.S. from the Milan, Indiana Flying School, where he was involved in the building and repairing of airplanes. He has twelve years experience building and repairing airplanes, locating and buying wrecked airplanes and parts.
- Harvey W. Kattelmann, A. & E. licensed mechanic & instructor, 1641 Rockford Place, Cincinnati, Ohio. He came to D.F.S. from Milan, Indiana Flying School.
- Gladys Gourley, Secretary, 1931-1932, She left to marry Walter Schmidt.

---

[4] Later became Gans' Furniture Store, then Murphy's 5 & 10c Store, and is now the "Good Samaritan" Store

[5] It later burned to the ground and the Madison Heritage Apartments was built on the site.

Mom and I finally got to see the "Madison Airport". We drove north on West Street to S.R. 29, up a hill with sharp curves, past several farms, 2.6 miles, until we came to a big old 3-story brick farm house facing the west side of S.R. 29. At the rear of the house were several very old and tall cedar trees and maple trees. Just south of the house was a small pond, a tobacco barn, and several other small buildings. This is the farm and home of John H. and Augusta Sachleben and three sons, Carl, Fred and Joe. It was an 80-acre farm located at the SW corner of S.R. 29 [6] and a county road[7]. The farm extends south along the west side of SR 29 to the old "Prenatt (Crowley) farm, and west to the "Shune Pike".[8] John H. and Augusta Sachleben leased 50 acres off the south end of the 80-acre farm for 5 years beginning on the 29th day of September 1931, to the D.F.S. for $300.00 a year.

Just behind and west of the house and trees were three small airplanes. (1) A "Waco 9" biplane, open cockpit, 8 cylinder in-line, 235 HP, water-cooled Hispano-Suiza engine, speed 138 mph. A biplane is one that has a wing mounted above the fuselage, and a wing mounted below the fuselage. This is Joe Vail's personal plane, "Peg." (2) a "Monoprep", "110 special"

OPEN AIR SCHOOL

Four Planes Acquired By Joe Vail and Associates for Madison Airport.

10-8- *1931*

Madison, Indiana has been fortunate to have enough air minded citizens who are planning for the future. The Dixie flying service has been organized and incorporated to operate on the John Sachleben farm, which is to be known as the Madison airport. The officers of the company are Harry L. Vail, of the Vail furniture company, John A. Naill of the Madison lumber company, Curtis Marshall, Richard Oglesby, Joe D. Vail and Dale Wilson.

Joe D. Vail, popular young man of Madison, who is very well known in aviation circles in this part of the country, is in charge of flying instruction, which will insure thoughtful care for students and passengers, which is so positively necessary. He is assisted by entirely capable ground school instructors. Richard Oglesby, who will be manager has had twelve years experience in building and handling air ships, which insures efficiency in this department.

The company now has four planes, a fleet biplane, a monoprep mohoplane, an Eagle Rock biplane, and a Waco biplane. The new hangar will be located directly fronting upon the Michigan road, state road 29, making it easy of access to patrons. The company will operate nationally and already have several out of town students signed up. Mr. Clyde Beyer of Pittsburgh, Pa. has already reported for training. Among the local students are Mr. Paul Wallace, who is now almost ready to solo. Mr. David Gentrup of North Madison, and Mr. J. P. Wilkinson the well known road contractor. Quite a lot of enthusiasm has been shown by the townspeople who have inspected the flying equipment now on the field.

parasol monoplane, 5 cylinder radial, air-cooled "Velie" engine. This is a 110 HP, 90 mph., two-seat, dual control, enclosed cockpit training plane, where the instructor sits beside the student. A parasol monoplane is one in which the wing is mounted above the fuselage. (3) A "Fleet" biplane, open cockpit, 7-cylinder radial air-cooled engine. It was an Army training

---

[6] Later changed to "Michigan Road", then changed to S.R. 421, and has now been changed back to "Michigan Road".

[7] Later became Hy 107, and is now Hy 62 and Clifty Drive

[8] Now N. Cragmont Street

plane that had been wrecked. A low wing monoplane is one in which the wing is mounted below the fuselage.

Beyond the planes looking west was a field of grass and weeds where tomatoes were grown, and a grove of tall trees was at the west end of the field. There was no runway, hangers, and not even a windsock that I could see. At first I thought we were at the wrong farm, but this had to be the place according to the directions Mr. Oglesby gave us.

I was really disappointed; I couldn't believe what I was seeing. The pamphlet they sent me showed a large hanger with a workshop, planes to be worked on, an office, and east-west and north-south runways. There was not even enough ground for a north-south runway. This was in sharp contrast to the Cleveland airport and school I attended a couple of weeks ago. They had several large modern hangers where planes were stored and repaired, classrooms and office, concrete runways and quite a few planes. We sat there awhile and then decided to go back to town. We stopped at Rogers Drug Store, (NE corner of Main and West Streets), for a coke and to decide what to do. I finally decided to go back to the office and ask for the $398.00 that I had paid for my tuition.

I learned another important lesson. Mr. Oglesby took me into his office and convinced me how lucky I was, that I was the first student and if I got my license I had a job with them. And as they were just starting I would grow with the company, rather than starting with a well-established company. As far as the pamphlet, that was just to show what it would be like in the near future. I could see right then that as long as he was in charge the sky was the limit and I thanked him for talking me into staying.

He gave me a letter containing instructions, to give to a doctor in Seymour, Indiana, who was to give me a physical exam, approved by the federal government, that all pilots had to pass. If I failed the exam I would get my $398.00 back. As this was only the second time I had been in Indiana he gave me detailed instructions on how to get to Seymour.[9] I took my mother to the railroad station down on First Street between Vine and Mill Street and put her on the train for home.

I took off up Highway 7 to North Vernon, turned west on Highway 50 to Seymour and finally ended up at the doctor's office. There were two doctors with the same last name; I didn't pay any attention to the initials so naturally I went first to the wrong doctor who happened to be a vet. I never knew whether he was any relation to the doctor I was to see for the exam. Being only 19, I didn't have much interest in, or knew much about doctors.

The exam was very tough, it took most of the afternoon and I was concerned whether I could pass it. When I was in grade school they said I should wear glasses so I wore them until about 1928 when I found out that pilots were not permitted to wear glasses. I quit wearing glasses and exercised my eyes everyday, straining to see how far I could see things clearly.

In addition to the usual physical exams, heart, lungs, etc., the test concentrated on endurance, push-ups, sit-ups, coordination, and eye and ear tests. I was surprised that I didn't have any problem reading the eye

---

9   In 1929 I drove straight through the middle of Indiana with four men
     in a 1927 Buick touring car headed for Los Angeles

charts. He brought out some cards with a lot of dots of different colors, and without much time you had to tell him what initials, numbers, or picture you thought you saw. This was supposed to tell him whether you were colorblind and how much.

The doctor took a pencil, telling me not to turn my head, started in front of my nose, and slowly moved the pencil to the left and the right, and I was to tell him when I could no longer see the pencil. This was to tell your side vision and whether you had any blind spots. He brought out a box about a foot square with just one side open, and it had two round pencil-like pegs, each one in a small track about 2" apart. Attached to each peg was a 10-ft string. He gave me the end of each string explaining that I was to line up the two pegs side by side with the strings, about 10ft from the box, after he moved them away from each other. He showed me what they would look like and then moved them apart. I moved the left string, then the right, moving the pegs back and forth. It was hard to tell much difference so after half-dozen tries I said that I thought there were lined up. The doctor said it was almost perfect, and was one of the most important tests. It was a depth perception test to see, for instance, if you would be able to tell the difference between a light pole and a fence post from the air.

In a hallway about fifteen feet long there was a white stripe running the length of the hall. I was to stare at the stripe for a minute or two, then close my eyes and walk the length of the hall. I did real good, but never knew the purpose of the test. The secret is not to hesitate, just start walking naturally as soon as you close your eyes as you can still see the stripe if you had stared hard enough.

There were several other tests such as the doctor holding a pocket watch behind each ear moving it away until you could no longer hear it ticking; hop down the hallway on the left and then right foot with no support; tapping each knee with a rubber hammer, checking reflexes; blood and urine samples, etc. Finally the doctor told me I passed the tests and to come back in six months. I was glad to get that over with, I was tired as if I had worked all day.

When I got back to Madison I met **Harvey Kattelmann,** the local mechanic; he was a big 6'2" and only 19 years old.

They got another plane in today while I was taking the test at Seymour. It is a "Monoprep", a monoplane-training plane just like the one they have, so that makes two training planes, Joe's Waco, and the "Fleet" Harvey was working on.

I got my flying equipment today, goggles - $2.10, pilot logbook - $1.25, helmet - $4.50.

So I am all prepared and anxious to start taking flying lessons tomorrow, I probably won't sleep much tonight and hope I get along ok.

Everyone I've met so far seems to be very nice. There are a million pretty girls in this town, or so it seems in comparison to the ones at home, our neighborhood was at least 90% foreign-born, Italian, Irish, English, German, Swedish, Polish, and Chinese, although they all seemed to be very nice, the girls weren't my type. Half the parents couldn't speak English.

I rented a room for $5.00 a week in a 2-story rooming house at 327 E. Main Street. Lemuel Richart and his wife owned the house. Across the street was the Jefferson County Courthouse. 329 E. Main Street was another rooming house owned by a Mr. Spillman. 331 E. Main Street, at the NW corner of Walnut Street, was a hardware store and tin shop owned and operated by William and Clarence Hoffman.[10]

I filled out an application for a pilot's license, which must be filed, under oath, with the Secretary of Commerce.

I found a swell place to eat just around the corner from my rooming house. At 417 N. Walnut Street there was a two-story brick house, Barbers Rooming and Boarding House owned and operated by a Mrs. Barber and daughter, Inez.[11] The food was home cooked and great, placed in the center of the table in large bowls and everyone had to serve himself. Now I know the meaning of the saying "he had a boarding house reach," and "heaven help the hindmost." Each meal cost 25c, all you can eat and included coffee and one piece of pie.

After eating there for a couple of weeks and getting to know everyone pretty well I would occasionally hide my pie plate under a stack of empty dishes and tell Inez she forgot to bring my piece of pie. It never failed; she

---

[10]   These three buildings were later demolished and replaced with a new hardware store owned by the Kelly Bros. Later Ace Hardware took over the business, and now the building was remodeled and is the home of the Salvation Army.

[11]   Mrs. Barber's husband worked at the Madison State Hospital (Cragmont). It is a Hospital for the mentally ill located on the hilltop.

always brought me a second piece. After the first time, I'm sure she knew what I was doing and must have felt sorry for me because she never said anything. Every chair was occupied at every meal. Mrs. Barber said I would have the first vacancy as she was working on a deal with D.F.S. to house and feed the students as they arrived.

October 3, 1931 – Saturday

I started to work today, at 8:00am, washing and waxing the Monoprep training plane. That was quite a job but it gave me a chance to become familiar with all the different parts of the plane and to watch for anything that might need replacing. I met one of Joe's personal students today, **Dave Gentrup**, a student at North Madison High School. He seems to be a nice guy and has 4 hours dual instruction so far; he has been practicing landings and takeoffs.

Dave Gentrup & Me at Hanger 3/10/1932

J.P. Wilkerson and Paul Wallace, the other personal students of Joe's, were next to take dual instructions. By this time it was getting pretty late. We went up at dusk for only 15 minutes flying time, the air was calm and we didn't get to do much except to become familiar with the plane in flight, but at least it was a start and my first lesson.

The closed cockpit is small; you sit so close side-by-side, that your shoulders touch the instructor's and you sure don't have any trouble hearing him, even though there is no muffler on the engine. The seats have seat belts and the only thing solid is the floor and dash. Everything else is rigid tubing covered on the outside with fabric. The plane is equipped with dual controls, throttles, rudder pedals, joysticks, which are 30" long rigid tubing with rubber tips. The joysticks are attached to brackets mounted on the floor, and the pilot's legs straddle it so it can be moved in any direction. It controls the elevators and ailerons. This arrangement permits either the instructor or the student to fly the plane.

There are just a few instruments on the dash; fuel level gauge, oil pressure gauge which should read 60 lbs. with engine running, compass, altimeter - it tells you how high up in the air you are (pretty accurate until you get close to the ground), air speed indicator, choke, gas shutoff switch, and tachometer gauge (revolutions per minute) - it indicates the speed of the propeller as you speed up or slow down the engine. With the engine throttled, the engine should be idling about 400 rpm. These two gauges play a big part in flying the plane. Then there is one on/off switch, which is connected to the magneto that provides the spark to start the engine.

Last, but not least, is a fire extinguisher mounted on the dash within easy reach, and Joe said, just half-jokingly I think, it was there in case a student froze on the controls in an emergency he could whack him one and make him let go of the controls in a hurry. I never forgot that explanation. There is no battery, starter, or brake on the plane.

Cockpit at 400 ft

I am going to try to get to the field first tomorrow; maybe I'll get in more flying time and begin to learn something.

This is some town on a Saturday night. All the farmers come to town, and all the town people congregate on Main Street sidewalks, so it's hard to find room to walk. All the stores, barber shops, etc. are full of customers, Main Street and all side streets are all parked up with every type of car, truck, or buggy.

I was surprised to see that the Boy Scouts directed the traffic. As usual there were some town "toughs" acting tough. I went to one of the two shows in town - 25c. People were waiting in line to get in both shows. I was surprised to find black people had to sit in the balcony. After the show I bought an alarm clock $1.00, watch 79c, pen $3.00, ink 15c, ointment 35c.

Following the flying lesson tomorrow, I'll have the rest of the day to loaf.

October 4, 1931 – Sunday

I got up too late for breakfast today, it's a good thing I have my own car. By the time I arrived at the field the wind was blowing stronger and was quite gusty (in this area it generally blows to the E from the W). In getting the plane ready to fly I uncovered the engine and checked it for loose bolts, oil or gas leaks, the hoses and wiring, and the engine mounting bolts. Then I filled the gas tanks with 30c a gallon aviation gasoline through a large funnel covered with a piece of chamois to prevent anything getting in the gas in the tank. The oil level was ok.

Next the landing gear was checked for cracks, loose bolts, also the tires and wheels. Then the wings, struts and wires were inspected, taking hold of the wing tips, I gave it a sudden jerk to see if there was any defect in the wings or struts. I looked over the stabilizers, elevators, rudder, and ailerons to see that they are solid and braced, and that the fabric is in good condition. Checked the tailskid, it consists of three layers of spring steel, resembling a Model A spring and is mounted on the rear of the fuselage to hold the tail off the ground and acts as a brake when landing. This inspection is done before every flight, you can't get out and walk if anything goes wrong up there.

After making sure the wheel chocks were in place so the plane couldn't roll and the switch was in the off position so it wouldn't start, Joe showed me how to crank the engine to start it. Standing facing the engine he turned the propeller three times clockwise to draw gas into the cylinders of the engine. He then positioned the prop so that it was shoulder high on the right side of the plane, and the engine was under compression. Standing directly in front of the prop he braced himself, took hold of the tip of the prop, swinging his right leg forward, he gave a sudden pull downward and his right foot swung, backward, helping him to step back away from the prop. He said that this maneuver would keep you from stumbling into the spinning prop.

After trying it I could see that swinging your right leg at the right time helped both in suddenly pulling the prop downward and in backing away from a spinning prop. It takes quite a bit of strength, skill, and practice, and it must be done suddenly or the engine won't start. If you ever get into the path of a spinning prop, you're mincemeat. If you have ever cranked an automobile engine to start it, it is the same theory; the difference being you might end up with a broken arm, which has happened several times. Engines too large to start in this manner have an "Inertia Starter", a flywheel is either hand cranked or electrically driven until it reaches the proper speed, and then a clutch engages it to the engine.

Following the prop cranking lesson, I opened the cockpit door and climbed in, fastened the seat belt, made sure the "joy stick" moved freely in all directions, that the ailerons moved with the stick and the control wires had no slack or play. Also that the rudder moves freely when the left or right pedal is moved.

Now we are ready to start the engine. I turned on the gas valve; made sure the ignition switch was in the off position and said to Joe "gas on, switch off." He stepped back away from the prop and said, "contact." I turned the ignition switch on; with my left hand holding the stick back against my seat and my right hand on the throttle, I replied "contact". He stepped forward, took hold of the prop and gave it a sudden pull downward. The engine started with a roar; it didn't have a muffler, as Joe moved back to safety. I pushed the throttle forward a little and let the engine run about 600-700 rpm until it warmed up, and checked the oil pressure. The oil pressure should be between 40 & 60 lb.

Holding the stick back against the seat, this raises the elevators and holds the tail firmly on the ground so the wheels don't climb over the chocks as I pushed the throttle wide open. As the engine ran smoothly I closed the throttle, Joe pulled away the chocks and climbed in beside me. He took over the controls and I kept my feet on the rudder pedals and my hands on the stick and throttle lightly so as not to interfere with him.

You always take off and land into the wind, or crosswind if necessary, never down-wind. The "wind-sock" tells you what direction the wind is blowing. Most of the time the wind blows from the west to the east in this area, so most of the time we will be taking off and landing toward the west. Once in a while the wind will be from the south, so we will either have to take off and land cross wind or take a chance on the short north-south runway.

As we were facing the wind and the west Joe suddenly opened the throttle until it was all the way open. As we started rolling forward rapidly, he moved the stick forward and the tail left the ground in flying position, and the horizon appeared on the nose of the plane. As we were going down the field, he used the rudder pedals to keep the plane going in a straight line as the plane had a tendency to go to the left or right rather that straight ahead. To turn the plane around on the ground, push the rudder pedal to the floor, push the stick forward a little, and open the throttle.

As the plane picked up speed, it began to feel lighter and Joe slowly pulled the stick back and the plane left the ground just in time to clear the trees on the west end of the field and we were flying, it is great. The field is really not long enough, if the wind is calm it takes the plane longer to get up flying speed than if there is a strong "head wind".

If the engine misfired and lost power you would end up in the trees.

In landing, if the wind is calm the plane needs a long runway, due to less air resistance, to set down and come to a stop, than if there is a strong head wind.

We flew straight ahead until we reached about 1,500 feet altitude. This gives us about two miles of safe gliding area in any direction to a safe landing if the engine should quit. Occasionally, Joe would move the stick slightly to the left or right to keep the wing level. The wind was quite rough and bounced us around but I soon got used to it.

Joe had me doing figure "8's" today. We picked a round barn and a smokestack as the turning points. Moving the stick slightly to the left, at the same time pushing lightly on the left rudder pedal and moving the stick forward or backward to hold the nose on the horizon, we turned left around the round barn, We headed toward the smoke stack where we moved the stick slightly to the right and at the same time pushed lightly on the right rudder pedal and moved the stick forward or backward to hold the nose on the horizon, we turned right around the smoke stack.

This is not as easy as it sounds. This maneuver teaches how to turn without losing or gaining altitude, too much "bank" can cause "side slipping", and the plane will slip sideways and downward losing altitude. Too little "bank" or too much rudder will cause the plane to "skid" and lose speed and could go into a nosedive. After making the turn you try to keep the wing level and nose on the horizon. After enough practice and experience you will automatically "feel" how steep to "bank" and how hard to push the rudder pedal. Actually if the plane is banked properly at the proper speed very little rudder control is needed.

After practicing figure "8's" Joe told me to find my way back to the field, but I was lost and had no idea where the landing field was, all the farms looked alike to me. He approached the field from the east into the wind, and when we reached the point where the plane will glide safely, he slowly closed the throttle until the engine was "idling" at a few hundred rpm. He moved the stick forward so the plane "nosed down" on the proper angle making sure the wing was level. If the gliding angle is too steep the plane will come in too fast to land, and if the gliding angle is too flat the plane will lose flying speed and fall. Facing us was the three-story brick farmhouse, the tall cedar and maple trees, and the electric high wires. If we come in too high over the trees we will "overshoot" the field and have to fly around and try again. If we come in too low we will land in the trees, so there is not much room for any error in judgment.

I imagined I felt the top of the trees scraping the wheels. We "leveled off" close to the ground, losing flying speed with the wheels barely clearing the ground. We slowly pulled back on the stick until the tailskid was level with the bottom of the wheels and the plane touched the ground in " a three-point landing". Holding the stick back against the seat so the tailskid drags the ground we rolled to a stop.

Perfect. A perfect landing is when the plane is an inch or two off the ground when both wheels and the tailskid touch the ground at the same time. We were up in the air thirty minutes today. All airtime must be entered in the "Pilots Log Book", both dual and solo time, ground time doesn't count. We turned the plane around by pushing the left rudder pedal to the floor, pushing the stick forward halfway, or neutral, and pushing the throttle open just enough to turn the plane. We taxied back to the trees, covered the engine, put chocks against the wheels and tied the plane down.

Before getting out of the "ship" Joe told me that there was a "Crozier Airport" started April 30, 1930, and owned by James E. Crozier. It is located at the NW corner of SR 7 and Clifty Road, later named SR 107, and now SR 62. He said no one from our field is to land over there at any time; in an emergency choose a farm field because both airports are feuding.

We are going to Cincinnati, Ohio tomorrow at 6:30am to pick up an "Eagle Rock" airplane. My working hours will be from 6:30am to 5:00 PM. I am anxious to get my next lesson.

October 5, 1931 – Monday

It rained this morning and was cloudy the rest of the day so I didn't get to go up today. I'll probably forget everything I learned yesterday by the time I go up again.

We left for Cincinnati at 7:30am in an old flat bed International truck, with racks and plenty of rags to protect the airplane we were going after. Harvey drove and it was a wild trip. I slept in the back on the rags going up, most of the time. We made the trip for nothing. The owner of the "ship" didn't pay the hanger rent so they wouldn't let us have it. It got chilly coming back so I borrowed Harvey's jacket and crawled under the rags and went to sleep again. We got home at 8:30pm.

I don't feel good, both of my legs are badly poisoned by poison oak and I can hardly walk. I sure hope I get better soon. I bought powder 25c, sugar of lead 10c, and Ungentine 50c.

October 6, 1931 - Tuesday

It rained "cats and dogs" this morning, and was cloudy the rest of the day.

We worked on the "Fleet" which had been wrecked, it's going to take a lot of work to get it flyable again, but they say it will really fly great. I think it is an Army training biplane. I was the first one up after making the pre-flight inspection this afternoon, and was up 45 minutes making a grand total of one hour and 30 minutes in the air. We practiced taking off keeping the "ship" from drifting, and how to find my directions. The "ship" veered to the left, I pushed the right rudder and it veered to the right, I was zigzagging down the field like a drunken sailor. Joe finally took over and straightened the ship up and we took off. The rudder is hard to get used to, you have to move it a lot on the ground, but barely move it in the air, as the airflow is much greater in the air. Joe brought us in on a 180-degree landing. It sure was great. I wonder if I will ever solo. Dave G. sure has his landings down pat, and he now has six hours in the air. All of Joe's students have more flying time than I have, but so far none have soloed, they started taking lessons long before I did.

A boy from Washington, PA hitchhiked to Madison in a day and a half to look the place over. Oops, it was an "American Eagle" that we were trying to get yesterday, not an "Eagle Rock". We will probably go after it in a day or two. They are working on our new field just south of the one we are now using, and are getting it in pretty good shape.

I was going to write some letters tonight, but my legs still hurt a lot. I hope they get better soon.

October 7, 1931 – Wednesday

I didn't get to go up today and it is getting to be a habit.

We went to Cincinnati, Ohio after that "American Eagle" at 7:30am and we had a great time. All three of us had to sit in the cab as it drizzled all day so I didn't get to sleep any this time. We had a pretty tough job dismantling the "ship", but finally got it apart. The main problem was trying to pack it, wings and all but the fuselage, on the small International truck. We tied the tail of the fuselage to the truck and towed it behind us. I got in the back of the truck to watch over the ship. It was so crowded that I could hardly move. We got home at 11:30pm. We were so tired that we left the truck and all on the street.

The plane is an awkward looking thing; the nose appears to be twice too long for the fuselage. It was a biplane - license number C620E, with a

Hispano-Suiza 235 HP, eight cylinder water-cooled engine. (My legs don't seem to get any better).

October 8, 1931 – Thursday

I didn't get to go up again today and it is getting monotonous.

Joe had a bad cold and the weather was damp, dark and cloudy. I guess by the time I'm a pilot I will be a better mechanic.

We unloaded the truck the first thing this morning upon the second floor of the Penn Maddox Ford garage, 101 E. Second Street.

It was quite a job as the ramp was narrow, steep, and made a right angle turn. This is where we will work on the plane until the hanger is built.[12]

We cleaned the fuselage of the Eagle then went out to the field and helped Harvey check the valves and spark plugs on the "Prep". They are having a little trouble starting the "Prep" so they are going to put a booster on it tomorrow. The more I work on the ship the better I like it. I guess all of Joe's students are about ready to solo while I still have only 1:30 hours in the air.

---

[12]   The building has since been torn down and was later the C&R Auto Parts Store

My legs are beginning to get better. I bought razor blades for 43c. Wrote a letter home today.

October 9, 1931 – Friday

This must have been an off day, I hope so, and Joe wanted to know if I had a bad dream last night. Actually I did, but it was primarily a sad dream. We were up an hour and I didn't do anything right. He tried to teach me how to take off and land, but it was no use. If I didn't overshoot the field (too high up to land), I undershot the field (too low to reach the field). Joe sure was mad. I couldn't even locate the field from the air, all the farms looked alike to me. Being from the city I was used to locating by streets, but locating the field was one thing I had to learn now. Taking off I would fishtail down the field, first heading left, then heading right, I could not seem to keep the ship going in a straight line. It sure was a heck of a day. The weather was fine, but the wind was awful bumpy up in the air. My total time is now 2 hours and thirty minutes. I sure hope I do better tomorrow.

I paid for my meals for the past week - $5.00. My legs are a little better.

October 10, 1931 – Saturday

The weather was fine, but the wind was pretty choppy.

I had another bad day. We were up an hour making my dual time in the air total 3:30. Usually you put in about 10 hours dual flying instruction before you solo. I am showing very little progress and it is getting to be a habit. I did fairly well on my takeoffs and hope I can do the same tomorrow. But boy those landings! One time I thought I was going to bounce sky high. When I level off for a landing I level off too high and consequently the plane bounces. It is considered a perfect landing when you level off, engine idling, wing level, and wheels and tail skid about 2" off the ground. When taking off, landing, climbing, descending, or flying straight ahead the wings are to be kept level and that is not always easy to do. If one of us got in the habit of flying with either the left or right wing lower than the other, we were nicknamed "One Wing Low". If you came in too high too land, you had to push the throttle all the way open hoping that the engine didn't misfire, and fly around and try again. We had a pretty stiff crosswind and that makes it much more difficult to land, especially when learning. Ideal landing condition is to land straight into the wind; this also holds true in taking off. I am beginning to get the feel of the ship and hope I have a good day tomorrow.

Dave Gentrup, Joe's other student, also had a bad day, probably due to the stiff crosswinds. (I think it is almost time for Dave to solo). In taking off, and especially landing in a stiff cross wind, it is necessary to fly "crab-wise", that is, very slightly flying the plane in the direction the wind is coming from and depending on how strong the cross wind is. Before the plane reaches flying speed the crosswind has the tendency to push the plane sideways. I got a letter from Mom today.

October 11, 1931 – Sunday

The weather was cool and cloudy with some showers. The wind was so rough we didn't get any flying time today.

Dave went up and when he landed he did a nice "ground loop", that is the tip of the wing hit the ground and the wind blew the plane around 180 degrees. I started to go up but Joe decided it was too bumpy to fly. We washed part of the American Eagle's engine with gas today.

There was plenty of excitement in town today. Last night a hypnotist hypnotized a local boy who was to stay asleep in a store window until Monday morning for $50.00. At 11:30 Saturday night he woke up but still under the influence of hypnotism he started to tear up the town. They had to get the Hypnotist to snap the boy out of it and was that boy scared! So the Hypnotist had to get his wife in the window as a substitute. They almost ran the guy out of town.

The American Legion had a parade today to dedicate a flagpole in the courthouse yard, right across the street from my rooming house. Some parade. Those Legionnaires sure are painting the town red. Some of them were drinking on the courthouse steps; the Mayor came up and told them they better quit that. They told him they didn't have time to go to the cellar every time they wanted a drink. What a bunch!

I haven't been using my car much, but it still runs good.

October 12, 1931 - Monday

No flying again today, the wind was very treacherous.

Dave went up at 7:00am this morning and as he was landing the wind lifted the plane about six feet off the ground, what a bounce. So that means no flying today for the rest of us. Dave now has about eight hours dual flying time. It is beginning to look to me like I'll never solo. Finished washing the "Eagle" engine with gas today.

I met a young fellow who works at night in the garage where our plane, the American Eagle, is stored on the second floor of Penn-Maddox Ford

Garage, 101 E. Second Street. He is quite a goofy guy and his panhandle is Clyde Weber. We will be storing the planes to be repaired here until the hanger is built.

They started on the runway and hanger today. The runway and hanger is to be built just south of what is now Crestwood Drive, on the west side of SR 29 (now Michigan Road). On the east side of SR 29 is the new home of William Hoffman[13] directly opposite the planned runway and hanger. The rest of the east side of SR 29 is farmland.

I bought 5 gallons of gas today for 69c. Think I'll write a letter to Herb Schwartz tonight. We might take a trip to Cincinnati tomorrow to get some plane parts.

October 13, 1931 – Tuesday

The weather is fine, cloudy, a few thunderstorms today and windy.

The 13th must be unlucky – we were up for about 15 minutes and as usual I got lost. By the time I found the field the engine started to act up, first it run like blazes, then it would slow down to idle. So we hurried up and landed. Checking the engine we found that oil was getting into the carburetor and loading it up.

Joe started teaching me to fly with my right hand controlling the "joy stick". Being left-handed it is hard enough learning to fly left-handed, let alone learning to fly right-handed, but then most "ships" are equipped for right-handed flying. We didn't go to Cincinnati today so I washed and waxed the tail surface of the "Eagle" instead. The "Prep" has been fixed (carburetor) so I hope to get to go up tomorrow; we will be looking for emergency landing fields. While flying you always look for a place to land in an emergency, such as engine failure, something goes wrong with the controls, or the plane itself.

There is a fair in town, and two of the dwarfs are staying in the boarding house next door. Boy, how the two boarding house owners do fight (Richerts, where I live, and Spillman). The other night they almost came to blows; each won't let the other's boarder's park in front of their place. It makes it very inconvenient for us, I have had to move my car more than once rather than getting into an argument.

Got rubber heels put on my shoes for 50c. I am going to write a couple of letters.

---

[13]   Now the "Madison Plaza",

October 14, 1931 – Wednesday

It rained like cats and dogs all day so there will be no flying today. I have just about forgot all I ever knew about flying.

I washed two of the "Eagle's" wings and waxed the tail surfaces today. It sure seems as though I'm doomed to be the chief scrubwoman.

Everyone calls me "Friday" as I am always with Harvey and am his helper. They are working on the new field just south of the field we are using now and have almost finished building the tool shed. I will be very glad when they get the hanger and runways ready to use. I am going to try to get to sleep at the hanger and save my room rent. At the present time it is really too dangerous coming in for a landing so close to the treetops, a slight miscalculation and you are in big trouble, or worse.

I wrote Pearl a letter last night and will try to write Mom a letter tomorrow. The mosquitoes have just about eaten me up, and my legs are bothering me again. Went to the show tonight and saw the four Marx Bros. in "Monkee Business", 25c. I bought some peroxide, 10c.

October 15, 1931 – Thursday

Well, it rained all day again today so naturally there was no flying. I hope the weather is fit for flying tomorrow.

Harvey is going to talk to Joe about giving me some more flying time each day. Washed two "Eagle" wings today, painted the license No. C630E on one wing and the step plate on another wing. We stripped the fabric off one wing. It was quite a job taking the fabric off the wing as it is sewed around every rib (and there are a lot of them) every six inches and also around the edge of the wing. The fabric was in bad shape, very brittle, and beginning to sag in spots. I got a piece of the wing spar from the "Fleet" as a souvenir.

A spider bit me last night and my right hand has swollen about ¾ of an inch, I hope it is not poisonous. It is bad enough to be bitten by mosquitoes so I think I am going to declare war on the insect world. I wrote a letter to Mom tonight after I had received a letter from Mom and Pearl today. I had to lock the door to my room today as the boarders from the fair are inclined to roam in other people's room. I'll probably write a couple of letters tomorrow.

October 16, 1931 – Friday

The weather is fair, but very cold and windy. You guessed it, no flying today.

This is a very poor time of the year to enroll in a flying school, but I wanted to take advantage of the job offer. I have to get up at quarter-to-five tomorrow morning to fly. The wind is usually calm early in the morning and late in the evening. I hope.

I painted the license number on the other wing and cut up the old fabric we stripped off the "Eagle". I kept a small piece of the old fabric and a piece of the untreated fabric from the same wing. A 235 HP, 8-cylinder, water-cooled, Hispano - Suiza engine powers the "Eagle".

Went to the library tonight but they don't have very many books on aviation. It is in a large two-story brick home at the corner of Main and Elm Streets.

I might get a chance to earn some money washing Oglesby's car, I hope. Paid my room - $3.00, and board - $5.00, tonight. It sure left a large dent in my pocketbook. My hand is a lot better and I found the spider that bit me and killed it. My legs are still bothering me. I wrote my sister Bert a letter.

October 17, 1931 – Saturday

Fine weather today, although the wind was quite tricky.

I got up at 5:00 o'clock this morning and we flew for thirty-five minutes practicing "shooting at the field". (Approaching the field and runway to land, but at the last minute pushing the throttle wide open, climbing straight ahead to a safe altitude, turning around and doing it all over again). This is the most important part of the training, to consistently land the plane safely, so you practice, practice, and practice until it becomes automatic. It is very simple; once you are in the air you have to come down someway, somehow, sooner or later. The tall trees are a constant threat, coming in too high over the tree tops you won't have enough runway to land, but by the same token coming in too low you will crash into the tree tops. So if you miscalculate, you better come in too high.

I finally got to wash Oglesby's Ford and started to wax it, will finish it tomorrow. Harvey and I brought thirty sacks of cement from town this morning on the old flatbed "International" truck for the hanger foundation.

I am going to do my flying real early in the mornings; that's going to be hard to do, as I am a "night person". I am feeling comfortable flying with my right hand.

Met a fellow tonight by the name of William Dixon. We swapped yarns about his trip and my trip to California. Might get a chance to go to Wisconsin. I paid 40¢ for laundry. Weber tightened the brakes on my car today.

October 18, 1931 - Sunday

The weather was fine, but cold. We went up for forty minutes this morning and practiced takeoffs and landings.

I took off like a drunken sailor. I have a total of five hours dual flying time. The wind was pretty choppy.

I went to the "United Evangelical Church" tonight with Mr. & Mrs. John Sachleben, and sons Joe and Carl. They are the family who own and live at the airport. It is like my church in Pittsburgh.

While there I met Ed Becht. I drove to Hanover College sightseeing, a beautiful place, just out of town on the hilltop. Then over to Cragmont Asylum, a place for very serious mental ill patients, and it is a beautiful place. Both places overlook the Ohio River. I paid 30c for film.

October 19, 1931 - Monday

The weather was cold, but nice, and the wind was very smooth today.

We went up for 10 minutes tonight, right before dark. Made one good take-off and one good landing. Landed "hot" (fast) because of a strong tailwind. Hit that ditch in the runway an awful whack. I am getting the "feel" of the ship.

Cragmont at 400 feet – May 19, 1932

Harvey and I hauled four loads of gravel from a creek, into the truck, and then shoveled it onto the field. I am worn to a frazzle. Found two horn corals and one piece of flint in the creek and am saving them.

Joe flew his "Pap" to Indianapolis today. I am supposed to get up at quarter to six tomorrow to fly, but I don't think I can do it – I am a night person. I wrote Mom a letter.

October 20, 1931 - Tuesday

The weather was fine today, and the wind pretty calm.

Made two landings and "takeoffs" by myself today. Almost landed in the treetops once. Still a little rough on my "take-offs" as I keep forgetting that you have to use a lot more rudder on the ground than in the air. I'll learn – I hope. Am getting to know what flying is all about by now, I was up 45 minutes practicing figure 8's, power turns, and climbing turns. My dual time is now around six hours. The average dual time before soloing is about 10 hours, so it won't be too long now. The tailskid wore out so I installed a new one.

I hauled two truckloads of creek gravel today to the new field. I received a letter from Pearl and Herby and wrote Looie a letter. I then went down to the garage on Second St. tonight to see Weber, we are going on a spree on Sunday. It is chilly tonight.

October 21, 1931- Wednesday

The weather was fine, warm, and the wind was mild.

I had a bad day today. Was up an hour, six flights, and it seemed as though everything went wrong. First my "takeoffs" were of the famous "corkscrew" method, and then twice I brushed the trees with my landing gear. Joe said he was going to tie a 10-foot flag to the tail of the "ship" when I solo.

I hauled two truckloads of creek gravel today to the new field.

I borrowed the complete "Dayton School of Aviation Correspondence Course" from Oglesby, so I can study up on aviation. This book, by Robert C. Rockwell, is really a collection of lessons and tests put out by the Dayton School of Aviation, Dayton, Ohio (a correspondence course).

Contents of the text is as follows:

Lesson No. 1 – History of the First Engine Driven Airplane
Lesson No. 2 – Aerodynamics
Lesson No. 3 – "
Lesson No. 4 – "
Lesson No. 5 – Flying
Lesson No. 6 – Airplane Construction – Materials
Lesson No. 7 – Airplane Construction – Fuselage
Lesson No. 8 – Airplane Construction – "

Lesson No. 9 – Airplane Construction - Wings
Lesson No. 10 – Airplane Construction – "
Lesson No. 11 – Airplane Construction – Control Surfaces
Lesson No. 12 – Airplane Construction – Dopes & Lacquers
Lesson No. 13 – Airplane Construction – Landing Gears
Lesson No. 14 – Airplane Construction – Evolution of the Seaplane
Lesson No. 15 – Airplane Construction – Seaplane Construction
Lesson No. 16 – Airplane Construction – Flying-boat Hulls
Lesson No. 17 – Airplane Construction – Control Systems
Lesson No. 18 – Airplane Construction – Balanced Controls – Wires & Cables
Lesson No. 19 – Airplane Construction – Wires and Cables
Lesson No. 20 – Airplane Construction – Rigging
Lesson No. 21 – Airplane Construction – Assembly & Rigging
Lesson No. 22 – Propellers
Lesson No. 23 – "
Lesson No. 24 – Importance of the Aircraft Engine
Lesson No. 25 – Aircraft Engine Types
Lesson No. 26 – Types of Aircraft Engines
Lesson No. 27 – Principles of the Engine
Lesson No. 28 – Carburetion
Lesson No. 29 – Ignition
Lesson No. 30 – Cooling of Aircraft Engines
Lesson No. 31 – Lubrication, Fuel Systems, Superchargers
Lesson No. 32 – Structural Details
Lesson No. 33 – Valve Units, Crankcases, Reduction Gears
Lesson No. 34 – Care and Operation of Aircraft Engines
Lesson No. 35 – Modern Aircraft Engines
Lesson No. 36 – Aircraft Engine Instruments
Lesson No. 37 – Aircraft Navigational Instruments
Lesson No. 38 – Aircraft Engine Instruments
Lesson No. 39 – Parachutes
Lesson No. 40 – Aerial Photography – Mapping
Lesson No. 41 – "
Lesson No. 42 – "
Lesson No. 43 – Meteorology
Lesson No. 44 – Aerial Navigation
Lesson No. 45 – "
Lesson No. 46 – Testing Aircraft Materials
Lesson No. 47 – "
Lesson No. 48 – Aircraft Radio

Lesson No. 49 – Radio Transmitters
Lesson No. 50 – Aerostats
Lesson No. 51 – "
Lesson No. 52 – Flying
Lesson No. 53 – Airport Rating Regulations
Lesson No. 54 – Air Commerce Rules & Regulations
Lesson No. 55 – Commercial Aviation
Supplement I – Nomenclature of Aeronautics; A-E
Supplement II – Nomenclature of Aeronautics; F-R
Supplement III – Nomenclature of Aeronautics; S-Z
Supplement IV – Blueprint Reading
Supplement V – Airship Factory & Dock

"Several Students"
March 10,1932

L to R – Dave Swett, Arthur Prosek, Jack Yellitz, Max C.F. Sittner
Kneeling – Marlin Christiansen E. Linwood Wright Jr. Frank J. Peternell

There was a big fire in town. Two houses burned down. I don't remember seeing houses burn down in Pittsburgh, although I know there were many. Received a letter from Mom and wrote Fred a letter.

October 22, 1931 - Thursday

The weather was fine, warm and the wind was pretty mild. Yesterday I had a bad day. Today was a good day. Was up thirty-five minutes and made three good "takeoffs" and two good landings although I came in "hot" once and had to use the whole length of the runway to set "her down"; but I DID get her down. Hope I can keep up the good work.

Boy, a dream "ship" landed at our field today - a new "BUHL PUP", a one-seater, low wing monoplane. It's a "pippin". The fuselage, fin, and ailerons are aluminum, and the wings, elevators, rudders, and stabilizers, are fabric. It is powered with a three cylinder air-cooled radial engine and cruises at 85 mph. It costs $1,250.00.

That book I borrowed from Oglesby yesterday is surely a wonderful book, and it has a world of valuable information in it. I learned more studying one chapter than I ever knew before.

I hauled two trucks of gravel from the creek and one truckload of water to the new airport, and then the truck broke down.

Mom sent a box of taffy and medicine. She thinks of everything. Can you beat it?

I have to report at 6:00 o'clock tomorrow morning.

October 23, 1931 – Friday

The weather was fine, warm, and a very strong crosswind.

Yesterday was a good day - today was a bad day. The "takeoffs" were ok, but the landings were awful.

Every time I would set the "ship" down the crosswind would lift it off the ground again making it bounce pretty darn hard. Dave is supposed to solo today.

Hauled two loads of lumber, one load of cement, and one load of water to the new field today. Helped shovel cement and gravel into the hanger foundation.

Weber gave me a picture of himself and the "Elaine" at the outboard motorboat race on the Ohio River at Madison. He also gave me a picture of the "Elaine" in the races. The "Elaine" is stored in the same garage as our planes.

October 24, 1931 – Saturday

It rained like thunder all day long with a very choppy crosswind. No flying today so I washed and waxed the "Fleet". I can't wait till the "Fleet" will be in shape to fly. What a ship!

Dave didn't get a chance to solo today. I took a ½ day off and listened to the Pitt – Notre Dame football game on the radio. Final score; ND 25–Pitt 12. Loafed the streets with Joe, Carl, and Becht this evening. Weber and I are going to have a great time tomorrow.

Fleet Aircraft

Clyde Weber and "Elaine" on the Ohio River

October 25, 1931 – Sunday

The weather was fine, cool, with a very mean crosswind. Consequently, no flying today. My total dual flying time is now 8 hours and 15 minutes.

Wonderful day, boy, a wonderful day! Went to Sunday school this morning at the Evangelical Congregational Church at the NE corner of Main and East Streets.

I then went with Weber to what proved to be a perfect day. We went to New Albany to get his girl, Gladys, then came back to Paris Crossing and picked up a girl for me, and she sure is keen! Her panhandle is Rosalee Darin. She is one of those "firey' French gals and a darn nice gal she is, by cracky, as they say. The locals sure talk different. We bummed around town a bit and then took Gladys back to New Albany. On our way back Weber was driving pretty fast so I gently reminded him a couple of times with a good hard jab in the ribs, he soon caught the idea and we proceeded home at a nice slow gait.

On the way back we saw three rabbits and Weber killed them all with three shots. I got two rabbit tails as a good luck mascot. These are the first real rabbit tails I ever saw. As a matter of fact, these were the first wild rabbits I ever saw. Also, I got me a "mascot" from Rosalee in the form of her picture so now I really will be able to fly. We, to my sorrow, dropped Rosalee off at 3 o'clock in the morning.

I am going back to Paris Crossing real soon. It is a small country town, so small that you drive through it before you know it is even there. New Albany is 67 miles from Madison, making our trip 240 miles of good times. Boy, what a day! "Hoop a la la"! Got in bed at 5:00am, just as the front door was unlocked. Date @ 80¢.

October 26, 1931 – Monday

The weather was cool, and a very strong; tricky crosswind blew hard all day so we didn't fly again.

Ho Hum! Got up at 7 o'clock this morning so I only had two hours sleep. Oh, well, it sure was worth it. At the garage, Harvey tore the 'Prep' engine down for an overhaul and inspection.

An "Arrow Sportster", a swell two-seater ship, powered by a 'Kinner' five-cylinder radial air-cooled engine, 90 hp at 1800 rpm, made in Lincoln, Nebraska, dropped in on us.

Shoveled gravel all day and it was a tough job keeping awake. I only went to sleep once on the truck and woke up just before the boss came up. They sure are razzing me about "ratting around", as they call it, all night long. They didn't think I would go on a blind date. Well, to be truthful, I didn't think I would myself.

Weber tore the engine out of his Ford tonight. I wrote Paul a letter. Oh, what a life!

October 27, 1931 – Tuesday

It rained "cats & dogs" and blew a miniature gale all day, so no flying.

Still sleepy, I will have to catch up on my sleep as I dozed a few times today and that's bad! Harvey and I worked on the "Prep" engine at the garage all day and it is almost ready to be installed in the plane. I kept five of the used piston rings as souvenirs as this was the first "ship" I flew.

Weber is talking about putting some extra equipment on my Cheve. They are still razzing me – that's ok, I razz right back.

October 28, 1931 Wednesday

The weather was fine, cool, and the wind was very tricky. No flying.

Harvey and I finished putting the "Prep" engine together and installed it in the "ship". Started the engine and let it run for about four hours to "break-in" the piston rings. It sounds great, doesn't miss a beat. Had to clean a two- cylinder radial air-cooled Aeronca engine this morning. I guess I'll have to learn how to fly all over again. It has been five days since my last lesson. It took me a month to get in eight hours; at that rate I'll never finish my course and get a license.

Took four pictures of the Ohio River and Kentucky from the top of Michigan Hill.

I am catching up on my sleep and I am planning on going to a party at church Friday night. Weber and I are planning on going on another party Saturday. I got a terrible haircut for 25¢.

October 29, 1931 – Thursday

It rained very hard all day. You guessed it, no flying for the sixth day. At least I am getting to be a pretty good mechanic, which is one of the requirements to get a license. We worked on the 0X5 engine, cleaned it and ground the valves. It is beginning to look like an airplane engine.

I received a letter from Paul and wrote Mom a letter.

October 30, 1931 – Friday

It was pretty cold with showers and the wind was very tricky so there was no flying for the seventh day.

A big day, first washed and waxed the Aeronca tail assembly this morning. Then went to Milan, Indiana with Joe in the flat bed International truck to get a mess of plane parts and two "Waco's" and one "Monoprep". We made several trips and finally got all done at 7:00am on 10/31/31. While at Milan I met Al Hotop, chief welder of the "Cloud Coupe" Corporation, and he told me to see them about a job when I graduated. I watched him welding a "Cloud Coupe" fuselage frame and he seemed to be doing a dandy job. I got an International turnbuckle as a souvenir.

I started to write Paul a letter but haven't finalized it.

October 31, 1931 – Saturday

I went to bed at 7:00am and got up at noon. It was cloudy, cold, and the wind was awful tricky, so there was no flying for the eighth day.

I went with Clyde Weber to New Albany to pick up his girl friend Gladys; came back to Paris Crossing but she wasn't available. Weber came up with another girl called "Red Head". I never did find out what her real name was but it's just as well, it turned out to be an awful date.

We went to a masquerade party at the high school gym, located on S. Broadway St. near the Ohio River; I believe it's called the "Brown Gym". The party was pretty comical. After taking Gladys back to New Albany and the "Red Head" to Paris Crossing we got home at 2:00am. As the rooming house rules states, among other things, that "the front door is locked at 11:00pm and stays locked until 5:00am"; I was locked out.

November 1, 1931 – Sunday

The weather was fair, but chilly. Finally, on the ninth day, we flew for 35 minutes.

Concentrated on takeoffs and landings. The five takeoffs were just fair, but the five landings were great. I brought the "ship" down in landing position about two inches off the ground, and that is considered better than perfect. I still have trouble taking off straight down the runway.

I went down to the garage on Second Street and visited Jim Robbins, the night mechanic. About 3:00am he got a telephone call requesting a wrecker. We went out and got back at 4:00am. Finally, at 6:00am I went back to the rooming house, washed, changed clothes, went around the corner and had my breakfast at Barber's Rooming and Boarding House. I haven't had any sleep since yesterday noon. Gas and lunch cost me $1.35.

November 2, 1931 – Monday

Fair and cool today.

We went to Milan, Indiana, at 6:00am to pick up a truckload of "junk", and airplane parts. While there I joked with Al as we were loading the truck and he gave a "Cloud Coupe" rib and a piece of rib fiber as a souvenir. I am getting quite a collection.

We flew for 25 minutes concentrating on landings and takeoffs, made four nice landings and four rotten takeoffs. It should be simple to takeoff – just head the ship straight down the runway, push the throttle wide open, push the joy stick forward enough to raise the tail so the plane and wings are level. When the plane reaches flying speed gradually pull the "stick" back and take off. My problem is I still can't seem to keep the ship from going towards the left edge of the runway and when I push the right rudder pedal to bring the ship back to the center of the runway it drifts to the right edge of the runway. So I zigzag down the field. My total dual time is now 9 hours 15 minutes and is getting close to the average 10 hours it takes to solo.

I went to Dave Gentrup's for supper and to meet and visit with his family. They live in a fine house on the property of the Cragmont State Hospital for the Insane, owned and operated by the State of Indiana. The food was great; the hospital raises all of its food, vegetables, fruit and meat. The patients do all of the farming under supervision as part of their therapy. Dave's dad is in charge of the boiler room and utilities. He said that he was as interested in the progress I was making in flying as he was in Dave's, and that made me feel good.

I wore my new flying coat today and everyone thought it was great. The people at the U.S. Rubber Co. in Pittsburgh gave it to me as a going away present. It is genuine horsehide, lined with sheepskin, with a sheepskin collar.[14]

Received a nice letter from Earl, and I have not finished the letter to Paul. Paid my weekly room rent ($3.00) and board ($5.00).

Haven't had any sleep whatsoever for eighty-one hours, but boy am I going to make up for it.

November 3, 1931- Tuesday

The weather was fine but cool and the wind was very wild and gusty, but flew for 30 minutes anyway.

---

[14] I wore it for many years, never could wear it out

It was the worse day I ever had as I made three fair take-offs and three rotten landings, besides coming within 3" of chewing a car to pieces with the propeller as I was taxiing off the runway. I should be able to fly by now as I have 9 hours 45 minutes dual time.

I cleaned two floors and took the dash out of the "Eagle" today. They took a picture of me in the "Prep" and Joe getting in the plane, for advertising purposes in an article in a pamphlet.

Joe giving final instructions before taking off with student for first lesson.
MAIL YOUR ENROLLMENT AT ONCE AND GET YOUR TWO WEEKS GROUND
TRAINING AND YOUR FIRST FLYING LESSON WITHOUT EXTRA COST.

Paid 40c for film, 4 pictures of the Ohio River, one of the foundation of the hanger, and one of all of our "ships".

Hisso Waco; 2 "Preps" 110 – 122 & 0X Waco in old field – October 24,1931

I finished Paul's letter and sent him "Rosie's" picture. Bought a set of boxing gloves in Milan, Ind.

Foundation of new hanger at Dixie Flying Service – October 24, 1931

Clyde H. Beyer

November 4, 1931 Wednesday

The weather was fine, but getting colder, and the wind was pretty rough.

We were up only 10 minutes today and I did ok. As I have only about 10 hours flying time in a month I asked Joe for more flying time and he said ok. It will take me forever to graduate at this rate.

A prospective student arrived yesterday and Oglesby asked me to let him bunk with me last night, never again.

Oglesby is up to his usual tricks; he took my "Coupe" rib so I took his "International" rib. I worked on the "Prep" and hauled two truckloads of creek gravel.

I got a letter from Mom. I might get to move to "Barber's Boarding & Rooming House, right around the corner at 417 N. Walnut St. on Friday, but am not sure yet. I have been eating all my meals there and Mrs. Barber had promised me the first opening.

November 5, 1931 - Thursday

The weather was fine, but cold – the wind was rough. After a fine breakfast went to the airport at 6:30am for some dual flying time.

When Joe showed up, late as usual, he told me to get the ship ready to fly. After the usual inspection of the plane, he got in the cockpit, I went to the front of the ship and standing facing the engine shouted "gas on, switch off", and he replies "gas on, switch off". After priming the engine and making sure the wheel chocks were in place, I turned the prop so that the left tip was about six feet above the ground and shouted, "switch on". I put both hands on the tip and shouted, "contact", he replied "contact". I gave a sudden pull downward, and the engine started. This method of starting the engine is always used so there is no chance the engine will accidentally start while you are turning the prop. After the engine warmed up we took off and I made three nice take-offs and landings.

As Joe was getting out of the cockpit and I was about to park the plane, he suddenly told me he felt it was too cold for him to fly, the cockpit is not heated, but if I wanted any flying time I would have to go up by myself. I was so surprised I didn't know what to say or do, finally with a silly grin he said, "GO ON! DO IT!

So I did. "I SOLOED"!

As I was about to take off he tied a piece of fabric about 3" wide and 5 feet long to the tail, as he had promised, saying it was the tradition; so that any plane flying in the area would know and stay well away, that a student pilot was flying solo for the first time. This was the moment I had waited for many a day.

At exactly 7:45am, November 5, 1931, I soloed after 10 hours and 40 minutes of dual flying time. Only dual and solo time in the air is entered in the logbook.

As he waved and walked away, I pushed the throttle all the way forward, with the engine running wide open, and the stick and rudder in neutral, the ship started down the runway. Slowly pushing the "stick" forward the tail left the ground and the ship leveled off in flying position with the nose on the horizon. As the plane picked up speed I used both rudder pedals to keep going straight down the center of the runway. Feeling the plane getting lighter and wanting to leave the ground, I gradually pulled the stick back past neutral and the ship took off at a safe angle and passed over the tall trees at the end of the runway.

Flying straight ahead until reaching a safe altitude at 1,500 feet, I leveled off and pulled the throttle back until the engine was running at cruising speed. Never! Never! Joe keeps preaching – attempt to turn right or left until reaching a safe altitude and flying speed, or you are sure to dive into the ground. Looking over at Joe to see what he thought about my takeoff; there was no Joe, just an empty seat; suddenly realizing there was no one there to tell me what to do or not to do, it was a strange feeling, alone at last, what a thrill! How lucky the birds are. The loud "cackling" of the engine (it doesn't have a muffler), was sweet music to my ears and very reassuring.

After practicing climbing, descending, keeping the nose on the horizon in level flight, and at the same time keeping the wings level, I flew over Madison and the Ohio River. The horizon is an accurate level gauge while in the air regardless of the altitude. While passing over the airport I noticed that they were waving at me to come down. Everything had been great so far, and all I had to do was get this "baby" on the ground. Circling the field, heading into the wind, lining up with the runway, I slowly pulled back on the throttle to slow the engine speed. I pushed the "stick" forward so the "ship" would glide at the proper angle and speed over the high lines and treetops without stalling, and not come in too fast or too high over the runway. I pulled the stick back to neutral and as the "ship" started to settle I pulled the stick all the way back and touched down in a nice three-point landing. After taxiing to the parking place, shutting the "ship" down, and covering the engine, I floated over to where Joe and Harvey were waiting before my feet ever touched the ground. Joe untied the 5 ft. banner and gave it to me as a souvenir. I kept it for many years.

I helped drag some tree stumps out of the new field and then cleaned Joe's Waco. Also hauled a load of coal and wood so I didn't have time to celebrate.

I had soles put on my shoes for $1.00. Loafed around town with Weber tonight looking at all the pretty girls then went to Ingles Drug Store and had a fountain coke with Nellie Harrod. Wrote Mom & Pyatt a letter.

November 6, 1931 – Friday

The weather was fine and cold. The wind was mildly rough.

I had a lot of rotten luck today. It all started this morning. After carefully checking the ship from propeller to rudder, filling the tank with gas I cranked the engine, let it warm up then took off heading west as usual. I overshot the runway 5 times in a row, using all 50 minutes trying to make a decent 3-point landing; it was terrible. I will have to do better tomorrow or bust, and not forget to carry my mascot again.

The tall trees and high lines at the edge of the field are beginning to intimidate me. If you come in over five feet above the trees you are sure to overshoot the field. If a pilot can land in this short field he can land anywhere.

Harvey & I worked on the "Eagle" the rest of the day. Then came more bad news. A piston broke, a connecting rod burned out, and the fourth cylinder wall had several deep scratches in the Cheve. Weber and I honed the cylinder wall until it was smooth as glass, and then installed the new piston and connecting rod. We refaced the valves, ground the valve seats, scraped off all the carbon, and put in all new gaskets and oil. We worked on the car from 9pm to 6am stopping once for midnight lunch, 3¢, and no sleep again. The car runs swell, but the engine is pretty tight and it will take a few miles to loosen up. It sure helps that Weber works at the garage. He doesn't charge for his labor and the parts are wholesale.

I wrote Maurice a letter after supper. He was the bookkeeper I worked for before quitting to take up flying.

November 7, 1931 – Saturday

The weather was fine, but a bit cooler and the wind was very tricky. I didn't get to do any flying today.

Went to bed for a couple hours this afternoon trying to catch up on the sleep I have been missing. I paid $3.00 to Mrs. Ritchert, my landlady and $5.00 to Mrs. Barber for my meals this past week. Probably will be moving to Mrs. Barber's Rooming and boarding house around the corner on Walnut Street next Friday. Wandered up and down Main Street all evening with several fellows watching all the pretty girls go by.

November 8, 1931 – Sunday

The weather was fine but quite cool, and the wind was quite "ragged". Didn't get a chance to fly today.

They began using the new runway today but it is not in very good shape yet. Harvey wiped the landing gear and prop off of the "Prep". He came in too high while landing and tried to sideslip in to keep from having to go around again, but failed to correct the drift. As the wheels touched the ground, the ship tilted, the left wing scrapped the ground and the prop hit the ground and broke.

The Cheve is running like a million bucks but no date today so I went down and spent the evening with Weber at the garage. He got a phone call about a serious car wreck so we jumped in the wrecker and took off. We found a brand new Plymouth wrapped around a tree with the two women and two girls who were seriously cut. One of the women went "nutty". It was a mess. If the car had been equipped with safety glass they probably would only have been in shock.

November 9, 1931 – Monday

The weather was fine but cool and the wind was slightly rough. No flying again today.

Harvey and I worked on the wrecked "Prep" most of the day removing the broken prop and twisted landing gear. It is going to take a lot of work to straighten the landing gear and make it safe to use again, and a new prop will have to be installed. It was lucky that it didn't damage the engine or frame. Lucky for me that I didn't wreck the ship. They would probably "ground" me for no telling how long.

November 10, 1931 – Tuesday

The weather was cloudy and it drizzled but the wind was not so bad. No flying again today.

49

We haven't finished working on the "Prep" as they had us painting the new hangar, and what I don't know about painting will fill a book.

Oglesby told me to quit loafing with Weber, and that staying out all night was affecting my flying. He also said that Rosalee was not the girl for me and I should drop her. Not a chance. He might be right about staying out all night affecting my flying. Weber said he was going to get married and there will be no more double dates. He borrowed my car so he could go hunting, and he very seldom comes home empty handed.

I got a letter from Earl Seitz, my high-school buddy. Mom sent me $20.00 to pay my room and board.

November 11, 1931 – Wednesday

Rained all day today and the wind was a little rough so there was no flying. We worked on the "Prep" landing gear and the streamliners most of the day. That error Harvey made landing sure is causing a lot of work.

I boxed with Weber for ten rounds and got whipped. Harvey borrowed my car so he could go on a date. The Cheve has an awful piston slap. Wrote Earl and Charles a letter.

November 12, 1931 – Thursday

It rained all day again and the wind was a little rough.

We finished repairing the "Prep" today. The field was a sea of mud so there was no flying.

There was a lot of excitement in town today. Eight men with "Sub-Thompson" machine guns drove into town in a new Buick and robbed the Madison Bank at 315 East Main Street at noon. They stole $22,000 and made a clean getaway. I drove into town for lunch just as the robbery was about to take place so I got to see the whole thing. After leaving the bank, they drove west on Main Street then turned right on Mulberry Street and that was the last time anyone saw them. The officers wanted Joe to follow them in a plane but he explained what the odds would be against ever finding them, all of our ships are out of commission and the field is a sea of mud. The robbers had planned it perfectly and were never caught.

We worked on the Cheve from 6:00pm to 4:00am and found three crack pistons. Bob Jones, the day mechanic at the garage gave me three used pistons. We installed the three pistons but one of the pistons got stuck in the cylinder and we had to smash it to get it out. It was almost daylight by the time we got the Cheve running again and hope we don't have to work on it for a while. One benefit of all this work is that I am learning all about engines, which will help me at test time. New connecting rod $1.00, midnight lunch 75¢, needle and thread 10¢ and 5¢ for postcard.

November 13, 1931 – Friday

It rained all day, very cool and the wind was rather rough, so there was no flying. It is just as well as it is Friday the 13th and every Friday the 13th reminds me of the time I almost drowned in the YMCA pool when I was about twelve years old. Fortunately my older brother Earl was with me and he saved my life. I was never as sick as when they brought me back to life.

Harvey and I put lacing on the cowling of the "Eagle" and gave it a thorough checkup. It was a quiet day and we goofed off quite a bit. The Cheve is still not running very good.

I paid my room ($2.50) and board ($5.00). I am going to move around the corner to Barber's Rooming and Boarding House in the morning. Boxed with Weber and ended up with a headache.

November 14, 1931 – Saturday

The weather was fine, warm, and the wind was slightly strong.

I soloed for 25 minutes getting ready for my private pilot's exam, practicing gentle figure 8s at 800-ft. altitude. The figure 8 is just a complete 360-degree turn with the throttle on and then reverse the turn and making another 360-degree turn in the opposite direction. I use Stephan's round barn as one end of the figure 8 and Sachleben's house as the other end of the figure 8.

Approaching the barn in level flight at about 800-ft. altitude, I moved the stick slightly to the left of neutral and pushed the left rudder pedal slightly forward. When the ship started to turn to the left I brought the stick back to neutral, but held the rudder pedal forward until the turn was completed. Then I moved the stick to the right of neutral and pushed the right rudder pedal slightly forward to come out of the turn. Again in level flight, I headed toward the Sachleben's house and made a right turn by moving the stick slightly to the right of neutral and pushing the right rudder petal slightly forward. When the ship started to turn to the right I brought the stick back to neutral, but held the rudder pedal forward until the turn was completed. Then I moved to stick to the left of neutral and pushed the left rudder pedal slightly forward to come out of the turn, keeping the nose of the ship on the horizon. While turning I moved the stick forward and backward at the same time I moved it left to right. It is really hard to keep

51

from gaining or losing altitude while turning. Practice, practice, practice. The beginning and ending of each turn should be at the same point.

It takes a lot of practice and Joe said that when you can make a perfect figure 8 consistently you have mastered one of the most important flying maneuvers. I made one sloppy take off, but three nice 3-point landings, and after parking the "Prep", I cleaned the inside as well as the outside.

Dave and I hauled two truckloads of lumber from <u>Naill's Lumber Company (West and Second Streets)</u> to the airport.

Former site of Naill's Lumber Yard.

SW corner of West & Second Streets.

I finally moved to Barber's Boarding and Rooming House on Walnut Street. It didn't take long, as I didn't have much to move, just clothes. It is a nice little room 6 ft. by 8 ft. on the second floor at the end of the hall; at the other end of the hallway is the bathroom. There's just enough room for a cot, very small dresser, and a small hall tree to hang my clothes on, but at least I will have privacy. Vail's

Furniture Store gave me two large pieces of cardboard, about the size of my cot, which I put under the mattress. I keep my pants between them while I sleep to keep them pressed. The Cheve is OK and I hope to have a big time tomorrow. I paid for my laundry - 50¢.

November 15, 1931 – Sunday

The weather was fair and cool, but the wind was awful tricky, so no flying today.

We picked up Weber's girl at New Albany at 2:00pm and went to get "R.D." and didn't get home until 5:00am so I missed another night of sleep but we sure had a swell time. We drove almost 300 miles and saw sixteen rabbits but didn't get a chance to shoot any. Paid 66¢ for gas and $1.00 for food.

November 16, 1931 – Monday

The weather was fine, warm, but the wind was awfully tricky. I am down in the dumps today, flew 50 minutes solo and did pretty darn bad, practicing steep figure 8's, landings and takeoffs. There is no time for pleasure flying, just keep practicing for the test. I'm beginning to think that Oglesby is right; I need more sleep, or at least some sleep. Practiced steep figure 8's at 1,000-ft. altitude using Stephan's round barn and Sachleben's house for the turns. Making steep figure 8's is the same as gentle figure 8's except the stick is moved further to the left or right of neutral, the left or right rudder pedal pushed further forward and the engine speed is increased. I always hate it when it's time to come down. Turns from 49 to 90 degrees are quite different; the rudder will take the place of the elevators and the elevators will act as the rudder. Between 45 degrees and 90 degrees the stick controls the direction of the ship and the rudder pedals control the elevation. It takes a lot of practice as the bank and speed of the turn is always different, even the right turn and left turn is different due to the engine torque. If too much rudder is used in proportion to the ailerons the ship will skid. If too much aileron is used in proportion to the rudder the ship will side slip downward. Joe will probably get into that the next time he goes up with me. We hauled Joe's Waco's tail assembly, horizontal adjustable stabilizer, vertical stabilizer, elevators, and rudder to the second floor above Oscar Bear's Jewelry Store, 206-208 East Main Street, which we use to repair tail assemblies and ailerons until the new hanger is completed[15].

---

[15]   After Oscar Sr. and Oscar Jr. retired, Manpower Temporary Services has occupied the first floor. After we quit using the second floor John Dickson used it as tailor shop, following him Claire Tingle used it as a seamstress shop.

We had to take the fabric off the tail assembly as it had deteriorated, lost it's life, and wouldn't pass government inspection which is coming up pretty soon. It was quite a job, as we had to cut all the lacing, remove the fabric, and then remove the cloth tape that had been wrapped around the metal frame. We sanded the frame smooth and painted it with "dope" proof paint and let it dry for over an hour so the covering fabric won't stick to it. The fuselage, and tail assembly, is usually a framework of welded steel tubing. We wrapped cotton strips around the tubing that comes in contact with the cover fabric and brushed one coat of "dope" on the strips, opened all the windows and doors, and went down the stairs and had a Coke at Inglis' Drugstore, southwest corner of Main and Mulberry.

After about an hour, giving the dope a chance to dry, we went back up and put the covering fabric on the rudder and elevators, lacing the fabric to the cotton strips we had wrapped around the frame tubing. This was a slow, tedious job, every stitch had to be perfect. After Harvey inspected every stitch and gave his OK, we were ready to paint them with dope. He also checked to see that the covering fabric was snug without wrinkles; but not so tight that it will prevent the dope from doing its job.

"Sea Island Cotton" is the recommended fabric to cover the plane, by two methods, blanket and the envelope method at 90 degrees. "Sea Island Cotton is used as reinforcing tape and finishing tape. The lacing cord is cotton or heavy linen thread. Heavy linen thread is used with a commercial sewing machine when making the blanket or envelope.

Dope is a solution of cellulose acetate or cellulose nitrate, dissolved in solutions such as acetyl, acetone ether, or naphtha. It makes the fabric-covering shrink and become taut, air tight, weatherproof, and leaves a hard smooth surface.

We painted the "dope" on the rudder and elevators as fast as we could as it dries fast. When we were about half finished we started to act silly and by the time we finished the first coat we were literally very drunk from inhaling the fumes. I don't know how we got down the steps or how we got home I am awful sick. We worked until 9:30pm.

November 17, 1931 – Tuesday

It rained all night. The weather was awful.

No flying so I hauled two truckloads of sand and gravel to the field. Harvey and I drove the International truck to Milan, Indiana and picked up a "Cloud Coupe" monoplane. Al gave me 2 wing ribs as a souvenir. It rained "cats and dogs" all the way home. We got home 8:30pm and I am still sick. Everyone is still talking about the bank robbery. I got a letter from Bert and Herbert.

November 18, 1931 – Wednesday

The weather was cloudy, cool, and the wind was calm for a change. It's strange, but the weather and wind always seems to calm down around sundown, even the sun tries to come out no matter how cloudy it has been all day. It sure was a very fine day today.

I flew for an hour today, practicing spirals in one direction from 2,000 ft. attitude, with the engine throttled (idling). I banked left, lowered the nose enough so the plane won't stall and circled the field gliding toward the runway and landed in a normal landing attitude, wings level, the wheels and tail skid touching the ground at the same time within 200 ft. of a line. I repeated the spiral, banking to the right. While spiraling with the engine throttled, each one of the five cylinders kept whispering 1-3-5-2-4, 1-3-5-2-4 in perfect rhythm and was sweet music to my ears knowing that if I had to suddenly open the throttle they would immediately respond with a loud cackle. This maneuver will be part of the flying exam and is to show that you are capable of landing a plane safely if the engine fails. I did fine.

We went back above Bear's Jewelry Store, and put finishing tape over the lacing on the rudder and elevators, "doped" the tape and got out in a hurry. We are going to bring some large fans when we put on the other three coats of dope. If the dope fumes get strong enough there is a good chance of an explosion so we don't smoke or turn light switches on or off. This dope is dangerous stuff.

I hauled two truckloads of sand and gravel to the field and picked up the wood and nails the carpenters left scattered around the hangar. Then I built three fires in the hangar to dry the place out.

Got a nice letter from Fred and wrote Mom & Herb a letter. Bought a heater for my car; an $11.00 heater for $1.75. It fits around the exhaust manifold and uses the heat from the exhaust to warm the inside of the car.

November 19, 1931 – Thursday

The weather was miserable, very foggy and it rained hard all day. No flying.

We went back to the second floor of Bear's Jewelry Store with three large fans, and opened all the windows. We put the covering fabric on the horizontal and vertical stabilizer, lacing the fabric to the cotton strips. We painted the stabilizers with dope, after turning on the fans, we could still smell the fumes but the fans did help a lot and we went down the stairs to wait for the dope to dry. After about an hour we went back up and put the finishing tape over the lacing and painted dope on the tape, left until the fumes had disappeared, returned and closed the windows, doors, and shut off the fans for the day.

I took a truckload of sand to the airport by myself. I had it pretty easy today, polished my car and of course it had to rain (polished it in the Maddox's garage) and am getting it in shape for the winter, putting weather stripping in all of the windows. I loaned Harvey my car so he could go on a date tonight. I found out tonight that Weber has been married for two months; it's a secret but I have my suspicions. I wrote Fred and Pyatt a letter.

November 20, 1931 – Friday

The weather was cold with some rain. The wind was like a violent hurricane. No flying.

A new student came in today. His name is Bob Parker, Box 362, Galax, Virginia.

I put pieces of tin over the knotholes on the hangar wall and then I built a fire in the office. We went back to Bear's Jewelry Store and sanded the elevators, stabilizers and rudder smooth and gave them the second coat of dope.

I tried to put the heater on my car but a bolt broke so I will try to put it on tomorrow. I went to a basketball game and saw Madison High beat Hanover High 26 to 19.

The hanger is going to be very nice when completed, an office, a large workshop, and room to store six airplanes. There are 25 windows. It is 60 ft. wide and 88 ft. long, and 12 ft. high with a 1/2 round roof.

I am still hauling gravel for the apron in front of the hanger. They graded the runway but it still will be grass and with a lot of rain we won't be able to take off or land. But the north - south runway is so short it will have to be perfect flying conditions to be able to use it; strong headwinds and dry ground.

10-23 *1931*

Plan Huge Hangar.

The new building for the Dixie flying service airport on the Michigan road will measure 60 by 88 feet and will be 12 feet high. Sufficient space is t obe provided for 6 airplanes. About twenty-five windows will provide plenty of light. The hangar is to be built at the edge of the Michigan road and a runway will be constructed for planes to taxi to and from the point of take off at the field.

November 21, 1931 – Saturday

The weather was cloudy, cool, and the wind was pretty mild. There was no flying.

The "Prep" didn't pass the Commerce Inspector's inspection. I don't know what the problem was. I finished putting the pieces of tin over the knotholes on the hangar wall. I wish I could get in some flying time.

We went back to Bear's Jewelry Store, sanded the stabilizers, elevators and rudder smooth, tinted the "dope" with a color Joe wants the finishing coat of spar varnish to be, and applied a third coat of dope. We left the windows open and the fans running and went down to Inglis's Drug Store for a Coke to kill an hour to give the dope a chance to dry.

While there I flirted with a very pretty girl, Nellie Harrod, who worked behind the soda fountain. There is one big advantage to being a flying student, if nothing else, and that is plenty of pretty girls to date. They seem attracted to the flying school guys and the local boys resent it.

We finally had to go back up, sand the tail assembly and give it the fourth coat of "dope". You can still smell the dope and it makes you kind of silly. They have to figure out a way to get rid of the fumes. I have had a headache since we started using the dope. Now after this coat dries we have to paint the tail assembly with a finish coat of spar varnish.

Polished the Cheve for the big day tomorrow. I paid my room and board for the week, $6.00.

November 22, 1931 – Sunday

What a day! The weather was warm, the sky's clear, and the wind was calm. I flew for one hour today.

I had great take offs, figure 8's and spirals. All of a sudden I feel comfortable with the ship, I get it started and climb in without a thought, just as though I was getting in my car. Not to brag, I made three perfect landings, sat the ship down 2 in. off the ground, now if I can just do it every time. I haven't had any problems taking off lately.

I went to Paris Crossing and picked up R.D. at 1:00pm and took her out to the airport. I introduced her to Joe and talked him into taking her up for a plane ride. As a student without a license, the only one who can go up with me is the instructor. I might have made a mistake, as she seemed to take a liking to Joe and he seemed attracted to her; Joe is single.

Following the plane ride we bummed around town until suppertime only to find out that she wasn't hungry so we went to the movies and saw "Penrod and Sam" and enjoyed it. We drove back to her house and sat around till 11:00pm. I did not lose any time driving home. This was the first time I had a single date with R.D. and it sure was great. Oh boy! Show $0.50 and gas $1.40.

November 23, 1931 – Monday

The weather was warm, clear and the wind was a strong cross wind. Flew for 30 minutes today.

The strong crosswind made the takeoff and landing more difficult. Instead of taking off straight down the center of the runway as usual I kinda "crabbed" a little into the wind, by pushing the right rudder pedals slightly and moving the stick slightly to the right of neutral to compensate for the crosswind. The takeoffs were ok. This was the first time I had to take off in such a strong crosswind and it was a good lesson, as I will have to takeoff and land in many different situations and conditions.

I practiced 180-degree turn and land with the engine throttled. Going away from the field with the wind, at 1,000 ft. altitude, I throttled the engine, started the glide and made a gentle 180-degree turn to come back to the field into the wind for a normal 3-point landing. I overshot the field four of the five times and had to push the throttle open, go around and try again. The crosswind may have had something to do with it but I have to learn to handle it. This maneuver is to show that a pilot can bring the ship down and land at a specific point without the use of the engine.

Coming in to land for the last time, I noticed Mr. Sachleben, who helped build the hangar and is the owner of the property, was working on the roof so I flew real low and slow over the hanger and waved at him. He must have thought I was going to crash into him because instead of waving he jumped off the roof, which is 15 or 20 ft. above the ground. He is in his late 40's, real tall, and thin with long arms and legs. With his arms and legs waving around he looked like a giant spider falling and I laughed so hard I almost forgot to push the throttle forward and go around again and land. After landing and parking the plane I went over to see if he was OK, he was but I got a real good bawling out until I convinced him I was only trying to be nice and wave at him and that I would never do anything like that again. After things calmed down I painted part of the hanger, then went out on several of the highways and put up signs giving directions on how to find the airport. Got a letter from Mom with good news. She is coming down Wednesday or Thursday for the dedication of the airport and hanger on Thanksgiving Day, November 26, 1931.

Another new student came in today, Andy G. Nufer, Jr. R.R. 3, Winamac, Indiana.

To Dedicate New Hangar on Thanksgiving Day

Several hundred cars visited the Madison airport yesterday to inspect the new hangar of the Dixie flying service. At 4:30 p. m. Pilot Joe D. Vail gave an exhibition of stunt flying for the amusement of the crowd.

The date of the dedication of the new hangar has been set for Thanksgiving day, at 2:00 p. m. A member of the dedication committee said today that admittance will be free and a large crowd of persons interested in aviation is expected.

The full announcement and program appears elsewhere in this paper today.

Four new students have enrolled since last Friday at the Dixie Flying Service, all of whom are to begin receiving instructions as soon as they arrive here.

The new students are: Al Thockle, Iowa; George Dykstra, Michigan; John Zelinz, Jr., of Pennsylvania and Clair L. Kirsch, of Ohio.

The Dixie Flying Service now owns eight planes, most of which are in flying condition. Students enrolled have a choice of several different makes, styles and designs of planes. A mechanic who keeps the planes in perfect condition at all times, is employed at the airport and after Thanksgiving the planes will be stored in the new hangar where office quarters and a repair department will be maintained.

The new hangar has been erected on the Sachleben farm, along state road 29, only a few miles north of this city.

November 24, 1931 – Tuesday

Today the weather was rotten. It rained all day, very cold, and the wind was extremely wild. No flying as the field is a lake.

I put up some more signs saying "To Dixie Flying Service at Madison Airport" and passed out circulars saying "Madison Airport Dedication Program Thanksgiving Day, Thursday, November 26, 1931 at 2:00pm, rain or shine", in Scottsburg, North Vernon, Osgood and Versailles.

Passing through Paris Crossing I threw a handful in R.D.'s yard. I am beginning to get familiar with this area, both in the air and on the ground and I have picked out several nice fields to use in case of an emergency landing. You should always have one in sight while flying. We sanded the tail assembly and gave it a coat of spar varnish. The fumes are not near as bad as the "dope". The assembly looks great and as soon as it dries we will have to put the elevators, stabilizers, and rudder back on Joe's Waco as he is going to fly it on Thanksgiving.

I put the heater on the Cheve but broke off another stud and will have to replace it. They are so corroded that they break off instead of unscrewing. I had my first flat tire today. Before leaving home I put four new tires and tubes on the car - $24.00.

I was sound asleep when Barber's phone rang, she called me and said my sister Bert and her husband Martin just got in town and as it was late they didn't have any place to stay. I got up, dressed, and got them a room at the Madison hotel.

November 25, 1931 – Wednesday

The weather was very cold, clear, and the wind was extremely rough. No flying today.

We put the tail assembly on Joe's Waco and it sure looks great so he is ready to fly her tomorrow. We still have planes and parts of planes we are working on at Maddox's garage.

A Curtiss-Wright Jr., parasol monoplane, flew in today for tomorrow's dedication and I took several pictures of it.

Front View of Major Bennett's Curtiss Wright Jr.
with Bob Parker & Andy Nufer
Joe Vail's Hisso Waco & Prop in rear

Rear view of Major Bennett's Plane

It is a queer looking craft with the radial air-cooled engine mounted on the rear of the wing behind and above the open cockpit. I can imagine what would happen if you nosed it into the ground. It is also called a "pusher" type, and seems to fly nice, lands nice and slow.

I took Bert and Martin to Maddox's garage and up to the airport to show them the planes in the new hangar. Then Mom and Ralph came into town about 2:00pm and we were sure glad to see them. They also got a room at the Madison hotel. I took them all sightseeing to Hanover College on the hilltop about 3 mi. west of Madison, and to the Cragmont State Hospital for the Insane. They both are beautiful places that I haven't been able to see either one from the air yet; practice, practice, practice. We then went for a ride out in the country and just happened to go through Paris Crossing and saw the girl friend out in the yard.

I will never forget the first time I met Martin. It was a long time ago and I thought it was really odd as he was wearing a suit and fancy looking gloves on a very warm day. While shaking hands with him I got a shock and found why he looked odd; his hands were hard as rocks. He had both arms cut off working for the Pennsylvania railroad in Pittsburgh. He transferred to Wellsburg, West Virginia and then to Steubenville, Ohio, so the railroad gave him a lifetime job. He was wearing the coat to hide the harness that went over his shoulders and his arms.

It was amazing what he could do, feed himself using forks and knives, can drink coffee etc. moving his thumb and first finger, and many other things such as shaking hands. The harness controlled his arms and hands by the movement of his shoulders, etc, and was very complicated and expensive. I had a chance to carry his harness and arms and it is hard to imagine that he could wear them all day. I guess they would weigh about 30 lbs. The gloves wore out pretty fast and the railroad replaced them regularly. I never heard him complain, gripe, whine, or feel sorry for himself over his misfortune and that had a lasting impression on me.

November 26, 1931 - Thursday - Thanksgiving Day

The weather was very cold, about 30 degrees, and the wind very rough. No flying as we had to get everything ready for the dedication.

I took Bert and Martin to the railroad station down on First Street between Vine and Mill Street at 8:30am. They get to use the railroad for free.

I hurried out to the airport to paint "Dixie Flying Service, Inc. Madison Airport" in real large letters on the front of the hanger. After getting everything ready for the dedication of the airport and hanger at 2:00pm, all the ships cleaned and lined up in a row we went downtown for dinner.

After dinner I got cleaned up and picked up Mom and Ralph and took them out to the airport to see the dedication. By the time the dedication began at 2:00pm there was a large crowd and over 200 cars; not bad considering the cold weather. I helped people find parking places, and then joined Mom and Ralph to watch the dedication.

MADISON AIRPORT
DEDICATION PROGRAM

2 P. M. Thanksgiving Day,

THURSDAY, NOV. 26, 1931

Honorable Frank J. Pritchard, Master of Ceremonies

Invocation — — —	Rev. J. C. Black
Music — — —	High School Band
Music — — — —	Drum Corps
Speaker — — —	Curtis Marshall
Music — — —	High School Band
Introduction — —	Of Proprietor's
Music — — — —	Drum Corps
Benediction — — —	Rev. J. C. Black

A Demonstration of stunt Flying by Mr. Joe D. Vail
Parachute Jump at 3:30 P. M. Weather Permitting
Passengers at $1.00 per person Thanksgiving Day only
—Morning and Afternoon—
Program to be held at the MADISON AIRPORT on the John
Sachleben farm, Thanksgiving Day, Nov. 26, starting at
2.00 P. M. regardless of weather.

On State Road 29

Every One Is Invited To Attend

A Dr. R.L. Crompton flew in from Osgood to see the airport dedication.

Joe put on a great exhibition of stunt flying, barrel rolls, figure 8's, loops, tail spins, whipstalls, and finally spiraled in for a great landing. The fellow never showed up to parachute out of the plane. Joe was busy the rest of day taking two passengers at $1.00 each up for a ride over the Ohio River and Clifty Falls Park. As his plane has open cockpits he loaned each passenger a pair of goggles. The pilot sits in the rear cockpit. After the crowd left we closed the airport and then went downtown.

It started snowing after supper, and a lot of hail the size of marbles fell, so we spent the evening sitting around talking. It was a real Thanksgiving, except "Pop" wasn't here, and I feel pretty blue. This is the first time I felt this way. Up till just now I've been very excited and happy having a great time.

No date on Sunday, oh my!!

November 27, 1931 – Friday

The weather was fine, but cloudy, cold, and the wind quite quiet. No flying as the field is too wet.

I drove the truck to the Madison Coal Company at the corner of Jefferson and Front Street, got a load of coal, and took it out to the hangar. Hauled three truckloads of airplane parts from Maddox garage to the airport. I just found out it's pilot was killed in the "Buhl Pup" that Harvey brought from Dayton, Ohio. It sure is an awful mess. I didn't think the plane could be smashed up like that one is.

Mom and Ralph took off for home about 8:00am.

I finally got the heater put on the Cheve permanently and cleaned part of the engine, in doing so broke a spark plug.

I stayed home tonight and wrote Earl Seitz & Harry a letter, also sent Dad a birthday card, only a week late, his birthday is November 21st. Finally catching up on my sleep.

November 28, 1931 – Saturday

The weather was awful, cold, rain, fog, but the wind was quiet. No flying as it is too muddy and wet.

Went down to Maddox garage and worked at taking the "Buhl Pup" apart. It is a small one-seater low wing monoplane with an air-cooled engine, covered with aluminum sheets riveted together instead of fabric. What a mess and what a smell, they didn't clean out the cockpit. Each rivet had to be chiseled off and there must be millions of them, or so it seems, as I worked on it all morning and can't see much progress. It's a cinch that I'll never stall a plane at 150 ft. like that pilot did. I'm going to come in at a steeper angle from now on.

I put on a dash control for the heater on the Cheve and fixed the flat tire, had to take the tire off the rim and put a patch on the inner tube. A nail had gone through the tire and put a hole in the inner tube.

After a swell supper as usual, Bob Parker, Andy Nufer, Harvey and I decided to study up on general aviation every Tuesday night and I think that's a swell idea as I am pretty dumb on the subject. Although Harvey is a licensed A & E mechanic (airframe and engines) he also lacks the knowledge.

Ed Becht, Bob Shucks, and I walked around town tonight looking for some excitement, and found none as usual so we settled for looking at the pretty girls as they walked by. I am down in the dumps as I don't have a date tomorrow.

Clyde H. Beyer

November 29,1931 – Sunday

The weather is bad, very cloudy, pretty cool, rained all day, but the wind is pretty quiet. The visibility was only about 50 ft. in the air, and a ceiling of about 200 feet. There was no flying so I didn't go out to the field today.

I was on my good behavior today. I got up early, had a great breakfast, washed, shaved, got my Sunday pants from beneath the mattress, put on my shirt, tie, and coat and went to Sunday school at the church on the northeast corner of Main & East Street.[16] During Sunday school the Sachleben brothers; Carl, Fred and Joe talked me into coming back to a play practice at the church after dinner. Following play practice I went out to the Sachleben's at the airport, and had a fine supper with them. They all treat me as if I am one of them, I think they have adopted me, and as I have no relatives here it makes me feel good. It sure was a dull day with nothing to do. Gee I wish I had a date, oh me! I hope I go to the country next Sunday and I'm dying to get some flying time.

Following supper at the Sachleben's we all went back to church for the evening services and I got home at 9:00pm.

I loaned Harvey my car tonight so he could go on a date. As Joe and I are the only ones who have a car, and no one borrows Joe's car. I bought three bars of Lifebuoy soap $0.25.

November 30, 1931 – Monday

The weather was cloudy and cold, and the wind was pretty rough, so as usual no flying, so I went back to chiseling the "Pup's" cockpit apart. I sure could use a mask. Bob Shuck and Ed Becht helped me finish taking it apart. I took it easy the rest of the day as Mr. Oglesby was out of town.

Ed Becht painted the D.F.S insignia on the back of my flying coat for $0.50. He is a great painter and Hawaiian guitar player. He's going to teach me how to play it.

I worked on the vibrator horn on the Chevy and now it sounds worse than ever.

I got my pictures today, 2 of Joe with Bob Parker and Andy Nufer, side and rear view of a Curtis Wright Jr. plane, and the picture of a "Prep" and two "Wacos" we got from Milan, Indiana. I paid 40c for the pictures.

---

[16]   It is now the United Church of Christ.

66

I went out to the Sachleben's after supper and had a great time playing euchre. I call Mrs. Sachleben "Mom" and Mr. Sachleben "Pop". Carl, Fred, and Joe introduce me as their "little brother".

After playing several games, Mr. Sachleben opened a bottle and gave each one of us a glass of "home brew"(homemade beer). He gave me a taste of "beer wine" which he made from the sediment at the bottom of the beer bottles. It had a sweet taste and I bet it wouldn't take much to make you tipsy. The brothers argued all evening over anything and everything.

When I got home I wrote R.D. a short note and a picture of Joe. Got a letter from Herb today.

December 1, 1931 – Tuesday

George Dykstra, Holland, MI

The weather was fair, cold, and the wind was a trifle rough.

I finally got to fly for 55 minutes, practicing 180-degree turns and land. I am still having the same problem I had the last time I flew, over shooting the field four of the five times I tried to land. The field was awful muddy but I didn't have too much trouble taking off although it took longer to take off as the mud slowed the plane down. I have an advantage over the other flyers; the plane doesn't have to lift as much as I only weight 140 lb. I am the only one able to take off from the north-south short runway when it is muddy or there it is no head wind. The ship was covered with mud so that means a wash and wax job.

Two new students came in today, **George Dykstra**, 304 West 15th Street, Holland, Michigan, and Max C.F. Sittner, Hudson, Kansas. I took George to the doctor in Seymour for his physical exam, and while there I was told I had to be checked by the eye doctor. This is a fine time to tell me as I have been taking flying lessons and getting ready to take the "private pilots test". You are not allowed to wear eyeglasses.

On the way home from Seymour, just east of Scottsburg, Weber passed us in a "Roosevelt", we chased him all the way home in George's old Dodge touring car, but could not catch him. Weber was coming home from his wedding trip.

I changed the oil in the Cheve $0.80 and bought a gallon of alcohol for the radiator. I got a letter from Mom and Bert.

After a great supper, Andy, Bob, Harvey and I drove out to the new hangar to learn about aviation in general, and what we will have to do to pass the L.C. pilot test. Harvey was the teacher and uses the Dayton School of Aviation Home Study Course as a guide.

Tonight's subject was *"Clouds"*.

"Clouds".

Cirrus- thin and wispy, composed of ice particles, height 25,000 to 30,000 ft.

Cumulus - detached fleecy clouds, lumps like cotton, wool or soapsuds with flat base, composed of water particles up to 25,000 ft.

Stratus - low flat, spread out clouds, resembling fog,but not resting on the ground, often merging into "Nimbus", height 4,000 ft to 6,000 ft.

Nimbus — thick, extensive layers of formless clouds from which rain or snow is falling.

Cirro Stratus- a high thin whitish cloud -height 25,000 to 30,000 ft.

Cirro Culumus - small flakes arranged in groups or lines, height 20,000 to 25,000 ft.

Alto Cumulus - rather large white masses or groups often spread into lines, height 15,000-20,000 feet.

Strato Cumulus - Dark twisted clouds with blue sky showing through, 5,000 to 10,000 feet.

Cumulo Nimbus - Similar to Cumulus but have "false cirrus" above and "Nimbus" beneath, height from 500-1,500 ft.

Alto Stratus — A thick gray cloud, curtain-like, that shows a bright spot shere sun or moon shines, height 10,000 to 15,000 ft.

This was a very interesting lesson and I have a lot of studying to do, as I know less than nothing about clouds and never realized the importance of this knowledge. We finally turned out the light, went back to town, and it wasn't long before I hit the sack.

December 2, 1931 – Wednesday

The weather was cold, practically no wind, no clouds, and the sky clear as crystal with the sun shining bright. What a wonderful day for flying but Joe had to get the "Prep" inspected again; this time the Government granted us a license. We tried to wash the "Prep" but as soon as we put the water on it froze. So we tore down the old tool shed instead.

After dinner we took the "Buhl Pup" wing apart and removed all the broken ribs. I kept a piece of the rib as a souvenir and am getting quite a collection. The "Pup" is a pile of junk and I can't imagine it ever flying again, but we are going to try to repair it.

After supper we "chiveried" Weber and his wife. That is, we took them up and downtown in the wrecker making all the noise we could. When the party broke up I went home and studied "clouds" again until I fell asleep. Sure had a swell time. Got three prints of Joe today -15c.

December 3, 1931 – Thursday

The weather was fine, cold, and the wind smooth.

I flew an hour today, practicing "360" turns and landings. Starting into the wind at the edge of the field, at an altitude, I throttled the engine (bringing it back to idle) made a sharp turn, going away from the field watching my flying speed, the strength of the wind and the gliding distance to the field. Gradually making a second turn so the ship was in the proper direction to come in to land. I made a straight glide for the runway and only overshot the field twice.

This is another maneuver that takes a lot of practice and experience to be successful, and is also part of the test. If the glide is short of the field never try to stretch it and lose flying speed and stall, instead push the throttle forward to be sure of landing with enough flying speed. If the glide is going to be too long or overshoot, the experienced pilot will "sideslip" his ship, losing altitude without gaining speed. I was warned not to try this until I practiced at a safe altitude; instead I pushed the throttle forward, gained flying speed, went around and tried it again. Sideslipping is not permitted during the flying test. After taking the test Joe is going to teach me how to "sideslip" the ship. This very important maneuver can save your life in an emergency.

The runway was frozen and made landing rather rough. I am learning to judge height, distance, and speed better every day.

We're moving out of the Maddox garage, and made several trips, with the truck loaded with parts of planes to the hanger. What a mess of stuff.

I had a fuss with Oglesby. He wanted me to sleep at the hanger as sort of a night watchman. I agreed to do it if he would pay my room and board. He wouldn't do it, so I wouldn't. Then he got mad and so did I.

I am down in the dumps. I got a card from Emma, R's sister, saying that R had to stay with her sick brother-in-law. Darn in-laws anyway! If you believe that, stand on your head. Well I guess that means I won't be doing anything on Sunday. Such is life. Wrote Herb a letter.

December 4, 1931 – Friday

The weather was cold, very foggy, and the wind was very rough, so no flying as the field was awful muddy.

David Swett & Rodney Pert

A new fellow student came in today, **Rodney Pert**, Bluehill, Maine.

I spent the day hauling plane parts, pieces of planes, and a mess of stuff from Maddox garage to the hanger. Still down in the "dumps". Wrote

Mom and Emma a letter. Went to a play practice at church with Ed Becht and Bob Shucks tonight and had a swell time.

December 5, 1931 – Saturday

The weather was cold, cloudy and the wind a little rough.

I flew for 30 minutes practicing spirals, 180 and 360-degree turns and lands, landing on the old field without "over shooting". I tried but couldn't find Paris Crossing from the air. It is too cold to stay up long. Today for the first time I "propped" the "prep" and it started. Cleaned the office and hauled ashes for the apron. We put a couple of wings up in the hanger rafters.

Loafed around the streets all evening with Becht, Shucks, Joe and Carl Sachleben. Got a letter from Earl Seitz and Harvey Pyatt.

December 6, 1931 – Sunday

The weather was warm, clear, and the wind was pretty darn rough. No flying today.

After a great breakfast, I washed, shaved, put on my only Sunday best and went to Sunday school. After dinner I went back to church and took part in a play practice. After supper, Ed Becht, Bob Shucks and I went for a ride out in the country and on the way home went through Paris Crossing. We had a fine time though I miss the girl friend a lot. Oh me! Maybe she doesn't like me and this is just a stall, so if she is foolin' with me, look out! Oh well! Flying is my first love anyway.

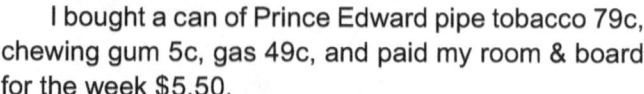

I bought a can of Prince Edward pipe tobacco 79c, chewing gum 5c, gas 49c, and paid my room & board for the week $5.50.

December 7, 1931 – Monday

The weather was fine, cold, clear, and the wind was pretty smooth. No flying again today. Two new students came in today, **Frank J. Peternell**, Box 185, Meadowlands, Washington Co., Pennsylvania., and **Lawrence Burklow**, 920 West Center Street, Park Ridge, Illinois, and I took them to Seymour for their physical examinations. While there I didn't find out anything new about my eye and so it must not be too serious or they wouldn't let me fly. On the road to Seymour a tire went flat.

Lawrence
Burklow

We got back to the airport in time to help the electricians put wiring and lights in the hanger. I don't know the first thing about wiring or lights but it sure is interesting work.

Fred loaned me his guitar and Ed Becht is going to teach me how to play it. Shucks gave me finger picks and a met. I bought a thumb pick and second string for the guitar for $0.20. Stayed home tonight and studied the lessons on clouds we had last Tuesday night.

Frank J. Peternell

December 8, 1931 – Tuesday

The weather was real cold, slightly foggy, and the wind blew a miniature gale. Flew for 30 minutes. Made one good and three "punk" landings. The "Prep" glided like a brick and kept me quite busy, but I liked it when bucking the wind the ship practically stood still and gained 1,000 feet altitude in about 50 ft., almost like an elevator.

After gaining plenty of altitude, I slowly brought the throttle back until the ship was not moving forward at all and still had plenty of flying speed. After hovering over the hangar for a couple minutes I wondered what would happen if I brought the throttle back a little more, I did and the ship was actually flying backward. It was strange, as though the ship was standing still and the ground was moving in front of the ship. The ship had plenty of lift and was air worthy as the head wind was strong enough with thrust of the prop, to safely support the plane. The weaker the head winds the more the throttle and prop speed was necessary to create the proper amount of airflow over and under the wing to keep it flying.

It sure was fun, but created a new problem I never had before. The wind was so strong when gliding in for a landing that I had to use a little throttle to make any headway; the ship just didn't want to settle down.

There was no danger of "overshooting" the field, as the forward speed was so slow that there was time to make the necessary corrections, which included keeping the wing from hitting the ground due to the gusty wind. In taking off the ship used only about ½ the usual distance of the runway, it felt like it couldn't wait to get up in the air.

Before flying I had to replace a broken tailskid with a new one. It sure pays to give the ship a thorough inspection before taking off.

We helped the electricians wire the hanger all day. Another new student came in today. No schooling tonight so Becht and I practiced on my guitar instead. I bought an oil pan gasket for the Cheve, $0.15.

December 9, 1931 – Wednesday

The weather was fine, pretty foggy, and the wind was fairly rough. It drizzled all day. The field was a sea of mud so there was no flying.

I washed the "Prep" cowling and hauled three truckloads of cinders from the Gas Company to the airport. The gas company uses coal to make gas at the corner of Walnut and First Street.

Hurrah! Finally had my eyes examined today and they are 20-20. I didn't get bawled out today so Dick (Oglesby) must have been in an extra good humor.

While waiting for supper I read the *"History of the First Engine-Driven Airplane"*, Lesson # 1 of the Dayton School of Aviation correspondence course. It was about Orville and Wilbur Wright, their lives, struggles, and successes in building the first engine driven airplane.

After supper I went on another blind date with Dave Gentrup Her last name was Lohrig and she was a pretty darn smooth girl, but I still like the one from the country a heap better. We went to the movies and then to Inglis' for cokes, following a joy ride. Gas $0.82, a movie show $0.70. I have to "pinch" every penny since Mom is paying all my living expenses.

December 10, 1931 – Thursday

The weather is cool, pretty cloudy, and the wind rather calm. The field was too wet to do any flying.

I loaded and hauled four truckloads of cinders from the Gas Company and unloaded the cinders on the runway. Also hauled a truckload of engine parts from Maddox garage to the hanger. It about broke my back. That was really a hard day's work. I got a letter from Mom and Paul.

After supper I studied *Lesson No. 2 - "Aerodynamics"*, which is simply the study of how the air affects the plane and how the plane affects the air. It explains how and why the plane flies: due to the speed of the wind over and under the wing whether by headwind or prop speed. The lift of a plane is determined by the angle of incidence of the wing, the velocity of the wing, the curvature of the wing, and the area of the wing. I now know why I flew backward the other day. Ground speed and air speed is relative, as motion is relative. If it takes 60 mph wind flowing over and under the wing to keep the plane flying and you have a head wind 30 mph and don't increase the prop spead, the actual ground speed of the plane is 30 mph. This is sure interesting and I hope I can remember it all as it is part of the written exam.

These are the questions to Lesson # 2

1. What is the meaning of the word "aerodynamics"?
2. Name the factors determining lift and state which of these is the most important?
3. What is meant by the camber of the wing?
4. How does camber affect lift?
5. Define angle of incidence.
6. What is meant by "positive stagger?" negative stagger "? And "gap"?
7. What effect does speed have upon the rising of the plane from the ground?
8. What effect has wing area upon the lift and in what proportion is area to lift?
9. Why does an airplane fly?

THE DAYTON SCHOOL OF AVIATION

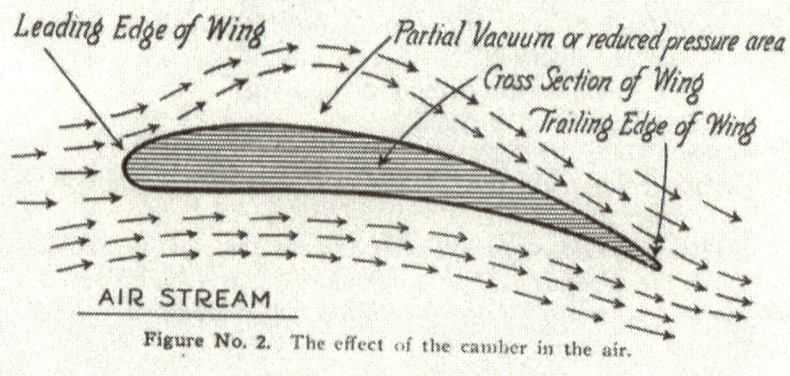

Figure No. 2. The effect of the camber in the air.

he is now greatly responsible for the lift of the airplane. With strong pressure beneath the wing, due to the speed with which the ship is being driven through the air, and with a partial vacuum or greatly reduced pressure above the wing, it is no longer hard to understand what holds the ship up and enables it to carry both passengers and freight.

Factors Determining Lift

We now understand the theory of lift, and know how the combination of engine and wings causes the plane to stay in the air. Next we are to learn of the four principal factors determining lift.

1. The angle of incidence of the wing.
2. The velocity of the wing.
3. The curvature of the wing.
4. The area of the wing.

Incidence

The angle of incidence of the wing is the acute angle between the plane of the wind chord and the line of thrust. The wing chord is the distance from the leading edge to the trailing edge of a wing and is commonly determined by the line of a straight edge touching the lower surface of the

8

THE DAYTON SCHOOL OF AVIATION

wings at two points. The line of thrust is usually the longitudinal axis of the plane or a line parallel to it. While these terms may seem somewhat hard to and confusing to you at this time, they will be fully explained and mentioned many times in the lessons to follow. For the purpose of this lesson you may simply consider that the angle of incidence is the number of degrees that the front or leading edge of the wing is elevated above the rear or trailing edge. This is not technically correct, but is all that you need to know or remember now.

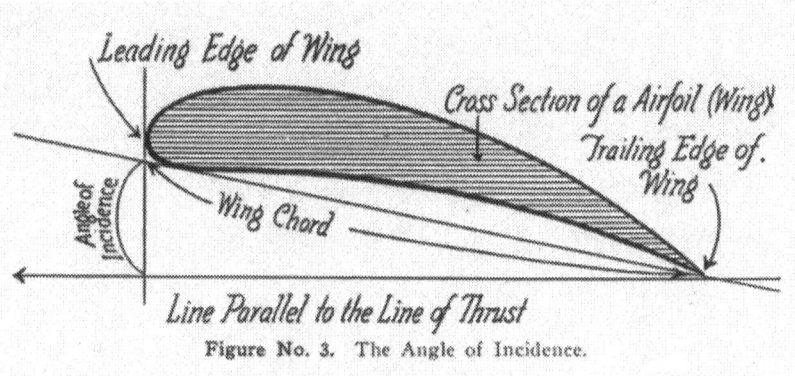

Figure No. 3. The Angle of Incidence.

Generally the angle of incidence is never more than 2 or 3 degrees, that is, the entering edge of the wing is seldom elevated more than 2 or 3 degrees above the trailing edge when the wing is put on the airplane.

Velocity of the Wing

The "thrust" or speed with which the wing is forced through the air is the principal factor affecting lift – just as the speed with which you run, holding the cord of your kite is the principal if its lift, if there is no wind blowing. And here again we must remember the first thing that we learned about Aerodynamics, namely, that motion is relative. It is just the same to your kite whether you do the

9

THE DAYTON SCHOOL OF AVIATION

work against the advantage of lift and cause a disturbed area behind the trailing edge of the wing, called drag or drift. This drag is the enemy to thrust or the speed forward just as lift is the enemy of weight or gravity.

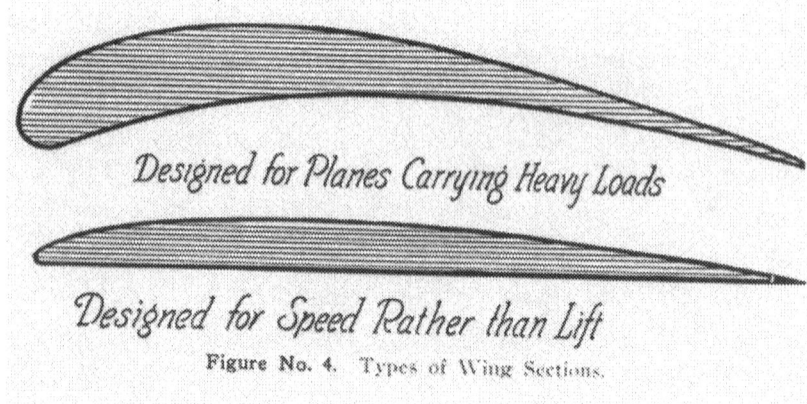

Designed for Planes Carrying Heavy Loads

Designed for Speed Rather than Lift

Figure No. 4. Types of Wing Sections.

The Area of the Wing

Of course you can readily appreciate that the area of the wing, (the span times the chord) as well as the camber, has much to do with its lifting force. For all practical purposes, we may say that the lift generated by the wing is in exact proportion to its area. That is, if we wish to double our lift, then we must double our area, and so on.

The Wings of a Biplane

A biplane is an airplane with two sets of wings, one above the other. The distance between the two sets of wings is known as the "gap".

Now the gap must be sufficiently great to keep the action of the air over the surface of the lower wing from interfering with the flow of air under the surface of the upper wing, and vice versa. Sometimes this is controlled by placing the upper wing in front of the lower wing, which is known as "positive stagger" of the wings. And in other

12

THE DAYTON SCHOOL OF AVIATION

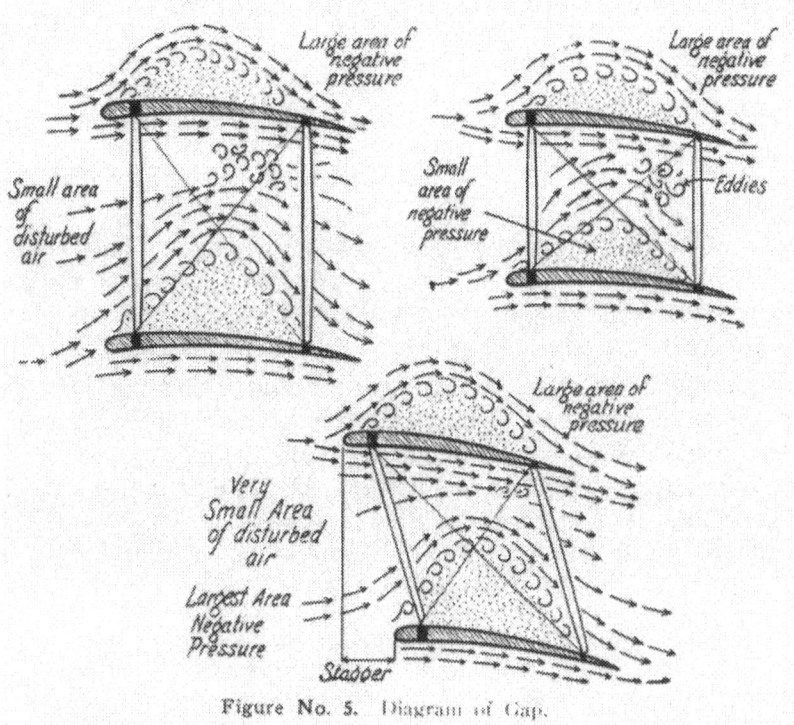

Figure No. 5. Diagram of Gap.

cases the lower wing is placed ahead of the upper, then it is called "negative stagger".

The principle of lift is exactly the same as in the case of a plane with but one set of wings, (a monoplane). The only new problem presented is that of correctly placing the wings so that the air on one set will not interfere with the action of the air on the other set. And we will discuss this in much greater detail when we begin to study about airplane rigging.

There are other characteristics of the airfoil (wing) besides those described above, which have considerable bearing on its effectiveness, but I am going to leave them for your next lesson. I believe in getting just one big idea at

13

Becht and Shucks came over and we played checkers the rest of the evening. Not to brag, I won but that is because I was taught by a pro when I was going to high school.

December 11, 1931 – Friday

The weather was warm, cloudy, raining and the wind was very rough.

No flying so I hauled three truckloads of cinders and spread them on the runway. I am almost dead.

I spent the rest of the day studying "Aerodynamics" and Lesson 3. This lesson studies all parts of the plane that sets up resistance when driven through the air, at high-speed, and the fuselage is the worst. Things affecting resistance are, the square of the speed, the size increases resistance, the density of the air increases resistance, and most important, the shape of the object. Streamline forms modeled after the birds and fish offer much less resistance than any other shapes.

This lesson also explains the importance of "The Center of Gravity, Center of Pressure, and the Three Axis of an Airplane".

These are the questions to Lesson #3

1. *What are the main factors affecting the resistance of an airplane? Which of these is the most important?*
2. *Why does a streamlined form offer the least resistance?*
3. *What is meant by the center of gravity?*
4. *How does the center of pressure differ from the center of gravity?*
5. *If the center pressure is too far behind the center of gravity, what will the airplane do?*
6. *Name the axis around which an airplane rotates.*
7. *What is meant by pitching and yawling?*

Harvey, George and Andy went somewhere in Kentucky to pick up a wrecked ship.

Becht, Shucks, and I went to play practice at church after supper. About this time next week I'll be starting home. I bought soles for my shoes $0.20.

December 12, 1931 – Saturday

The weather was fair, cloudy, and the wind was pretty smooth. No flying today as Joe Vail was awful sick.

One of these days I'm going to figure out how many days that I did not fly. They are hiring a transport pilot from Cincinnati, **Bud Guild**, who used to fly at Milan, to help Joe teach the students. Seems to be a swell fellow. He is short and stocky; Joe Vail is tall and slender.

Fellas having fun with Bud G.

I put up signs out on the highway warning drivers that planes will be flying low over the highway when taking off and landing. Then I loaded, hauled, and unloaded two more truckloads of cinders. I don't know what this has to do with learning to fly; it's just wearing me down.

I finished the day studying *"Aerodynamics" Lesson No. 4*. This lesson studies the three units of the plane controlling the movement around the three axis. The elevators, which control the pitching of the ship, or its rotation around the lateral axis. The rudder, which controls the yawing of the ship, or its rotation around the vertical axis. The ailerons, which controls the rolling of the ship or its rotation around the longitudinal axis.

These are the questions to Lesson No. 4

1. *Describe the horizontal and vertical stabilizers.*
2. *Why are the stabilizers used?*
3. *Name the controls of an airplane.*
4. *What is the purpose of the elevators, the rudder, the ailerons?*
5. *How does the pilot operate the elevators, the rudder, the ailerons?*

Since I have been seriously studying, I am beginning to understand why I do what I do when I am flying. Before, I just did what Joe told me to do and never knew why.

December 13, 1931 – Sunday

The weather was warm, drizzly, foggy, and the wind was smooth. I flew for 20 minutes.

I took off and when I reached the attitude of 500 ft. I was in the fog and couldn't see the ground, actually couldn't see anything so I slowly pushed the stick slightly forward hoping to see something and found myself flying over the top of some tall trees so I banked toward the runway. Instead of landing I pushed the throttle forward and climbed back up into the fog just for a thrill and almost didn't get down. The fog had dropped below 50 ft., I could barely see the tips of the wings, and didn't know whether the plane had drifted to the right or left, whether it was slowly climbing or descending as the altimeter is not accurate close to the ground. Without the horizon as a guide I had to guess the position of the plane so I tried to keep circling what I thought was the field looking for the hanger. If the plane had drifted out of the range of the field I was in trouble, if the plane was too low we were going to crash. I was beginning to sweat, so I asked the "Supreme Pilot" in the sky to get the ship safely down to earth.

The saying "flying by the seat of your pants" is true with these types of planes, there are no instruments telling you when the wings are level, when the ship is climbing or descending, or flying level. If the plane is climbing you have a tendency to lean into the back of the seat, diving you lean away from the back, banking to the left you lean right, banking to the right you lean left. A pilot with a lot of flying hours and experience is very good at it and swears by it.

Watching the gas gauge, which was 1/4 full, it was time to come down one way or another, reminding myself not to forget if we crashed to shut off the ignition switch immediately. Slowly pushing the stick forward, with the altimeter showing less than 40 ft., I finally caught a glimpse of the orange hanger. As luck would have it, it was west and to my right, making a steep right bank keeping my eyes on the orange spot, and when the hanger appeared in front of me I leveled off and could see the electric poles I had to fly over. I throttled the engine, and after crossing over the electric high lines, the ship was about 20 ft. in the air, too high to land.

By this time the fog was almost on the ground so I had no choice but to land somehow. Joe never taught me how to sideslip in landing but I have studied the maneuver several times. I had planned to practice it at a safe altitude but never had a chance. Now was the time to do it so I banked the ship to the left pushing the stick forward and to the left, at the same time pushing the right rudder pedal forward. The ship lost altitude fast so I brought the stick and rudder pedal back to neutral to put the ship in a normal glide for a landing, making sure there was no drift left in the ship when the wheels touched the ground. The plane bounced a couple of times. At least we were on the ground with no damage to the ship.

After wiping the sweat off my goggles I said a quick thank you to the "Supreme Pilot". This was the first time I didn't have to be flagged down.

A sideslip - this is the best way to lose altitude without gaining speed. When landing if altitude must be lost such as in an emergency landing in a small area. A sideslip is made with the engine throttled. In slipping it is important to make a proper bank so as not to stall and fall to the ground, and to recover properly so there's no drift left in the ship once the wheels touched the ground. It is easily noticed when the pilot has recovered from a sideslip and observes the ground a short distance ahead of the nose of the ship. A failure to correct all drifts before landing might cause a wheel to break, or "wash" out the landing gear. Sideslips occur in a very short time and very close to the ground leaving no time or room for error.

I went to Sunday school and church with the Sachlebens. After dinner I tightened the connecting rods on the Cheve so it will be in good shape for the trip home. This has been a dead day, no date again. I suppose she threw me down, sure am very blue, and will have to get over it.

I wrote Mom, Pearl, and Bert a letter. Bought oil for the Cheve $0.25.

December 14, 1931 – Monday

The weather was cold and cloudy, and the wind was strong. There was no flying today. I spent the day loading, hauling, and spreading 4 truckloads of cinders on the runway. I got an awful bad cold. I bought a pair of shoes $2.69 and a pair of socks $0.20.

Earl Seitz sent me some church magazines today and I feel sort of down in the dumps.

Yesterday a "Waco" from Michigan flew in for a visit. It is powered with a J5 Wright Whirlwind engine. It sure is a peach of a job.

Some dumb fool ran into my car yesterday and it is a good thing I didn't see him. He dented the left rear fender and body so that is another job for me to fix. I'm spending a lot of time working on the Cheve and I am gaining a lot of experience.

December 15, 1931 – Tuesday

The weather was clear, very cold, and the wind was pretty mild. Flew for 40 minutes.

Climbing up to 2,200 ft. I practiced side slipping a few times. I will be well prepared for the next emergency. I practiced climbing turns. This is the same as a normal turn except the nose of the ship is raised above the horizon. This maneuver should also be practiced many times at a safe altitude, and not done at low altitude because of the dangers of too sharp a turn, or too steep a climb, which will result in the controls becoming sloppy, a sign that the ship is about to stall. Also practiced spirals. It sure was cold so I spent the rest of the time practicing 180 degree and 360-degree turn and land, making one rotten landing right in front of Bud.

I loaded, hauled, and spread three truckloads of cinders on the runway. I'm becoming an expert at this. I still have a real bad cold.

We all went to the first regular class at the hanger after supper. Bud G. was the instructor and the subject was *Lesson #5, "Flying"*.

To be a successful pilot you must have complete confidence in yourself and the ability to fly and control your ship. The first job, an important one, is a complete inspection of the ship before taking off, but you must rely to some extent on your mechanic to keep the ship in perfect flying condition. Starting at the "prop" check the engine thoroughly, bolts, nuts, hose wires, leaks, etc., and then the landing gear, wings, fuselage, controls, stabilizers, elevators, rudder, and tail skid.

Following inspection, the engine is started with the assistance of your mechanic. Making sure the ignition switch is off, turn on the gas valve, and say "gas on, switch off" to your mechanic. He will make sure the "chocks" are in front of the wheels, then turn the "prop" around three or four times to draw gas into the cylinders of the engine. He will step back and say "contact", you turn on the ignition switch, and with one hand holding the stick back to your seat and the other hand on the throttle reply "contact".

The mechanic moves forward, braces himself, takes hold the prop and gives it a sudden pull downward; and the engine roars into life as the mechanic moves back to safety. You push the throttle forward so the engine is running at the rate of about 600-700 rpm until it is well warmed up, glancing at the oil gauge which should show 40 to 60 lbs. pressure. During this time holding the stick all the way back to raise the elevators and thus hold the tail of the ship firmly on the ground so the wheels won't climb over the chocks. If the engine is running smoothly with the throttle wide open, close the throttle, fasten your safety belt, nod to the mechanic and he will pull away the chocks.

Taxiing to the takeoff point, the engine speed should be fast enough to move the plane slowly over the ground. Ordinarily the controls will be left neutral, except the rudder that is used to steer the ship on the ground.

Checking the windsock to see which way the wind is blowing, and heading the ship into the wind, you open the throttle gradually until it is all the way open. The ship rolls forward gaining speed rapidly, move the stick forward soon as the tail leaves the ground and the horizon appears off the nose and the ship is in a flying position. Now the lift on the wing is enough to raise the ship and you are flying. Let up slightly on the stick, this will cause the elevators to rise, and the ship will climb. When reaching full flying speed, pull the stick back further to increase the angle of climb.

Reaching the altitude of 1,500 ft., put the stick in neutral and fly straight ahead, then to turn left you move the stick and the rudder pedal slightly to the left. Having made your turn, bring the stick and rudder just slightly past neutral. When the ship is pointed directly at the runway bring all controls to neutral. When reaching the point where the ship will glide safely to the runway, heading into the wind, throttle down until the engine is idling and push the stick forward so the ship will "nose down" on the proper angle. When very close to the ground, slowly pull the stick back until the wheels and tailskid are about 2 in. off the runway and as the ship loses its flying speed it will settle down in a neat "three-point landing".

These are the questions to Lesson No. 5

1. Describe in your own words, how you, as a pilot, would inspect your ship before taking off.
2. How would you start the engine?

3. *How would you make a flight?*
4. *Describe how you would make a right turn (the lesson describes a left turn).*
5. *What is considered a perfect landing?*
6. *Tell how you would go about landing your ship.*
7. *What is meant by "taxi"?*
8. *Why is altitude important?*

This lesson was for the new students as I have been through all this quite a few times.

Got a letter from Earl S. and Harry P. Bought two pairs of socks $0.20.

December 16, 1931 – Wednesday

The weather was cold, clear, and the wind was very smooth. 40 minutes of flying time.

Bud G. flew with me 10 minutes check time and it would seem as though everything I did was wrong. As an instructor he is entirely different than Joe Vail. We went up to 2,600 ft. for safety and he showed me how to stall the ship. This is the highest I have ever been in a plane. After stalling the ship a couple of times to his satisfaction, we went down and landed.

After Bud got out, I went back up to 2,600 ft. and put in 30 minutes solo flying time practicing stalls. I pulled the stick back until the ship was climbing so steeply that the critical angle was reached, and the sloppy controls due to loss of flying speed was a sign that the ship was about to stall. I immediately pushed the stick forward to drop the nose into a dive, pushed the throttle wide open, to regain sufficient flying speed and used the ailerons and rudder to keep the ship trimmed. (Sloppy controls mean moving the controls have little or no effect upon the ship). I stalled the ship two more times and had a lot of fun doing it. It was a nice feeling losing control of the ship and then being able to regain control of it. But it can be extremely dangerous and even deadly at low altitudes. I finally had to come down as I was out of time.

I loaded, hauled, and spread two truckloads of cinders on the runway, and then hauled two truckloads of office equipment from 216 East 2nd St. to the airport hangar.

I was tired, so after supper I stayed in and studied *Lesson No.6, "Airplane Construction, Materials"*.

The principal woods used in airplanes are; ash, balsa, basswood, birch, cherry, elm, maple, oak, pine, spruce, walnut, and plywood.

Four classes of steel are used in airplane construction:
- *Sheet metal, is cold-rolled mild carbon steel*
- *Bar form steel, is plain carbon steel*
- *Low carbon, steel tubing, medium carbon steel tubing, and chrome-molybdenum steel tubing*
- *Hard wire, stranded cable, and streamline wire*
- *Aluminum and Duralumin*
- *Silk, linen, and cotton are the fabrics used*
- *"Dopes", a solution of cellulose acetate or cellulose nitrate, dissolved in some solvent such as acetyl, acetone, ether, naptha, etc. used on the fabric*
- *Glue -animal, casein, & blood albumin*
- *Propellers are either made of wood micarta or aluminum alloy.*

These are the questions to Lesson #6

1. *Name five of the principal materials used in airplane construction.*
2. *How is plywood made?*
3. *What is duralumin?*
4. *What material is principally used for fuselage, wings, and tail surface covering?*
5. *For what purpose is dope used?*
6. *Of what are most modern fuselage frames constructed?*

If I can remember half of this lesson I'll be surprised.

Clyde H. Beyer

but this does not necessarily mean that all material tested for compression and tensile strength had one square inch of area. Material is calibrated and tested so that knowing the percentage or part of one square inch the unit strength may be figured.

The principal wire products used are hard wire, stranded cable and streamline wire. Hard wire is used internally for drift wires and stays, which are tension members only. Flexible stranded cable is used for controls, and non-flexible streamline wire for rigging. Streamline wires are made from (S. A. E. 1045) steel, cold drawn and heat treated; they are used externally to reduce resistance.

Wire products depend for a large part of their physical strength on cold drawing or rolling operations, and should never be welded or brazed, as these operations heat the steel beyond the critical temperature and remove the effects of cold drawing which cannot be recovered by heat treatment. Care should also be exercised in soldering cable and wire, as too long an application of hot solder will have a similar effect.

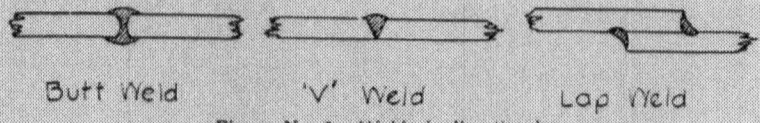

Butt Weld   'V' Weld   Lap Weld

Figure No. 3. Welds in Bar Stock.

Aluminum

Aluminum, like iron, is a natural metal ore, first smelted and then cast in molds. With the one exception of magnesium, it is the lightest of all metals used for commercial purposes. While pure sheet aluminum is used for cowling (cover for the engine compartment), gas tanks and the fairing (streamlining) of exposed parts, aluminum alloys are used far more extensively. By far the most important aluminum alloy is duralumin (called "dural," for short).

10

Figure No. 4. A Wing Spar of Duralumin under Construction, for a Ford Tri-motor All Metal Plane.

December 17, 1931 – Thursday

The weather was warm, clear, and the wind was pretty mild. I flew 30 minutes check time with Bud.

He showed me how to make vertical power turns, but they were the same turns that I have been practicing all the time.

The vertical power turn is made with the throttle wide-open, wing pointed straight down toward the ground started in a specific direction and completing it in the same direction as it is started. It is very hard to keep from losing or gaining altitude in the three complete turns. This is part of the limited commercial flying exam, and is made first to the right and then to the left.

Bud Guild got me all up in the air as he flies altogether different than Joe Vail and I insist on flying Joe's way. I'm having a terrible time learning to fly as Joe teaches me one way and Bud teaches me just the opposite. I would like to have the same instructor all the time, it sure would make it a lot easier if I would have just one bawling me out instead of two.

I have a strange feeling and I am a little suspicious that Bud is a little leery of the ship. He flew transport planes before coming here and this is a little two-seat monoplane with a five-cylinder engine that probably can't fly over 70 or 80 mph.

We went down to drop Bud off and I must have been still upset cause I made a rotten landing. The plane bounced about six feet high so I pushed the throttle wide open and flew around to try again. This time I landed but it was still a rotten landing. What a relief, to go back up for fifteen minutes solo time alone to practice vertical power turns.

I started and completed the three turns facing the Ohio River, as it is easy to see as you come around. Pushing the throttle wide open, moving the stick left of neutral and at the same time pushing the left rudder pedal forward until the turn was over 45 degrees. Then I pushed the rudder pedal (acting as the elevators) to keep the nose on the horizon, and to keep the ship from gaining or losing altitude. At the same time I pulled the stick (acting as the rudder pedals) back to keep the ship in a tight turn.

After three turns and facing the Ohio River I started reversing the controls, pushing the stick forward and pushing the rudder pedals back and forth until the ship was less than 45 degrees. Then moving the stick right of neutral, pushing the right rudder pedal forward and as the ship returned to level flight I put the stick and rudder in neutral and pulled the throttle back to cruising speed.

I then made the vertical power turns to the right. I am not comfortable making right turns - I'm more comfortable with making left turns. I wonder if this is because I'm left-handed. The turns were great, losing just a few feet. During the test you can't lose or gain 30 ft. in three turns. I sure get a kick out of the turns, and am going to do them every chance I get. The turns are very fast and anchor you firmly in the seat of the ship as it is on its side. The steady roar of the engine is sweet music to my ears.

Time to go down; with no one to rile me up I made a very nice three-point landing.

As usual, following the flying lessons I went to the Gas Company and loaded four truckloads of cinders and spread the cinders on the runway. I am beginning to wonder who got the best deal, the flying school or me. I flew 45 minutes today and spent the rest the day (about seven hours) hauling and spreading cinders. I am supposed to work on airplanes when I am not flying, but if it means I have a job when I get my license I'll do anything. I got a letter from Mom today.

After supper I looked over several lessons and then went out to the Sachleben's for the evening. We played euchre and the brothers argued as usual. We ended up the evening having a tin cup of "home brew".

I have been trying to save a little money by buying a bag of "Bull Durham" tobacco 5c and a pack of cigarette papers 5c and rolling my own cigarettes. I am getting pretty good at it but of course they are not as neat as the bought cigarettes. I feel guilty that I have to depend on my mother for all my money.

They were selling 12 in. long packs containing five cigarettes for $0.10 that you could cut into three making 20 cigarettes to the pack but it was never successful. I've tried smoking one without cutting it but soon ran out of air. They were so many types of cigarettes I can't remember the names of them all, "Bugle" was another brand of roll- your-own tobacco that I liked because it was pretty mild.

December 18, 1931 – Friday

The weather was cool, very foggy, and the wind was pretty smooth. There was no flying.

The fog was so thick you couldn't see the tops of the trees or 50 ft. in front of you. It refused to lift. Hauled two more loads of cinders and spread them on the runway and then took the rest of the day off.

Joe Sachleben gave me some "long green" tobacco to smoke. It is tobacco that they have been hanging in the barn to "cure" so that they can strip the leaves off the stocks, or stems, and take the tobacco leaf to market to be auctioned off to the tobacco companies. A lot of the farmers chew this tobacco rather than buying chewing tobacco at the store.

The first time I smoked this raw tobacco, it curled my hair, made my eyes pop out and took me five minutes to get my breath, wow! Rather than throwing it away I mixed a little bit of it with the "Bugle" tobacco making it go farther.

One day while I was still going to grade school we had corn on the cob for dinner. I took some of the brown dry corn silk, some tissue paper, climbed up on the third floor roof as no one was allowed to smoke. Rolling some corn silk in the paper I lit it and tried to smoke. The result was a lot of coughing and watery eyes. I never tried to smoke again until I started going to high school where everyone smoked. Instead of smoking cigarettes I took to smoking a pipe, but not at home. No one was to smoke, drink, or play cards, not even "Old Maid" at home. We had plenty of games to play and build.

I am going to start for home tonight so I checked the Cheve and added some water to the battery. After saying so long to all my friends and eating a big supper I got everything ready to hit the road. Bob Parker, a student, is going as far Cincinnati, Ohio with me. We left at 9:30 pm after saying goodbye to everyone. We went north on Michigan Road to Versailles, Indiana in dense fog. We had to crawl along as you could only see about 5 ft. in front of the car. We turned right on State Road 50 and at 10:00 pm a tire went flat. In the dense fog we jacked the car up, took the tire off the wheel. Then took the tire and tube off the rim, and took the tube out of the tire and patched it. Before replacing the tube we removed the nail from the tire. After replacing the tube in the tire and the tire on the rim we took turns filling the tire with air with a hand pump. Then we put the tire and rim on the wheel, let the jack down and we were ready to go, and after only a half-hour. We drove east on State Road 50 and arrived in Cincinnati at 12:30am.

After dropping Bob off, I drove through Cincinnati and got on Highway 22 at Norwood and headed east. Crawling along, as I could barely see the edge of the highway, I finally reached Wilmington, then Washington Courthouse, Circleville, Lancaster, and Zanesville, Ohio. Luckily there were practically no other cars on the highway so I had the road to myself. Even in the towns there were very few cars. The fog began to disappear and the further east I drove the less fog there was so I drove a lot faster, about 40 mph.

Continuing east on Highway 22 through Cambridge and on to Steubenville, Ohio, then crossing the Ohio River after paying a 25c toll into West Virginia at Weirton. 13 miles later I was in Pennsylvania driving east on Highway 22 until I crossed the Ohio River at Pittsburgh just below the "Point" where the Allegheny and the Monongahela rivers meet to form the Ohio River.

Driving on East Ohio Street past where I use to work and crossing the 40th St. Bridge (Washington Crossing) over the Allegheny River, I drove up the cobblestone hill to Pennsylvania Avenue and Main Street (just a couple blocks from Steven Foster's birthplace). I arrived home at 10:30 am Saturday morning. I used 14 gallons of gas and 3 quarts of oil costing $3.60.

December 19, 1931 – Saturday

The weather was warm, cloudy, and the wind was pretty smooth.

After dinner I slept for a couple of hours I hadn't had any sleep for over 30 hours.

Later Dad and I drove out to the Baum Boulevard in East Liberty, called "Gasoline Alley" as the majority of new car dealerships were along this boulevard. We stopped in the Marmon showroom, as Dad was interested in trading in his Marmon 6 cylinder 4-door, 7-passenger sedan (2 seats were attached to the back of the front seat and folded up when not in use) for a new car. The salesman showed us a Marmon eight-cylinder, four-door sedan that Dad seemed to like. Then he showed us a Marmon that he said was the first of its kind as it had a sixteen-cylinder engine. We took a demonstration ride in the eight-cylinder car and it was great.

The salesman, seeing that Dad was really interested in buying a car and would be trading in a Marmon, he suggested taking us for a ride in the sixteen-cylinder Marmon. He took us out into the country and when I kept bragging on the car he said I could drive it. I was in heaven, it rode so smooth and quiet, no vibration, the engine ran as smooth as an electric sewing machine. As I was going along at 80 mph he finally said to put it into high gear. I didn't know it had four forward gears. We were soon going over 110 mph when he said it was time to go back to the store.

Dad seemed interested, although it cost a lot more than the eight-cylinder car, until he found that it got 5 mi. to a gallon of gas. Dad told him he would have to think it over and would let him know what he decided to buy. I believe the Marmon factory is in Indianapolis, Indiana.

Something happened to the brakes on my car so I had to drive all the way home without brakes except for the hand brake (emergency or parking brake). The car only had brakes on the two rear wheels, (like most cars), and they were connected to the foot pedal with metal rods, which had to be adjusted periodically.

During my grade school years Dad and I would take a walk every Saturday evening after supper while my mother took care the store. As soon as we turned the corner onto Penn Avenue he would light up a stogie, a cheap cigar, and buy me an ice cream cone. We would walk out Penn Avenue, turn right on Negley Avenue, walk a couple of blocks, then turn right on Baum Boulevard and admire all the new cars along "Gasoline Alley". Finally we would turn right on Liberty Avenue and head home. I guess it would be about a 5-mile walk.

My dad really loved cars and we never grew tired of looking at the new cars. He always had to own the most expensive car in the neighborhood.

The only other time I ever saw my dad smoke was when he took a break in the bakery shop and I don't know if my mother knew he smoked. I'll never forget one brand he smoked, "Crooks"; it had an odd crooked shape. Years later I smoked them; they were a cheap mild cigar.

I still regret that I didn't ask him about his boyhood in Germany or his early life in America. Instead I jabbered about myself during our walk. I did learn a little. His sister living in Pittsburgh sent his older brother ship tickets to come to America. He didn't want to leave Germany at that time so my dad jumped at the chance to go to America. When he arrived in Pittsburgh he had to work to pay off the tickets.

Dad and his brothers were living in Eschwege, near Kassel, in the region of Hesse, of which the capital is Frankfurt. He was anxious to leave as both his parents had died and the money they left for the children was soon gone.

My brother Paul said that Dad came to America alone at the age of fourteen. But I have a letter written by Dad's nephew Henry Beyer to his cousin William Fisher on June 3, 1948, telling him that Martin came over alone at the age of sixteen in 1888.

He joined the First German Protestant Lutheran church in downtown Pittsburgh. He and another member of the church signed up to be missionaries among the Indians out west; and maybe get a chance to hunt for gold. He got along fine with the Indians, probably because he knew all about horses from A to Z. They sure didn't want him to leave. His parents owned a prosperous trucking business in Eschwege going as far as Muhlhausen in Thuringia. As there were no trucks, cars, or railroads at that time, horses were used in transportation, so Dad dealt with horses all his young life.

He stayed with the Indians for two years. I don't know if he converted any of them or if they converted him. I can't imagine anyone living with the Indians for two years and not being influenced by it. My Dad came to Pittsburgh with a few gold nuggets, sang German in the choir, met my mother and the rest is history.[17]

December 20, 1931 – Sunday

The weather was warm, cloudy, and the wind was calm.

After breakfast I got all dressed up and went to Sunday school at the First Evangelical Church at the corner of Center and Graham Street. All my classmates seemed extra glad to see me, even the ones that never noticed me before, and I haven't even gotten my pilots license yet. After church I went home for dinner then drove out to Harry Pyatt's house and talked over old times for a while. Harry and Earl Seitz made me a shortwave radio that could get stations from all over the world.

Following the visit with Harry I drove out to Donald Bicht's house and had a nice supper and visit with the family. I then went back to church for a Social League meeting. After the meeting I drove home and was ready to hit the sack.

---

[17] When telling this to the family during my 80th birthday party they acted as though it was a fairy tale. But I still have the letter my oldest brother, Paul, wrote to me about it in 1946. Then when my mother died, I got and still have, a letter Henry Beyer, my dad's nephew, wrote to his cousin William Fisher about it on June 30, 1948.

December 21, 1931 – Monday

The weather was warm, cloudy, and the wind pretty smooth.

Earl, Ralph, and I put on our "Sunday best" and Dad took us over to the Syria mosque, near Pitt football stadium, to take part in the "Shrine" annual Christmas program. It was very impressive and we had a swell time. Dad was made a "Noble" in the Syria Grotto Shrine in Pittsburgh on November 20, 1922.

I changed clothes and went to see Fred Moss. Fred and I went to grade school and high school together. They live on Mathilde Street, a real steep cobblestone street we used to sled ride. We drove across the 40th Street Bridge to the north side to see if the "buckeye" tree was still in the Catholic cemetery.

As little kids we rode our bikes over there, climbed over a tall rock wall, filled our bags with buckeyes and got out of there in a hurry. We had many a buckeye fight with those buckeyes. We drilled a small hole in the buckeyes and put a 2 ft. piece of strong string through the hole and knotted one end. The object was to swing the buckeye and hit the other fellow's buckeye trying to break his, each taking turns swinging. We discovered that the tree was gone.

But we did find the old "Indian Tobby Tree". It has seedpods about 6 in. long that the Indians were supposed to have smoked as cigars. We tried it a couple of times but it was not good.

Fred's dad worked at the Cadillac dealership eliminating all body noise in the new cars. I tried to talk Dad into buying a Marmon sixteen cylinder.

December 22, 1931 – Tuesday

The weather was warm, cloudy, and the wind was pretty mild.

My ear has been bothering me so I went to the doctor and he said it was in bad shape. I couldn't get the Marmon opened and had to get Earl to open it.

I loafed in the bakery shop all afternoon helping Dad bake German Christmas cookies and cakes. It reminded me of the time I took some to my German teacher in high school. He accused me of trying to bribe him and I do think it worked. I kept up my effort to talk Dad into buying the "16".

December 23, 1931 – Wednesday

The weather was warm, cloudy, and the wind was pretty mild.

95

After a late breakfast I drove over to the north side to Kress Brothers Company and visited with the old gang. They all seemed real glad to see me and asked me about the flying school. I quit working there to enroll in Dixie Flying School. After a couple hours at Kress's I went over to Paul's and helped him clean and polish his car. Following supper I ran into Red Sandstrom and Harry. Their house faces the alley behind our shop. We had a swell "woofing" party.

December 24, 1931 – Thursday

The weather was cool, cloudy and the wind was pretty smooth.

Ralph and I spent the whole day working on the car. He is going to a technical high school learning to be a mechanic. We removed the cylinder head, oil pan, and took the pistons and connecting rods out of the engine. The number one piston pin had come loose and scored the cylinder wall. We put new rings on the pistons, inserted the pistons and rods in the block, and tightened the rod bearings. After putting the cylinder head on the engine we adjusted the valve tappets, installed the oil pan and filled it with oil. Ralph accidentally adjusted the brakes looser instead of tighter. Then we test drove it and the car acted pretty good.

I bought a cigarette machine and a couple packs of rice paper and made a supply of cigarettes using "Bugle" tobacco mixed with "long green".

December 25, 1931 – Friday

The weather was cold, it snowed a little, and the wind was very rough.

Christmas was very good to me. I got a swell lounging robe, two pairs of socks, and a tie. After a swell Christmas dinner, Ralph and I went for a ride and the car works OK, so I went and visited Pyatt for a while.

We all went to church tonight to see the Christmas entertainment, which proved to be the worst I ever saw.

December 26, 1931 - Saturday

The weather was cold, cloudy, and the wind was pretty rough.

I am still trying to get Dad to buy the Marmon "16". I loafed around the house all morning, and after dinner drove to Fred's house. We went across the Ohio River to see the City-County airport on the south side. It sure is a humdinger; I never saw anything like it. After driving around town we went back to his house and had a swell turkey supper. Finally went home and listened to the radio for a while then hit the sack.

December 27, 1931 – Sunday

The weather was cool, cloudy, and the wind was pretty fair.

After breakfast I washed, shaved, put on my Sunday best and went to Sunday school. My Sunday school class was more interested in my experiences at the flying school than they were in the Sunday school lesson. I was tempted to exaggerate a bit, but on second thought, realizing that I was in church; decided that was a bad idea.

Earl Seitz picked me up at Sunday school and we took a ride out in the country where his family owned a small piece of ground, probably just a few acres. There was a small one-room shack that you could see the light through the cracks in the walls. When it rained the roof leaked like a sieve. During high school vacations, Earl, a friend of his, and I would spend a week at a time at the shack, but we had to be home by Saturday night so we could go to church on Sunday. There was no water or electricity and we had to dig holes as our "out house".

One week Earl brought three 6 ft. pieces of "4x4" lumber and made bows by using knives and pieces of broken glass. It took most of the week, off and on, and when finished the bows shot arrows we made about as good as bought bows. We made arm protectors and became quite good at target shooting.

Another week we decided to go rabbit hunting although we knew nothing about hunting. Earl had an old 45-caliber pistol that he was going to use if we saw a rabbit. Not having any luck we returned to the shack and there sat a real young rabbit in plain sight. Without thinking I decided to try to catch it and just as I ran in front of Earl he pulled the trigger. Either the bullets were no good or there was something wrong with the gun; or I wouldn't be here. It sure scared us and Earl took the bullets out of the gun and hid it.

We finally caught the rabbit and I took it home to raise but I had to take it back out in the country because it wouldn't eat. Later that week, while walking through the fields we saw a lot of chicken feathers near a large hole close to the bottom of the hill at the base of a big tree. Wondering what might be in the hole, if anything, we decided to go back to the shack and build a good-sized cage. We also took a sheet off of the bed hoping to throw it over whatever came out of the hole.

We built a fire with wet grass, leaves, and twigs so the wind would blow the smoke into the hole and make the critter come out. Earl and I held the sheet at the ready and his friend fanned the smoke into the hole. We got more than we bargained for as something big exploded out and almost got past us. But the three of us fell on it with the sheet trying to wrestle it into the to cage. I felt a sharp pain in my right side so after getting it, which turned out to be a large collie-type dog, into the cage, I looked at my side and saw where the dog had torn open my shirt and sliced open my right side. Through the blood I could see my bare ribs and it almost made me sick. We jumped into the car and they took me to a doctor who sewed me up. I still have the scar.

I tried to keep this from my mother but she caught me and said "no more of this monkey business". That was the last time I was at the shack until today.

The dog? Earl took it home but was unable to tame it and had to take it to the animal shelter. They said it probably had lived on its own since it was a puppy.

Earl dropped me off at home, and after dinner I drove over to the north side to West View to visit Pearl. They live at the end of the street, the house sits on the edge of a real steep hillside, almost too steep to walk up or down and if you fell it would be two or three blocks before you hit bottom. They had peach and apple trees and a large garden. Al, her husband, was a jeweler and worked in downtown Pittsburgh. He made me a swell green, gold ring with my initials on it for my graduation. Just a couple blocks down the road is the West View Amusement park where we went quite a few times down through the years.

After spending the afternoon at Pearl's I drove home, ate supper, then cleaned up and went to Social League at church.

December 28, 1931 – Monday

The weather was cool, cloudy, and the wind was smooth.

After breakfast I spent most of the day working on the car, only taking time out to eat dinner and go to the Western Auto store for parts. I put new exhaust pipes, muffler, horn, and cigarette lighter on the car. I also put glycerin in the radiator to keep it from freezing. Total costs for parts $2.37.

I took the car for a test drive down to Paul's and after we had some refreshments he gave me a lot of candy and some pictures. We said so long to each other as I was driving back to Madison tomorrow.

December 29, 1931 – Tuesday

The weather was cool, cloudy and the wind was pretty rough.

Dad decided to buy the 8-cylinder Marmon as he didn't own a gas station.

After breakfast I cleaned the car, filled the car with gas (Dad gave me 6 gal. actually he paid for everything as I didn't have any money) and I went around to tell everyone "so long." By the time I had dinner, loaded the car, and picked up Fred Moss, it was 2:30pm as we headed out of town.

We arrived at Wellsburg, W. Va. where my sister Bert lives, at 4:30pm. We stayed there about an hour while she fixed us a swell lunch. After lunch we drove down to Wheeling, W. Va. Then got on Hwy 40 and headed west through Cambridge, Zanesville, Columbus, Springfield, and Vandalia.

We turned onto Hwy 48 and headed south through Dayton to Cincinnati, Ohio. Then headed west on Hwy 50 to Versailles,

Fred Moss with my Cheve

Ind. We next headed south on Michigan Road arriving in Madison at 8:30 am, Dec. 30th.

This trip was a little longer than the one I took going to Pittsburgh. It was about 550 miles. We used 15 gal. of gas and 1 ½ qts. of oil - $3.06. Fred drove half way and we drove slowly as I wasn't satisfied with the way the engine sounded.

December 30, 1931 – Wednesday

The weather was cool, rainy, and the wind was pretty rough.

We got into Madison at 8:30am and I was sure glad to be back. We went up to bed at 8:30am. and got up just in time for dinner. After dinner Fred and I went around town to say "hi" to all the fellows and then found him a room.

I bought my driver's license and license plates for my car - $6.50. Following supper Fred and I loafed around town until it was time to hit the sack.

December 31, 1931 – Thursday

The weather was cool, cloudy, rainy, and the wind was rough.

We took the Curtis 0X5-V water-cooled engine out of the Waco and washed the engine with gas. I spent the rest of the day helping Joe fix the landing gear on his Waco. We packed the wheel bearings with grease, tightened all the bolts and nuts, checked the struts for cracks and alignment.

A new student came in today. His name is Chester T. McKinley, from Ravalli, Montana.

I loaned Fred my car while I studied *Lesson no. 7, Airplane Construction, Fuselages.*

There are three main types of fuselages being built and in use today; the tube fabric covered type, the all metal, and the Monocoque.

The most popular type of fuselage used today is made of a framework of steel tubes covered over with cotton fabric, which in turn is treated with dope. Three different types of wire are used for bracing; aviation wire, (or piano wire), streamline wire, and hard wire.

The engine mounting for a radial engine is a ring mount of tubing or sheet metal bolted or welded to the front end of the frame.

A V-type engine requires two horizontal bars, held in place by suitable bracing.

The all-metal type is made of a framework of steel tubes covered over with duralumin sheets instead of fabric. Duralumin can't be welded satisfactorily so all parts are riveted with duralumin rivets.

The Monocoque type has no frame at all. It is built of molded plywood fastened to molded bulkheads.

The questions for Lesson No. 7 are:

1. *Give a brief description in your own language of the three types of modern fuselage construction*
    a. *The Steel Tube Fabric Covered Type.*
    b. *The All Metal Type*
    c. *The Monocoque Type*

2. *Which of the three types do you prefer? Why?*
3. *Name three modern ships the fuselages of which are good examples of these three types of construction, stating which type of construction each is made.*
4. *What is an amphibian? How does it differ from a float seaplane or a flying boat?*
5. *Name one modern commercial amphibian, and tell of what its hull is constructed.*

Becht and Shucks came by and we practiced on the guitar.

Lesson 7 – Fuselages

THE DAYTON SCHOOL OF AVIATION

The engine mounting is made, depending on the type of engine used. A radial engine is usually carried on a ring mount of tubing or sheet metal, bolted or welded to the longerons at the front end of the frame; while a V-type engine requires two horizontal bars, held in place by suitable bracing.

Figure No. 5. Fuselage of Fairchild 21, the "training plane," under construction. Note fire wall, with oil tank placed in front of it. Also note the sturdy framework and ideal engine mount.

Other good examples of this type of all steel tube fuselage construction are the Keystone "Patrician", and the Fairchild.

The Keystone "Patrician" is a handsome twenty-passenger cabin airliner, propelled by three Wight Cyclone engines of 525 H.P. each. In writing of it, the company says "the all metal, fabric covered type of construction has been chosen as the most suitable. Fuselage, wing spars, and tail framework are of chrome molybdenum steel tubing, with welded joints throughout the structure."

6

THE DAYTON SCHOOL OF AVIATION

The Fokker F10 is now in regular passenger service from Chicago to St. Louis and Cleveland. It is a veritable "air Pullman," carrying 12 passengers and attaining a cruising speed of from 125 to 150 miles an hour.

The Fairchild, however, is more typical of the standard construction of the great majority of new planes. Most of them use an all steel fuselage, made entirely (longerons, struts, and diagonals) of chrome molybdenum steel tubing, welded in truss form. This fuselage is light, strong, and solid, but must be constructed and welded with great care.

Figure No. 6. Wright J-5 Engine Mount for a Travel-Air.

7

Clyde H. Beyer

January 1, 1932 – Friday

The weather was cold, cloudy, and the wind was fairly rough.

I flew for thirty-five minutes today practicing 180 degree and 360 degree turns, spirals, verticals, and figure 8's. It sure was great to be up in the air again. All things considered I did pretty good job. It was snowing pretty hard above 1,600 ft. but it never reached the ground.

At 1,800 ft. while doing verticals I ran into some clouds so I dove under them, ran into snow and had to go still lower.

The landings were nothing to brag about, one good, one fair, and one very poor. The "Prep" sure takes a beating on landings, and I am starting the New Year as usual.

We went to Milan to see about some airplane parts, equipment, and a "Prep." When it came time to leave I got to fly the "Prep" that we came after, from Milan to our airport and Joe didn't even "check time" me. This was the first time I flew off a strange field. I studied a road map and just followed the highway back to Madison and made a nice three-point landing, which surprised me.

We took the "Eagle" wings down from the rafters to check, repair, refinish, and re-cover with fabric. The fabric had been taken off way back in October.

The wing consists of a main spar, rear spar, bracing wires, metal compression tubes, a leading edge, trailing edge, false spar, and ribs.

The main spar, which is located just behind the leading edge, is made of a single piece of spruce, as is the rear spar, which is located just in front of the trailing edge and runs along the length of the wing.

A piece of plywood covering the length of the wing from just behind the leading edge to the main spar gives the leading edge additional strength.

The false spar is located at the outer rear of the wing in front of the trailing edge and supports the aileron; it is made of a single piece of spruce.

Ribs are made of plywood and have cap strips to prevent cutting the covering fabric and, are attached to the leading and trailing edges. They determine the shape of the wings.

Ailerons are made of wood similar to the wing that is covered in the same manner.

We were divided into groups, each group to a wing. After placing the wings on saw horses and inspecting the wings inch by inch for cracks, splits, loose glue joints (Casein is the preferred glue) or any other damage, we found everything OK. Harvey then measured and re-measured the wing, as the covering fabric must fit accurately because the fabric is sewed on a commercial sewing machine with heavy linen thread.

While Harvey was sewing the covering fabric together we sanded the wing smooth, then brushed on a coat of varnish to protect the wing from moisture. Shellac and linseed oil can also be used. After the varnish dried we sanded it lightly to remove the rough spots.

By this time Harvey had completed sewing the covering fabric together and we helped him place it on the wing. We each took the reinforcing tape, (Sea Island Cotton), and placed it over each rib on the covering fabric.

After being told how important rib stitching was, we were given a long needle and heavy liner thread. Then Harvey stitched the first rib emphasizing that "each stitch must be perfect".

I passed long stitches of heavy linen thread completely through the wing and around each rib ny running a long needle through the reinforcing tape and fabric on one side of the rib near the trailing surface. Then bringing it back through on the other side close to the rib. I tied a square knot along the top of the wing.

After Harvey inspected that first stitch and knot he gave his okay so I advanced the thread about 6 in. along the line of the rib toward the leading edge without breaking the thread. I repeated this stitch with a square knot, every 6 in. until the rib was completely stitched. Each rib is stitched in this manner. This was a slow, tedious job and I wouldn't want to do it for a living.

After all the ribs were stitched, and there were a lot of them every 6 in., Harvey made sure that the covering fabric fit properly, not too tight or loose, without wrinkles, so when the dope is applied it can do its job of waterproofing and strengthening the wing. He started at the inner rear corner of the wing lacing on the trailing edge, around the aileron gap and wing tip. He handed me the needle and I continued on the leading edge and then to the inner rear corner.

Harvey is responsible for everything we do, as he is the only one who has an A. & E. mechanics license. Being satisfied that the cover fabric was sewed and fit properly we got the ok to give the wing a coat of "dope". But this time we had every window and the hangar doors wide open with two great big fans running top speed. When we finished brushing on the first coat of dope (we could still smell it), we placed the finishing tape, Sea Island Cotton, over each seam, rib, and the lacing around the edge of the wing.

After supper, and that is always a battle at the table with hungry fellows at a boarding house, I fooled around town with Becht and Shucks for a little while then went out to the Sachleben's and played euchre all evening. We ended the card playing with the usual 1/2 tin cup of "home brew". The first time I tasted the homemade beer I thought it tasted like soapsuds but I am getting used to it.

January 2, 1932 – Saturday

The weather was cold, cloudy, rainy, and the wind was rough, so no flying today. Bob Parker came back last night.

As I had the experience of sanding, doping and vanishing covering fabric I spent the morning cleaning a bunch of airplane parts so the rest of the students could get the experience.

They sanded the first coat of "dope" smooth, then brushed on the second coat. After drying at least one hour they sanded the second coat smooth then applied a third coat. The fourth and fifth coats applied were tinted the same color as the spar varnish.

Finally, the finish coat of spar varnish was applied, left to dry thoroughly, then sanded smooth as glass. It was as tight as a drum, a great job.

After dinner I took the oil pan off the Cheve engine again and had Bob Jones tighten all the connecting rods and main bearings for $1.25. I can't seem to keep the bearings tight, or maybe I just think they are loose. It might be that it is loose pistons that are making the "slapping" sound. It sure would be a relief to have them fixed. I put the oil pan back on, replaced a bolt on the generator, fixed the radiator cap, and put on the new license plates.

After supper I walked down Main Street to the movie house and saw "Touchdown". It was a swell movie.

January 3, 1932 – Sunday

The weather was cool, cloudy, and the wind was a pretty steady breeze, but no flying.

After a great breakfast of coffee, two eggs and toast, and some pancakes, I went to Sunday school and church with the Sachleben's, came back to the boarding house and had a swell dinner.

Fred and I went out to the airport and took some pictures. Fred took a picture of the hanger, Chet, the fellow from Montana and me.

We drove over to Clifty Falls State Park, just west of Madison, and took some pictures of the scenery. It sure is a wonderful sight.

At 3:45pm we drove down to the bus station and Fred got on a Greyhound bus headed for Pittsburgh.

We got a little Aronca in today. It is a one-seater with a two cylinder air-cooled engine and looks just like a grasshopper. Hope I get a chance to solo it.

I stayed in tonight to study *Lesson No. 8, Airplane Construction, Fuselages.*

The steel tube fuselage is streamlined by placing a streamline "former" from the rear cockpit to the horizontal stabilizer and a smaller "former" on each side of the fuselage, running from the engine section in front to the tail post. Streamline "formers" or simply little frames of very light strips of wood (usually balsa) designed solely to give shape to the fuselage.

Units located in the fuselage: power plant, oil tanks, gas tanks, cockpit and cabins, instrument board and controls and the tail skid. The engine usually located in the forward part of the fuselage, it being bolted or riveted to the engine mount, which is bolted or riveted to the fuselage frame.

Back of the engine compartment is placed a firewall.

Cockpits and cabins are usually bordered by a cowling made of sheet aluminum that runs along the top of the fuselage from the engine compartment to the rear of the last cockpit. Celluloid windshields are placed around the forward side of the cockpit to break the wind. Seats are usually of the bucket type, which is simply a stamped sheet of aluminum made to fit, with a plywood seat covered with a cushion. Closed cabins have two doors, one on each side.

107

Clyde H. Beyer

Gas tanks are usually located in the front part of the fuselage, between the front cockpit and the engine compartment. Some are placed in the wing, or in the upper wing if a biplane.

Tailskid - supports the rear end of the fuselage when it rests on the ground. It is placed near the rear of the fuselage and prevents the rear of the fuselage from dragging on the ground, it also help to absorb the shock of landing. Many types use a leafed steel spring equipped with a steel shoe.

Instruments and Controls. The instrument board is located in the cockpit directly in front of the pilot. The instruments are usually circular and placed close together so the pilot's eyes don't have to wander over the instrument board.

The rudder pedals are located in the cockpit and are supported in a pivot at the pilot's feet.

The "Joy" stick is located so as to be practically in the pilot's lap and is supported in a pivot at the pilot's feet.

These are the questions to Lesson No. 8.

1. How is the fuselage streamlined?
2. What units are located in the fuselage?
3. How are the various type power plants installed in the fuselage?
4. How are the cockpits and cabins equipped?
5. Described 2 gas tank installations
6. Tell what you know about instrument boards.

I sure didn't know it was going to be this hard and I am just getting started. I thought all you had to do was to jump in the ship and fly, just like learning to drive.

THE DAYTON SCHOOL OF AVIATION

realize that while an open cockpit will give much more resistance than a closed one, it has the big advantage of giving the pilot a much better range of vision.

Figure No. 1. Welded Steel Fuselage of the Fairchild 41.
Note fittings for wing and landing gear attachment; also fairing about cabin.

Because of these contradictory requirements we soon discover that it is impossible to design a perfectly streamlined fuselage, one that will ideal in every respect – that the final design of the fuselage, like all the other parts of the ship, is a matter of compromise, of solving many different problems in a way that seems best for all. And even though our design does not give the perfect streamline effect, there are simple and inexpensive ways of correcting it, without in any way changing the framework or method of construction.

The steel tube fuselage, for example, has comparatively sharp corners and flat sides. But the designer overcomes

3

January 4, 1932 – Monday

The weather was cold, cloudy, and the wind was pretty mild. I flew the "122" "Prep" that I flew from Milan last Friday, for 30 minutes and it was simply awful. I undershot the field once, and overshot the field twice. All the maneuvers were sloppy and I couldn't seem to get used to this new "Prep". I asked Joe to keep me on this ship until I can fly it right.

We made a line about 12" wide across the runway from the south edge to the north edge of the runway with crushed limestone. It is a short distance west of the hanger apron. This line is used during the Private Pilot and Limited Commercial Pilot flying tests as the Pilot has to touch down in front of and within 200 ft. of the line. So when we land over 200 ft. beyond the line it is called "overshooting" the field. And when we land over 200 ft. in front of the line it is called "undershooting" the field. If you do that during the tests you flunk.

The next time I fly the ship I am going to "trim" the ship by adjusting the horizontal stabilizer. The former owner may have trimmed it for a heavier pilot than I am thus changing the center of gravity that affects its flying characteristics.

We ground the valves on the "0X5V" engine we took out of the Waco last Thursday. After supper I went back to the hanger and helped Harvey put the engine together. I cut my finger pretty bad with a screwdriver.

The car is running pretty good. Dave Gentrup fixed me up with a blind date for tomorrow. Dallas Layton moved last week.

January 5, 1932 – Tuesday

The weather was cold, cloudy, rained hard, and the wind was very brisk, so no flying.

We all worked on assembling the "OX5V" and the Sechley engine this morning.

After dinner I took **John Yellitz**, Mountain Springs, PA, a new student, to Seymour for his physical exam. We came back from Seymour, 41 mi., in exactly one hour. The Cheve is running pretty good.

I bought the "Dayton School of Aviation Correspondence Course" in book form from Dick O. for $5.00.

Dave G. and his girlfriend picked me up, and then picked up Melva Lohrig as my date. We went up to Cragmont State Hospital and saw a free movie for patients and attendants only. But we got in because Dave's dad is a big shot there. After the movie they served refreshments. I had a real nice time.

This was the first time I was so close to so many "cookoo" people. They sure are strange. Some are very dangerous and are not allowed to attend these types of affairs. A lot of them have the minds and actions of little children. The ones I saw were living in a world all of their own. It was a new experience for me and I left with a strange and sad feeling. They may not have "all their marbles" but they seem to have special talents in special areas. I met one man that didn't pay any attention to what was going on around him. I bought a 12" silver watch chain and heart from him for $0.25. He spends all this time making them by hand, using silver spoons for the material. I examined mine with a magnifying glass and each link is 1/4 in. long and is exactly alike, absolutely perfect, and unbelievable.[18]

Since then I have never had any desire to make fun of or belittle people with mental problems, politicians excepted.

One night about ten years later in the middle of the winter, a stark naked woman knocked at our door. We let her in and as she thought she was dressed we realized she must be from Cragmont. After putting a blanket around her we called Cragmont and a couple of attendants came and picked her up. The attendants told us she does this every chance she gets. At the time, Jane, I, and baby Diane were the only ones living in the hanging rock addition which was across the highway from the Cragmont grounds.

There is something wrong with the starter on my Cheve. It is a good thing a have a hand crank. I drained the alcohol out of the radiator.

A letter came from Rosalee today and it sure was swell. She's working in Anderson, Indiana so I guess I'll go see her Sunday.

Following supper, I ate too much as usual; we all drove out to the hangar to continue studying *Lesson No. 27 – Principles of the Engine.*

Compressing the air and fuel within the cylinder results in greater power and efficiency. The compression pressures in aircraft engines run from 100 to 125 lbs. per square inch. The compression ratio is the volume above the piston at its two extreme positions, and the average ratio is 5 to 1.

---

[18] I still treasure it and have it my display case in the living room.

Clyde H. Beyer

Pre-ignition – Self ignition of the fuel is not normal, hot valves, pistons, sparkplugs, or carbon deposits may cause the fuel to ignited prematurely and causes a drop in power and is harmful to the engine.

Displacement is the total volume swept by the piston. It is the area of the head of the piston, the stroke, and the number of the cylinders, and is specified in cubic inches. This is the means of classifying engines as to size. Aircraft engines develop one horsepower for every 1.5 to 4 cubic inches of displacement.

Horsepower is the work required to lift 33,000 lbs. a distance of 1 ft. in one minute. The work of an engine, that is, it's horsepower, is delivered to the member to be driven in the form torque. The latter is the turning movement of the crankshaft, and is designated in foot-pounds; the product of the load in pounds and the distance in feet from the centerline of the engines at which this reading is taken.

Brake Mean Effective Pressure – The average pressure on the piston throughout a revolution of the crankshaft of the power stroke is called the indicated Mean Effective Pressure (MEP). The brake MEP agrees with the delivered horsepower, which takes into account the mechanical or frictional losses of the engine.

The MEP, with the displacement and speed, are the three factors that determine the engine horsepower. In automobile engines the brake mean effective pressure ranges from 65 to 100 lbs. per square inch, while aircraft engines varies from 100 to 140 lbs. per square inch.

Engine speed. The engine horsepower varies directly with speed. We can expect an increase in horsepower at higher speeds provided the mean effective pressure does not fall off too rapidly. The mean effective pressure decreases beyond a certain speed. Aircraft engines must be operated normally at speeds well below that at which the power peaks and when tested at sea levels.

Propeller efficiency. There is a loss in propeller efficiency with increased engine speed. There is a certain ratio of the propeller tip velocity and forward speed of the airplane for maximum efficiency. The propeller is designed to absorb a given horsepower at a certain predetermined speed, at which the engine will deliver the horsepower, and produce a given forward speed for the airplane in which it is installed.

Mechanical efficiency is the ratio of power developed in an engine and the power delivered to the propeller. This difference is referred to as friction horsepower, that is, the power required to overcome the friction of all the moving parts of the engine. The friction horsepower of aircraft engines is from 10-15% of total power developed. The mechanical efficiency is from 90 to 85 percent, these figures decreasing as the speed of the engine is increased.

Thermal efficiency - A rather small percentage of the heat units applied to an internal combustion engine in the form of fuel is converted into useful work. The ratio of the heat units supplied to an engine to the heat units converted into work is called the thermal efficiency.

Distribution of losses – The cooling losses amount to about 25 or 30 percent of the heat supplied. The greater part goes to the heat given off in cooling the cylinders. Taking into account both the thermal and mechanical efficiency of an engine, we can see that only about 25% of the heat supplied in the fuel is actually converted into useful work or horsepower to the propeller.

Gas mixture. The air and gasoline are first mixed in the carburetor, but the process really continues to the inlet valve where the charge enters the cylinder. To obtain best power and operation from an engine the mixture must be correct at all times. The correct proportions are about fifteen parts of air to one part of gasoline by weight. It is a difficult problem to maintain the correct mixture due to sudden changes in the atmosphere, density and temperatures that occur with sudden changes in altitude. The heaters employed on aircraft carburetors are to prevent ice formation, but the mixture should be as cool as possible.

Detonation – It is always desirable to employ the highest possible compression ratio in an engine, but beyond a given point the temperature rise resulting from increased compression will cause detonation on ignition. Detonation acts like a sledgehammer blow, giving a metallic sound and if allowed to continue it increases in severity which will damage the engine. The type of fuel and the timing of the ignition play an important part in preventing detonation.

These are the questions to Lesson No. 27

1A. What is a fuel engine?
1B. Describe the operation of a four cycle engine.
1C. Describe the operation of a two cycle engine.
2A. Define Compression
2B. What advantage is compression?
2C. Define compression ratio.
3. On what does the compression ratio depend?
4. Define: Preignition, displacement, horsepower, torque, power curve peak, propeller, mechanical and thermal efficiency.
5. What is the difference between the Brake Mean Effective pressure and the Indicated Mean Effective pressure?
6. What factors determine horsepower?
7. What proportion of fuel and air are in a perfect mixture?
8. What are the advantages and disadvantages of a lean and rich mixture?
9. Why are heaters used on an aircraft carburetor?
10. What are some ways of obtaining maximum power output?

It is going to take a lot of concentration to absorb all this information. Right now I'm going to sleep on it. These night classes are ruining my "catting" around.

Lesson 27 - Principles of the Engine

THE DAYTON SCHOOL OF AVIATION

piston, or in one revolution of the crankshaft, the four-cycle engine in four strokes or two revolutions of the crankshaft, and the six-cycle engine in six strokes or three revolutions of the crankshaft. Most engines operate on the four-cycle principle. It is essential to understand these cycles in order to properly time the valves and ignition.

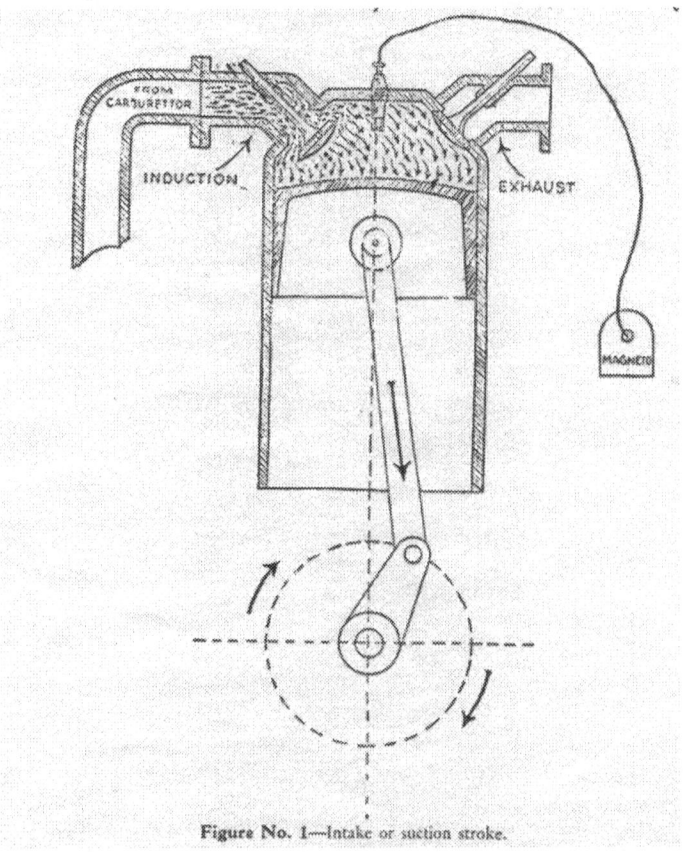

Figure No. 1—Intake or suction stroke.

FOUR-CYCLE ENGINES: As explained above, the operations are completed in four strokes of the piston, or two revolutions of the crankshaft in the four-cycle engine.

4

THE DAYTON SCHOOL OF AVIATION

During the first stroke, which is generally called the intake or suction stroke, the piston moves toward the crankshaft and sucks in an explosive or combustible mixture through an inlet valve that is held open for that purpose.

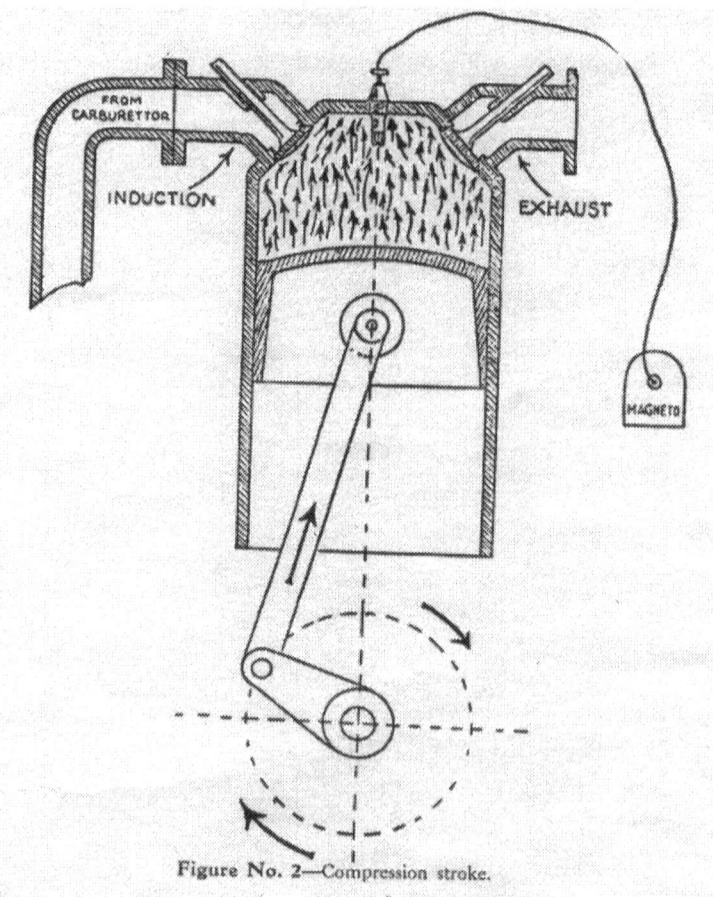

Figure No. 2—Compression stroke.

On the next stroke, which is referred to as the compression stroke, the piston moves away from the crankshaft to compress the mixture which was introduced into the cylinder during the first stroke. Both inlet and exhaust valves remain closed during the compression stroke. At the end of this stroke a spark occurs. Thus igniting the compressed gaseous mixture.

5

THE DAYTON SCHOOL OF AVIATION

During the third stroke, which is known as the expansion or power stroke, the piston moves toward the crankshaft under the pressure created by the burning gas.

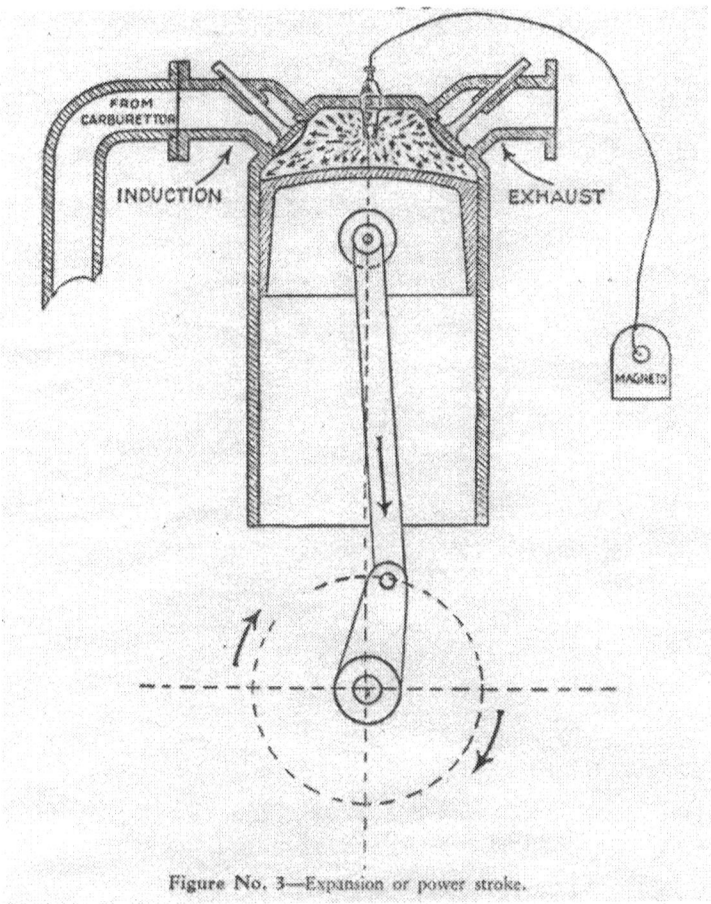

Figure No. 3—Expansion or power stroke.

Near the end of this stroke when the pressure is greatly reduced by expansion, the exhaust valve opens, permitting the burned gas to rush out into the atmosphere.

The fourth stroke is called the exhaust stroke or scavenging stroke, because the piston moves away from the crankshaft or toward

6

THE DAYTON SCHOOL OF AVIATION

The head of the cylinder, pushing out the remaining burned gases. At the end of this stroke the piston starts down again to begin another suction stroke and another cycle of operations.

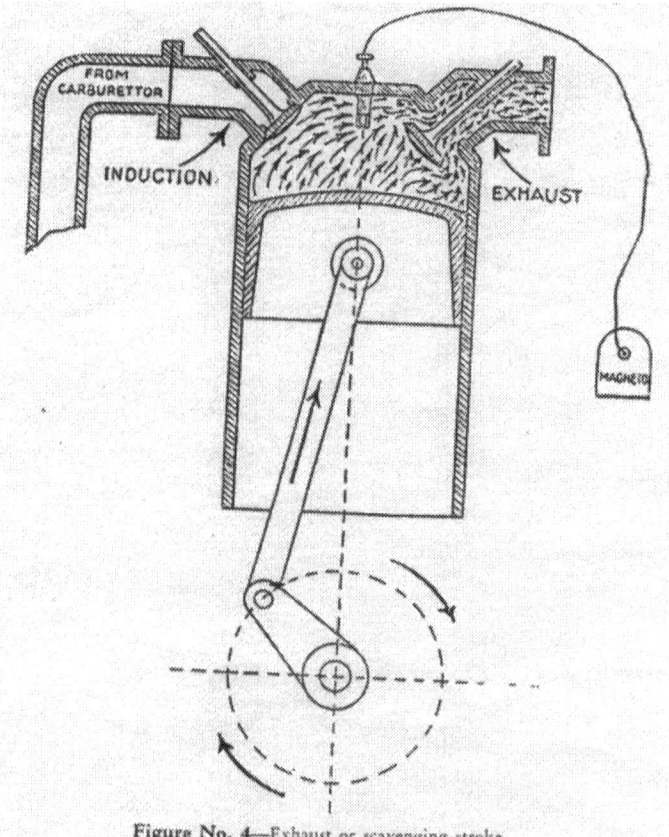

Figure No. 4—Exhaust or scavenging stroke.

An automatic switch known as the timer or breaker controls the spark, being driven at a speed in definite relationship with the crankshaft, and depending on the number of cylinders, type of ignition system, and the cycle of operations. The spark actually occurs a few degrees ahead of the end of the compression stroke to allow time for complete inflammability of the charge.

7

118

January 8, 1932 – Friday

The weather was cool, fine, and the wind was pretty smooth.

I flew for one hour and fifteen minutes after Joe check timed me for ten minutes. We flew the "Prep" that we got from Milan Indiana. They decided to keep trying to "break-in" the rings on the NC 122K. While Joe was with me he said I was doing OK. Coming from him that was a compliment.

After letting Joe out I went up in the old "Prep" NC 110K, and practiced all the maneuvers, climbing turns, spirals, stall, 180 degrees, 360 degrees, and figure 8's. The landings were rotten. I undershot the field two times and overshot the field three times. On the final landing I had to sideslip the ship in from 60 ft. up. It was surprising Joe didn't bawl me out, as you can't sideslip while taking the test for your license.

We got another ship in today from Cincinnati, Ohio, It is an International.

I painted the undercarriage and wheels of the Eagle, and am getting to be an expert at painting.

We went out to the hanger after supper and Harvey conducted the class in *Lesson no. 28 "Carburetion"*.

The function of an aircraft engine carburetor is to mix for delivery to the engine cylinders, gasoline and air in such proportions as to provide the most complete combustion with maximum power under various conditions.

Early types of carburetors - There were three general types, the simplest being the one in which the air was passed over the surface of the fuel and the mixture ratio depended upon the degree of fuel evaporation. Another method was to pass the air through wicks which had picked up the fuel, and a third method was to pass air through the fuel, permitting it to bubble to the top before going to the engine cylinder.

Simple Types of Carburetors. All carburetors in use at the present time are of the proportional flow type. They require a constant level chamber so that the fuel level does not rise above the jet. There are three methods to maintain constant level of fuel in a chamber. One is the overflow system; the second is the diaphragm system, while the third, which is in general use, is the float valve type.

There are two types of float valves, the concentric type in which the valve stem passes through the axis of the float, and the eccentric type in which the valve and float have different centers and the valve is operated by an arm.

The venturi is the best type of carburetor air passage. It aids the atomization of the fuel, and also produces the necessary pressure drop at the spray nozzle to lift the fuel from a float chamber.

Altitude controls. Aircraft carburetors are fitted with a device to regulate the gasoline supply for the changes atmospheric conditions encountered at higher altitudes. This device in the float chamber reduces pressure and would hinder the flow of gasoline to the jets and thus make the mixture more lean.

The Stromberg idling device is really a miniature carburetor with its discharge just above the closed throttle. An idling tube is directly connected with the main jet passage. It has a fuel nozzle which discharges into a mixing chamber where it meets the air entering through holes controlled by a needle valve. A slot, which becomes more exposed as the throttle is opened, forms the discharge nozzle into the main barrel. The suction increases as the slot opening becomes greater, thereby sucking up more fuel, as the throttle is opened.

These are the questions to Lesson No. 28

1. *What is the function of an aircraft engine carburetor?*
2. *Describe the early types of carburetors.*
3. *What three methods can be employed to maintain a constant level of fuel in a float chamber?*
4. *Describe 3 types of float valves.*
5a. *What are altitude controls?*
5b. *Why are altitude controls used?*
6. *How does the Stromberg idling device work?*
7. *Describe the modern aircraft carburetor*
8. *What advantage is a venturi in a carburetor?*

By the time class was finally over I was really confused. I had a lot of trouble understanding all the different types and makes of aircraft carburetors that are more complicated than automobile carburetors. The air gradually gets thinner the higher the plane flies, so the carburetor has to be able to adjust for that.

My Cheve carburetor is simple. Whenever trouble occurs I just take it apart, clean out all the jets, adjust the float level, and put in a new filter and gaskets. It has an adjustable screw to regulate the mixture of gas and air.

THE DAYTON SCHOOL OF AVIATION

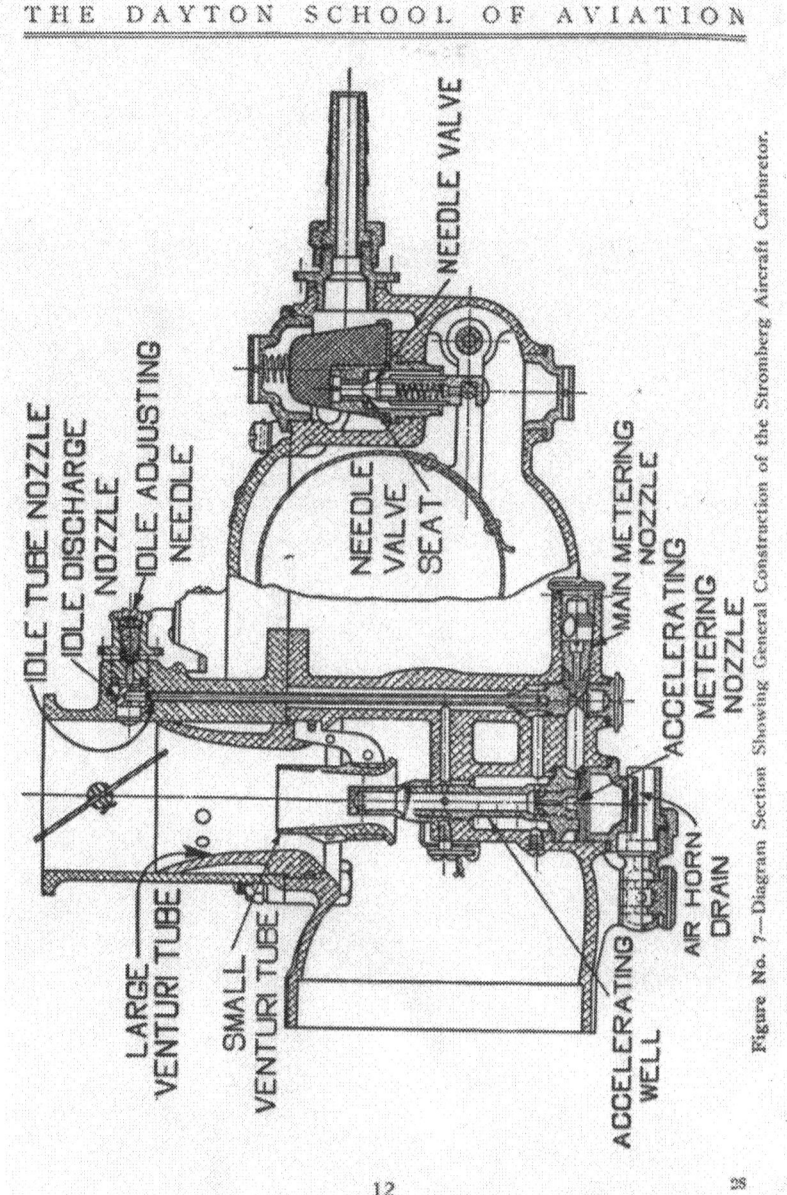

NEEDLE VALVE

IDLE TUBE NOZZLE

IDLE DISCHARGE NOZZLE

IDLE ADJUSTING NEEDLE

NEEDLE VALVE SEAT

MAIN METERING NOZZLE

ACCELERATING METERING NOZZLE

LARGE VENTURI TUBE

SMALL VENTURI TUBE

AIR HORN DRAIN

ACCELERATING WELL

Figure No. 7—Diagram Section Showing General Construction of the Stromberg Aircraft Carburetor.

12   28

121

THE DAYTON SCHOOL OF AVIATION

inverted and Vee types to meet the requirements of a variety of engines.

The float mechanism of the Stromberg carburetors are designed to operate between one and four pounds pressure, although they will generally feed sufficient fuel at eighteen inches gravity head, and withstand six pounds pressure

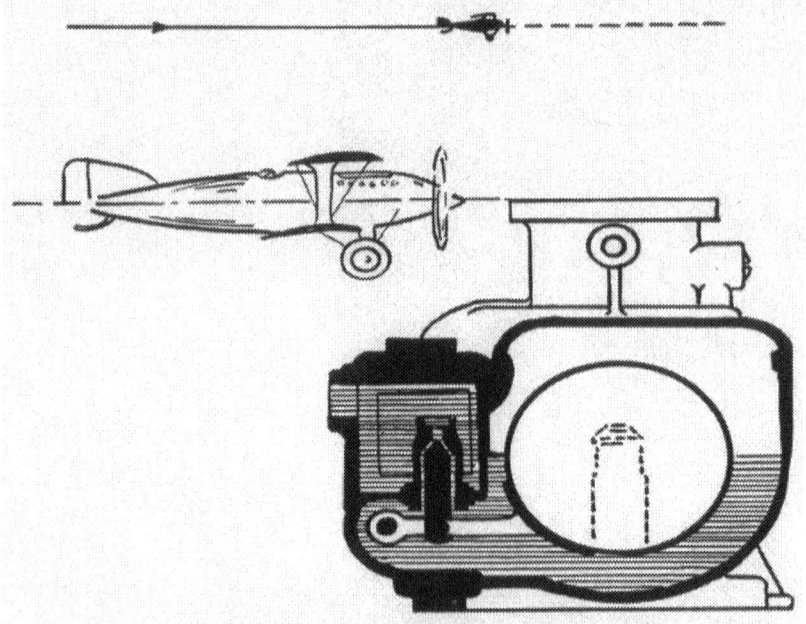

Figure No. 8—Level Flight, Normal Feed to Jet.

without flooding without the engine not running, and usually nine to ten pounds pressure without flooding sufficiently to affect the operation of the engine while running.

The floats are generally spherical or cylindrical in shape, and hinged with the pivot toward the tail of the ship. This arrangement provides feed to the jet during most airplane

13

All the other students have a big advantage. They have all completed the "Dayton School of Aviation Correspondence Course". These students get two weeks ground training and the first flying lesson free. So while I stay in and study all this stuff they are out chasing girls all over town.

January 9, 1932 – Saturday

The weather was very cold, cloudy, and the wind was pretty rough.

The fellow that bought one of our Waco's we rebuilt came and got it this morning. We watched him roll down the runway and take off with the engine sounding great. He was soon out of sight. It was a pretty picture. This proves that the planes we rebuild will fly again.

I put the struts and wires on the Eagle's tail surface then painted all the struts. By the time I got finished it was dinnertime.

I spent the afternoon working on the Cheve fixing the brakes, horn, license plates, and starter. The starter spring was broke. I want the Cheve to be in great shape for the trip to Indianapolis and Anderson tomorrow.

Becht, Shucks and I bummed around town tonight. The streets were crowded with cars, wagons, and buggies, and the sidewalks were crowded with farmers and their families. The stores seem to be doing a lot of business; I bought a pipe for $0.25 and two rolls of film for $0.50. Saturday is the big day for the stores and it is a job to walk up Main Street but we managed to watch a lot of pretty girls. There is a boy scout at each street corner directing traffic.

January 10, 1932 – Sunday

The weather was warm, clear and the wind was mild. A perfect day for a trip so I didn't fly today. After an early breakfast I filled the Cheve with gas and a quart of oil I got at the airport, as it doesn't cost me anything. By that time Becht showed up so we took off for Indianapolis. We drove north on highway 7 at 35 mph to Columbus, then took Highway 31 north to Indianapolis. While in the city I took a picture of the courthouse and the Capital Building. We went over to the airport and it sure is a swell "port". I took a picture of Commerce's Stinson-Detroiter and one of the Rearwing Jr., also a picture of the hanger.

From there we drove northeast to Anderson. We drove all around the city before finding the restaurant where Rosalee works. She seemed glad to see me and we had a great time. She looks 3 times as good as she used to.

Ed Becht with Stinson-Detroiter at Indianapolis – January 10, 1932

Rearwing at Indianapolis – January 10, 1932

About halfway home, a loud hammering noise began in the fourth cylinder. We drove real slow; practically crawled the rest of the way home hoping the engine wouldn't break down before we got there. It might be a broken piston or the connecting rod "burned", I will find out Monday.

I had a great supper, put on my Sunday best and went to church. It was a long day.

January 11, 1932 – Monday

The weather was cool, cloudy, and the wind was rough. No flying.

I put some exhaust stacks on the Eagle Rock engine and fixed the windshield on the Waco. Also painted the engine struts on the Eagle. While I was doing this Harvey, George, and Art put a coat of "dope" on the wing.

There has been a lot of stealing going on around here so Joe decided that each student would take turns staying at the hanger overnight. He asked me to stay tonight after I had supper at Barber's. Joe showed me a 45-caliber "Colt" revolver in the desk drawer that I should use to protect myself. He showed me how to use it, as I had never handled a gun. It would be just my luck that I would end up shooting myself in the foot. I sure hope that I never have to use it.

Following supper with the guys I took sheets, blankets, and pillow for the cot and my nightclothes out to the hanger. As it was early I wrote Earl, Fred, Herb, and Paul a letter.

We are going to have another class on Carburetion tomorrow night. Bud, the pilot instructor from Cincinnati, is going to conduct the class this time.

With nothing else to do I studied the lesson we had last Friday. I was surprised to learn that in two maneuvers the carburetor doesn't supply fuel to the cylinders. One maneuver is when the plane is flying continuously upside down and the other time is when a plane hits a violent downward air gust (or air pocket). If a "loop" is made properly the centrifugal force will keep the proper amount of gas in the carburetor and normal fuel supplied to the cylinders. The pilot also will be held hard against the seat.

January 12, 1932 – Tuesday

The weather was fine, cloudy, and the wind was stiff. I flew for 30 minutes today and it was kinda exciting and different.

The headwind was so strong that the ship took off two-thirds of the way down the runway. Usually you need almost the entire runway to take off. Once safely in the air I had to use more throttle than normal to make headway, and when I made my first turn the wind blew me sideways. But once I got turned around down wind the ship really flew. My figure eight's were different than usual, as were my vertical power turns, climbing and gentle turns.

As it takes only 40 mph flying speed to keep the ship in the air I had more fun than a barrel of monkeys. After banking into the wind I slowly pulled back on the throttle until the ship was standing still in the air, as the ship still had safe flying speed I pulled the throttle back until the plane was flying backward. I handled the controls just the same as if I was flying forward. It sure was a queer sensation.

While practicing the "180 degree turn and land" I cut the "guns" (pulled the throttle back to idle) as usual as I was approaching the runway, but the wind blew me away from the field. I had to push the throttle forward and fly the ship down under power, almost like an elevator.

With a very strong head wind, and no sudden gusts, it is easier to land as the ship is really landing slower according to ground speed. There is more time to make any corrections and you shouldn't run out of runway. It would be a different story if the engine quit and I haven't figured that one out yet. Except maybe you shouldn't be up flying in that type of wind, but it sure was fun.

I took a student, Ted Kirk, Toledo, Ohio, to Seymour for his physical exam. The trip and an exam take about two-thirds of the day.

Well, we had another class on Carburetion tonight. This was a lengthy and complicated subject and will take quite a bit of studying. I didn't learn any more than I already knew, as Bud didn't seem to know much about it. I stayed at the hanger again tonight.

January 13, 1932 – Wednesday

The weather was warm, clear, and the wind was awfully rough. No one flew today.

There was a hurricane last night accompanied by a violent rainstorm with thunder and lightning. It blew both of our outhouses (toilets) and two crossing signs across the field, and blew the windsock out of sight. It blew rain in the hanger and got me all wet.

The hanger creaked and groaned, and the big hanger doors shook so hard I thought they would break off. Joe rushed out to the hanger during the middle of the storm to see if I was okay. That's the kind of fellow Joe is; he got soaked. The electricity went off so we had to use flashlights he brought. It was a wild night and I didn't sleep a wink. They said this was the worst storm this area ever had, and I can believe it.

I drove the truck down to Naill's Lumber Yard twice and picked up lumber to repair the storm damage. We all pitched in and cleaned up the mess. We rebuilt the "out-houses" and put them back in place, found the crossing signs and put them back up.

I spent the rest of the day putting on the last coat of paint on the Eagle's struts, and finished the clamps for the "Waco" windshield.

Ted Kirk quit us and went home. I paid my weekly room and board $5.50, then went up to my room and continued studying "Carburetion".

January 14, 1932 – Thursday

The weather was cool, cloudy, and the wind was very violent, so there was no flying today.

I took the vertical stabilizer, horizontal stabilizer, elevators, and rudder off the "Eagle Rock". Then cleaned the Kinner 5, radial air-cooled, 90 HP at 1800 rpm engine.

It started storming something awful while we were eating supper and rained very hard. When it came time to drive out to the hangar for the night I had to drive the old International truck, as there is something wrong with my car's engine. The truck has a windshield and cloth top but no doors so I was sure to get wet.

It was pitch dark as I drove up the Michigan hill and the rain was coming down faster than the windshield wipers could clean the windshield (it was a vacuum wiper and not very efficient). It thundered and lightening constantly, and the lightning was so bright it blinded you, but I managed to reach the top of the hill and was driving slowly toward the airport when the truck lights went out just as a lightning flash lit up the sky and blinded me and I drove into a ditch. Luckily for me I stopped the truck a foot from a culvert.

I got wet waiting for someone to drive by but I was the only one on the road. Finally I saw a headlight coming toward me from the north so I got out of the truck and stood in the driving rain. As the car approached I flagged the guy hoping he would stop or I could be there all night. There were no houses nearby as this was all farmland.

The farmer stopped and as I opened the door I tried to tell him how wet I was, that I would get the seat all wet. He said "come on, get the hell in". I was so cold, wet, shivering, and with my teeth chattering I had a hard time trying to tell him what happened. He dropped me off at Harry Vail's fine home. I never knew his name.

When Mrs. Vail opened the door I started to tell her what happened but before I could finish she made me go in the house and over to the fireplace. I left a trail of muddy footprints on the beautiful carpet in the parlor. She made a fire and dried me out, and then loaned me some of Joe's clothes and insisted on keeping mine till she had a chance to wash them. While waiting for Joe to come home Mrs. Vail made me a cup of hot tea. I finally thawed out and told her the whole story. They sure are a mighty fine family.

Joe came home and after hearing my story, (he was surprised to see me wearing his clothes) we got in his Buick and drove out to the truck. After several attempts he pulled me out of the ditch and I followed him to the hanger, as I had no lights on the truck. We got soaked. After building a roaring fire in the "pot-bellied" stove we stood close to it and tried to get dried out; first with our backs to it and then our front.

By this time Dave Gentrup stopped by Joe talked to us about the Private Pilot's Test, given by an Inspector of the Department of Commerce, that we were going to take in the near future. He explained that the Private Pilot's written examination covered air traffic rules and those portions of the Air Commerce Regulations pertaining to pilot's privileges and limitations, and the inspection and operation of aircraft. He said the Practical Flight Test would be a series of 5 gentle and 3 steep figure eight turns from 800 to 1,000 ft. respectively. Spiral in one direction from 2,000 ft., with the engine throttled, and land in normal landing attitude by wheels touching ground in front of and within 200 ft. of a line designated by the examiner, and 3 satisfactory landings to a full stop.

The Private Pilot license has to be renewed once a year with a physical exam and ten hours flying time. Every pilot must keep all flying time in a logbook.

Joe loaned us a list of questions to study that are usually a part of the written exam. They finally left at 10:30PM. I sure learned a lot tonight. This was really the first time that Joe has sat down and talked at length about all aspects of flying. I won't have to be rocked to sleep tonight. I am "pooped".

January 15, 1932 – Friday

The weather was cool, the wind was rough, and it rained all day.

I took the cowling off the International ship and spent the day removing the paint. The other students worked on the other planes we are rebuilding. Some of the ships were:

Stinson-Detroiter, parasol monoplane with a radial air-cooled Wright Whirlwind NJ5, 9-cylinder engine.

Eaglerock Biplane, with a water-cooled HISPANO SUIZA engine, 8 cylinder, 235 hp, 140 mph.

Curtis-Robin, parasol monoplane, with a radial air-cooled, 90 hp, OX5 engine.

Fleet Biplane, 5 cylinder Radial air-cooled engine.

Monocoupe, parasol monoplane, with a radial air-cooled, Velie 5, 110, special, 110 hp, 90 mph engine.

Travelaire, Biplane, 90 hp, OX5 engine.

American Eagle Biplane, with a water-cooled Hispano-Suiza engine, 8 cyclinder, 235 HP, 140 mph.

American Eagle Biplane, with a Kinner K5 air-cooled radial engine.

Waco Biplane, with a water-cooled Hispano-Suiza engine, 8 cyclinder 235 HP, 140 mph.

International, OX5 engine

Cloud Coupe, Mfg. By Cloud Coupe Corp., Milan, Indiana.

We have a Packard Diesel radial air-cooled engine on a stand used in the classroom.

The fellows came out to the hanger after supper and Harvey taught us Lesson No. 29, Ignitions.

The most convenient and effective way of igniting the compressed mixture of gas and air in the cylinder is by an electric spark, which is not only positive in action but it can be produced to meet the requirements and readily controlled.

The several forms of electric ignition systems are the low tension or the make-and-break system, the low-tension magneto system, the high-tension magneto system, and the battery-generator ignition system.

For aircraft engines, the high-tension magneto and the battery-generator systems are employed almost exclusively.

The ignition system must essentially be simple. The timing apparatus must be such as to cause the spark to occur at the proper interval in the cycle of operations, and the current must be conveyed to the spark plug in the cylinder by suitable wiring or other apparatus.

Magnetism is the property possessed by certain substances to attract or repel other substances susceptible to its effects, and when manifested by passing an electric current through a wire or conductor it is called electro-magnetism.

When separating 2 connected wires through which a current of electricity is flowing, the tendency is for the current to continue flowing in the same direction. Therefore, at the instant the wires are separated, the current will break down the insulating air gap and a spark will occur. But *such a spark is weak and it cannot jump the gap unless the ends of the wire are close together.*

This action is similar to the action occurring in a primary ignition circuit, the breaker or contact points being the points of interruption. The primary circuit alone does not produce a sufficiently strong spark, so the voltage is transformed to a secondary current that is a more practical method than increasing the primary voltage. An induction coil is the means employed to produce the spark that actually ignites the compressed mixture in the cylinder.

The mechanism that produces the break in the primary circuit to cause a current induction in the secondary circuit is called a breaker. It consists of 2 metal contact points that are alternately brought together and separated by a cam on a rotating shaft. At the moment the contact points are separated, a high voltage current is induced in the secondary winding and a spark will jump across the gap of the spark plug. A distributor rotor and the wires connecting it with the various sparkplugs in the cylinders send the current to each cylinder at the correct moment.

A condenser must be placed across the contact points if an absolute interruption of current is expected, to prevent the high voltage arc from burning the points. Its function is to prevent the differences in pressure at the contact points that is responsible for the arc.

The spark plug is primarily a wire so insulated that the only path for the electric current is across the points which projects inside of the combustion chamber of the cylinder.

Two kinds of spark plugs are in general use; porcelain, and mica. Porcelain plugs are more liable to crack or break under high temperature and high compressions. Mica plugs readily foul from oil at idling and cruising speed making their use prohibitive. A spark plug should be selected to suit the particular conditions in each engine.

The Scintilla Magneto is more extensively used at the present time than any other type. Instead of the conventional magneto, it employs a bell-shaped rotating magnet that produces the reversals of magnetic flux through the core of the coil. The contact breaker assembly and the core are stationary. The condenser is incorporated within the coil between the primary and secondary windings. All of our ships use the Scintilla Magneto.

The Delco Ignition System. In this system a generator furnishes the current with a regulator to keep the electrical output practically constant over a wide range of engine speeds. A special battery supplies ignition

Clyde H. Beyer

current for starting and for operation in case the generator fails. A switch permits the engine to operate on either one set of spark plugs or both.

These are the questions to Lesson No. 29.

1. What is magnetism and electromagnetism?
2. Describe the simplest manner in which a spark can be produced.
3. Describe the method of increasing the spark so that it can be used for ignition.
4. What are the forms of electric ignition systems? (b) What systems are used most exclusively for aircraft engines?
5. What is a rotor? (b) How does it function?
6. What is a breaker? (b) How does it function?
7. What is a distributor? (b) How does it function?What is a condenser? (b) How does it function?
8. Describe the advantages and disadvantages of the various types of spark plugs.
9. Describe fully one type of magneto
10. What is the Delco Ignition System?

Lesson 29 – Ignition Systems

THE DAYTON SCHOOL OF AVIATION

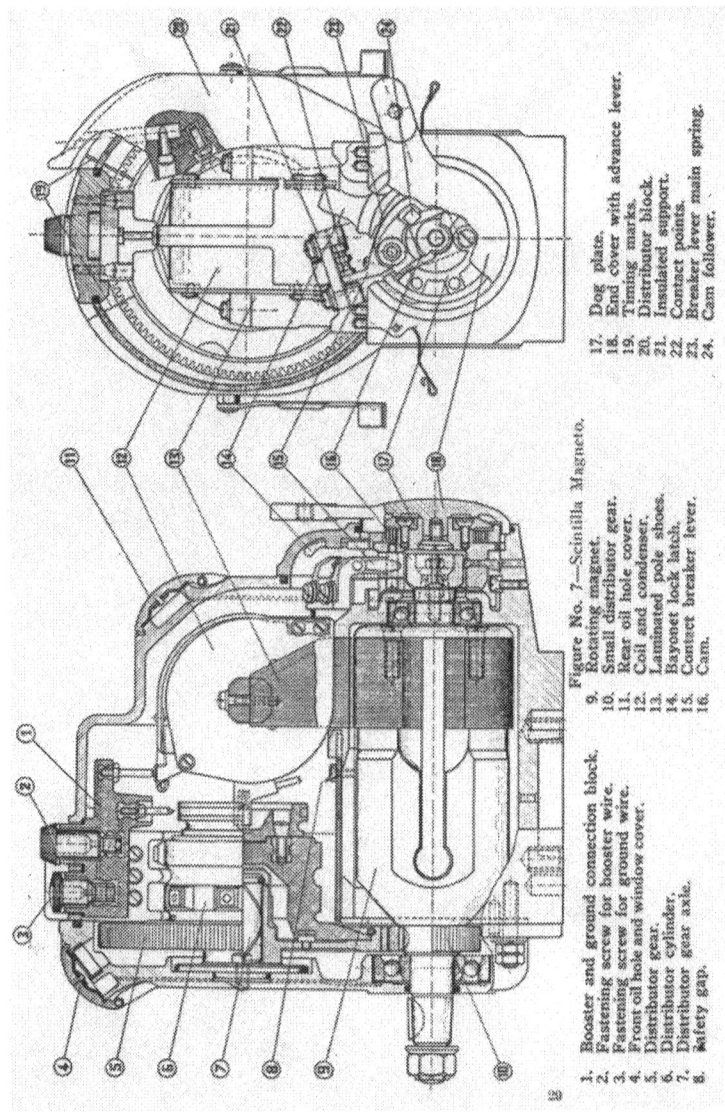

Figure No. 7—Scintilla Magneto.

1. Booster and ground connection block.
2. Fastening screw for booster wire.
3. Fastening screw for ground wire.
4. Front oil hole and window cover.
5. Distributor gear.
6. Distributor cylinder.
7. Distributor gear axle.
8. Safety gap.
9. Rotating magnet.
10. Small distributor gear.
11. Rear oil hole cover.
12. Coil and condenser.
13. Laminated pole shoes.
14. Bayonet lock latch.
15. Contact breaker lever.
16. Cam.
17. Dog plate.
18. End cover with advance lever.
19. Timing marks.
20. Distributor block.
21. Insulated support.
22. Contact points.
23. Breaker lever main spring.
24. Cam follower.

THE DAYTON SCHOOL OF AVIATION

iridium. When the letter D is added to the model number, it signifies that the magneto is provided with a connection for booster starting. This connection permits introducing a high tension current from an external source, thus facilitating the starting of the engine.

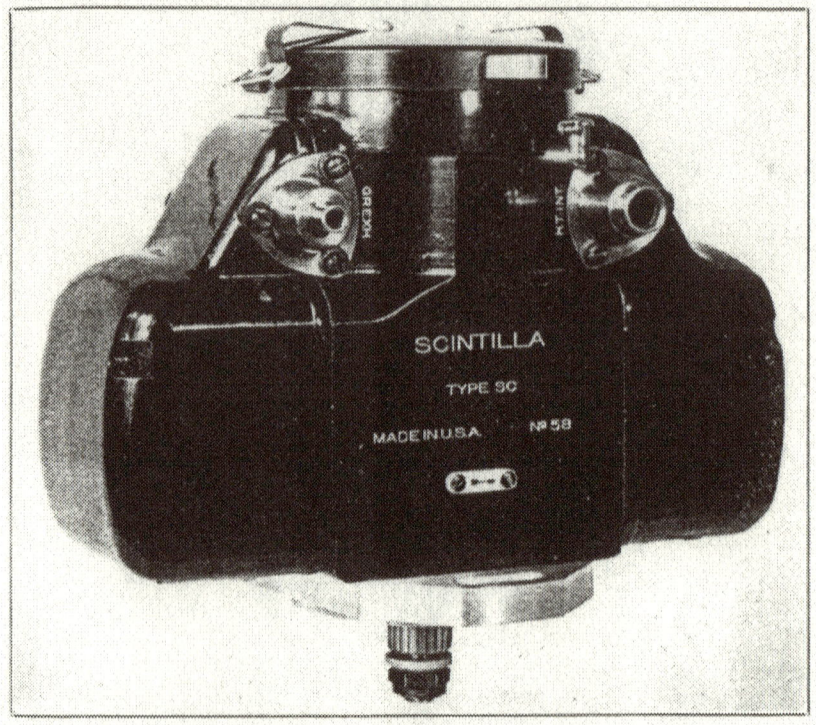

Figure No. 8—A Scintilla Vertical Double Type SC Magneto.

Scintilla also builds a double vertical magneto which is known as Type SC. This magneto has one rotating magnet producing two independent sparks every 90 degrees of its rotation. The sparks may be synchronized or staggered as required by a simple adjustment of the breakers. The magnet is rotated between two pairs of pole shoes which

19

THE DAYTON SCHOOL OF AVIATION

The principles of operation can be understood by referring to Fig. 12. In this position both switches are open. Suppose we close the left switch, then current travels from the battery through the ammeter in the switch and then to the left distributor. If the right switch is closed, then the current goes to the right distributor. The contacts in the switches are different, the four segments in the right switch being in one piece and in the left switch they are

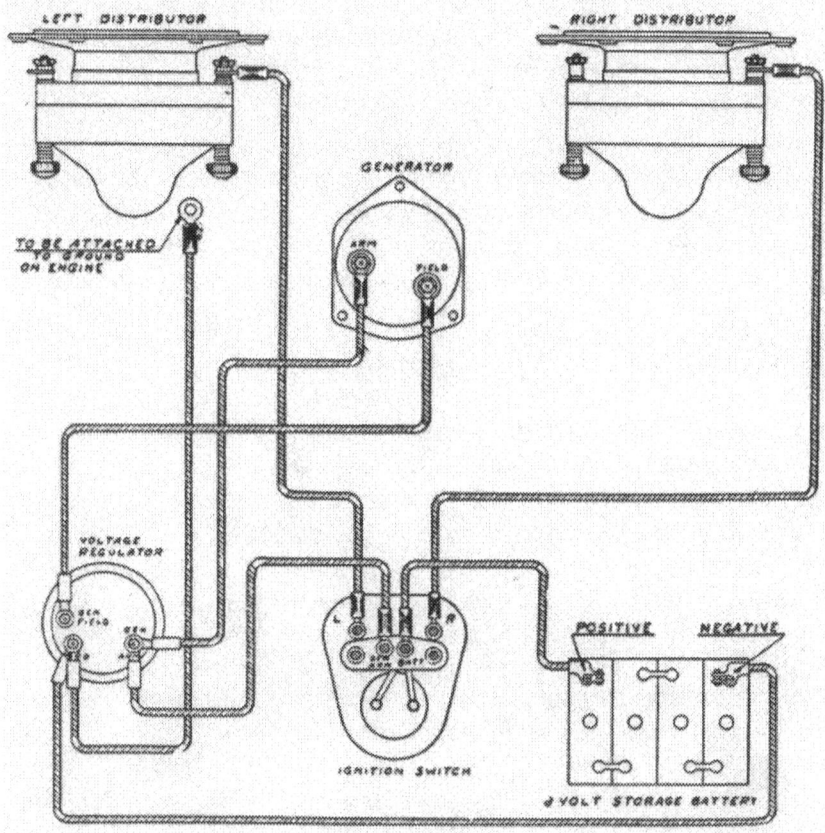

Figure No. 12—Wiring Diagram of a Delco Ignition System.

24

We finally called it quits as it was after midnight. I learned a lot about magnetos tonight.

I got a letter from Paul. It was still raining very hard as I went to sleep.

January 16, 1932 – Saturday

It rained real hard last night. The weather was cool and the wind was rough. It rained hard all day and the field is a good size lake. No flying.

I am still trying to get all the old paint off the cowling and worked at it until dinnertime. After dinner I worked on the Cheve till suppertime and found the No. 3 piston was broken. I took the head and oil pan off and removed the broken piston and rod. I replaced the broken piston with a used piston and rings, put the head and oil pan back on, and added the oil. The rest of the fellers had to work on the ships all afternoon.

Fred, Joe, and Carl Sachleben came over to the hanger after supper and we played euchre for a little while, as I had to study for the written exam.

Air Traffic Rules and Air Commerce Regulations.

It shall be unlawful to navigate any aircraft in interstate or foreign commerce unless such aircraft is registered as an aircraft of the United States. To navigate any aircraft registered as an aircraft of the United States without an aircraft certificate is in violation of the terms of such certificate. Any person who violates any provision shall be subject to a civil penalty of $500.00. Aircraft must be licensed before engaging in:

(1) Carrying persons or property for hire, or the U.S. mail
(2) Between 2 or more states, or to or from foreign countries
(3) Within the air space over the District of Columbia or any territory or possession of the United States.

Aircraft used solely for the pleasure or non-commercial purposes need not be licensed, but all aircraft must display the assigned identification mark.

An aircraft to be entitled to license and registry must be airworthy and equipped in accordance with the requirements of the Secretary of Commerce.

Registration means entry of licensed aircraft in an official license registry of the Secretary of Commerce.

Airworthiness means a condition meeting the minimum aircraft requirements –

(1) *The structural strength of wings, ailerons, tail surfaces, fuselage, including the engine mount fittings, control system, and landing gear.*

(2) *Cockpit, cabin, and control arrangements*

(3) *Power plant and power plant installation*

(4) *Equipment and instruments*

(5) *Propellers*

(6) *Design of fittings*

(7) *Materials and workmanship*

(8) *Flying characteristics and qualities*

Performance Requirements. *General flight, which includes a half hour flight test with full load to determine stability.*

Manufacturer's identification data. *The date of manufacture, serial number, and type of engine must b permanently affixed by a metal plate in the cockpit.*

Before an aircraft license will be issued, the owner must file under oath with the Secretary of Commerce an application for the license. Aircraft licenses will be issued for a period of not exceeding one year.

Identification mark *consists of the license number of the aircraft proceeded by the roman capital letter C. Located on the lower surface of the lower left wing and the upper surface of the upper right wing, the top of the letters or figures to be toward the leading edge. The mark shall also appear on both sides of the rudder. Private Pilots may pilot licensed aircraft but shall not carry persons or property for hire in licensed or unlicensed aircraft.*

An applicant for a pilot's license must be of good moral character. The minimum age requirements are 16 years for private pilots, and 18 years for industrial, limited commercial, and transport pilot.

Flying Experience Requirements. *Private Pilots, 10 hours of solo flying, of which at least two hours must have been within the last preceding 60 days prior to the filing of the application.*

Got a letter from Mom today.

January 17, 1932 – Sunday

The weather was cool and the wind was real rough. It rained hard all last night and finally quit about noon today. At last! There was no flying, as we don't have any seaplanes.

After a big breakfast I washed, shaved, and got my Sunday suit out from under the mattress. As it was nicely pressed I put it on and walked to Sunday school, which was only a block east on Main Street.

Following a great Sunday dinner Dave Gentrup and I spent the afternoon loafing all around town looking for pretty girls to date. Not having any luck Dave went home for supper and I went back to the boarding house. After eating as much as I could hold I went out to the hanger for the night.

Dave came out to the hanger and we studied for the exam. We studied so late that Dave decided to stay all night.

Ever since yesterday I have had something wrong with my stomach.

January 18, 1932 – Monday

The weather was cool, clear, and the wind was rough. No flying as the field is a sea of mud.

I worked on the "International" cowling again today, trying to straighten out the dents and fittings so it can be painted. We spent the rest of the day rearranging the ships in the hanger and cleaning up the workshop.

Joe, Dave, and I went out to the hanger after supper to iron out anything we might not understand about the Private Pilot exam. Joe gave us some more stuff to study.

__Accident Reports__. Where serious injury to persons or property is suffered or death results in operating licensed aircraft, the owner shall immediately report by telegraph or telephone to the Secretary of Commerce.

__Height__. Aircraft shall not be flown over cities, towns, or settlements except at a height sufficient to permit a reasonably safe emergency landing, which in no case shall be less than 1000 ft. Elsewhere aircraft shall not be flown at height less than 500 ft.

As Joe was getting ready to leave after a long "bull session", he surprised us by saying: "Get a good nights sleep because tomorrow is the big day. The inspector will be here bright and early in the morning to give us the written and flying exams". As he was going out the door he turned and said, "It might be a good idea to pray for a dry field and a calm wind". That wasn't the only things I prayed for, I kinda asked for a little help in passing the tests.

Before hitting the "sack" I looked out the door and the field still looked awfully wet. The wind was blowing really hard so maybe it will dry out some.

Earl sent me a letter with my license, and a check. I still don't feel well.

January 19, 1932 – Tuesday

The weather was cool, clear, and a stiff breeze was blowing from the south. That was a very bad sign. The field didn't look too bad but was probably soft.

I got up early, went down to the boarding house, changed clothes, washed, shaved, and had a nice breakfast of eggs, pancakes, and coffee. I went back out to the airport to get ready for the written and flying exam for my Private Pilot license. I was excited and a little nervous, as I was never very good at taking written tests. But when I met Joe Schumate, the Department of commerce Inspector who was going to give the tests, I felt a lot better as he seemed to be a regular "Joe".

Dave and I went in the office and the inspector gave us the written test to fill out. The written test is given first and if you fail you don't take the flying test. When we finished the test we gave the papers to the Inspector and waited outside with the rest of the students while he graded them. I thought the test was easy but I was still nervous because I sure wanted to take the flying test. He finally came out and told Dave and I that we passed, in fact that we did very well. That was a big relief and it made me more confident about taking the flying test. Homework paid off.

I had hoped Dave would be the first to take the flying test; that would give me a good idea about what to expect. But of course, I was first. After we pushed my favorite "Prep" out of the hanger and on the apron facing west in takeoff position I concentrated on not missing one little item of the pre-flight inspection. I was not going to give him any excuse to flunk me.

I checked the propeller for cracks or any other defects, and made sure the propeller bolts and nuts were tight and in good shape. Next came the engine to see if there were any gas or oil leaks, and to make sure that all the bolts and nuts were tight and in good condition. All the hoses, tubing, control cable, and rods were o.k. Next I examined the landing gear, struts, axles, wheels, bearings, and tires for defects. All the bolts and nuts were tight and in good condition. Then I examined the fuselage from "prop to tail" to make sure the covering was alive, tight as a drum, with no wrinkles, cuts, or scrapes. I made sure the surfaces of the horizontal and vertical stabilizers, elevators, and rudder were in good condition and that they were securely fastened to the fuselage. The tailskid checked out okay, the spring steel sections were firm yet flexible and were tightly bolted to the fuselage.

I went back to the wings. After checking the fabric, struts, and their fasteners, I grabbed the tip of the left wing and gave it a sharp jerk downward, and by doing the same thing to the right wing, you could determine whether there is any great defect in the wings or wires. Finding all the exposed control cables and fittings, as well as the doors okay, I climbed in the cockpit.

I glanced sideways at Joe and the Inspector, who was watching me, and they looked like they were getting tired waiting for me to finish.

The instrument panel looked okay, the seats, seatbelts, "joy stick", and rudder pedals were securely fastened. I moved the stick in all directions, the ailerons and elevators responded normally. The rudder moved freely as I pushed the rudder pedals.

Last but not least, I climbed out of the cockpit, checked the oil level and filled the gas tank using a funnel covered with a chamois to prevent any dirt getting in the tank. As the wheel chocks had been placed in front of and behind the wheels I was, ready or not, to go.

I walked over to both Joes and told them everything was okay. The wind blew pretty darn strong from the south, which made taking off and landing in a stiff crosswind a problem I didn't need. But I discovered I had a different problem. The inspector told me to start the engine and let it warm up.

Joe climbed in the cockpit and I stood in front of the propeller. Joe left the door open so we could hear each other. I told him "switch off, gas and choke on". He repeated, "switch off and gas and choke on". I turned the prop a couple of times to suck gas vapor into the cylinders. Then I stepped back from the prop and told Joe "switch on" and he replied, "Switch on".

I stepped forward, braced myself with both feet firmly on the ground, and took hold of the tip of the prop, which was about shoulder high. I called out "contact" and he replied "contact". I swung my right leg forward and gave a sudden pull downward as my right foot swung backward and helped me to step back away from the spinning prop.

The engine sputtered, sneezed, coughed, and spit black smoke out of the exhaust pipe as Joe adjusted the choke it settled into a steady roar. He set the throttle at idling speed and the engine cackled a steady, sweet,

1-3-5-2-4. After tying the stick back against the seat to prevent the ship from accidentally climbing over the chocks if the engine speeded up, he climbed out.

While I waited for the engine to warm up I noticed the inspector had written quite a bit on the pad he was carrying. I sure would like to know what he wrote. The inspector said it was time to go. As I got in the cockpit he told me to taxi to the north end of north-south runway. That was the different problem I had. I had never used the north-south runway; as a matter of fact I had never seen anyone use it. As short as the east-west runway is N-S runway is a good one-fifth shorter.

I didn't think I could get the ship off the ground before I ran out of runway due to the soft condition of the field but I guess I had to try. We never spent much time maintaining it. There were a couple of minor advantages. At each end of the runway there were ordinary wire farm fences about 3 ft. tall instead of those tall trees and high lines. There was however, a telephone wire across the south end of the runway above the fence. I sure was glad there was a very strong wind.

Joe pulled the chocks away from the wheels and I pushed the throttle forward just enough to slowly "taxi" to the north end of the runway. I turned the ship south as close to the fence as possible, and the runway looked shorter than ever. As I taxied I noticed the ship didn't seem to roll very freely and that is just what I had been afraid of.

I made sure the seat belt was fastened, then pushed the throttle wide open, and I pulled the stick back to the seat until the engine was running at top speed.

Then I pushed the "joy stick" slightly forward to raise the tail off the ground and the ship started to roll forward but way too slow to suit me. Suddenly, just a short distance down the runway as the ship was picking up speed, but not enough to become airborne; the wheels hit a ditch that was about 2 ft deep. We bounced about 15 ft in the air and as we didn't quite have flying speed the ship started to wobble back to earth. I tried with all my strength as though I could physically keep the ship in the air, and I did manage to keep it from touching the ground.

With the engine running wide open the ship straightened up just a few inches off the ground. With the wheels no longer dragging along the ground the ship picked up speed as we headed toward the fence. I was sure we, the ship and I, were in deep trouble, as it was too late to try to stop. I had to get the ship at least 3 or 4 ft off the ground to pass over the fence. It was impossible to get high enough to pass over the telephone wires so all I could hope for was to get over the fence and under the wires.

141

The strong head wind gave me a little extra lift but I was still just a few inches off the ground. At the last minute I pulled the stick back and as the ship flew over the fence I pushed the stick forward because the ship felt like it was about to stall. Flying straight ahead a foot off the ground I gradually gained normal flying speed. Lucky for me the field was a pasture with no trees.

After reaching 1,000 ft altitude I turned back to the field, taking my time to relax and prepare for the maneuver. As cold as it was, I felt damp but it was just sweat. I made sure the inspector could see me and made 2 nice gentle figure 8's without losing or gaining altitude, which was a surprise. Now all I had to do was to make two normal landings and one spiral landing.

I thought this was the end of the test because I couldn't land and take off three times across that ditch.

Finishing the maneuver I looked down and saw the students filling the ditch with gravel. I circled the field until the inspector flagged me down. Not knowing whether the landing gear had been damaged I decided to try to land with the tail touching the ground an instant before the wheels. I hopped, in that way it would help keep the ship from nosing over.

Thankfully the wind was still blowing very hard. I approached the runway, crossed the fence, and sat the ship down where the ditch had been, in a nice tail first landing. Nothing happened. After a quick inspection he waved me off. This take-off was a struggle but not near as bad as the first, and the ship barely cleared the fence. I made a normal 3-point landing and took off for the last time and made a fine spiral landing.

I passed the test and the inspector told me the first take-off wasn't pretty but I did great. I felt bad because Dave failed the test. I had to get my picture taken yesterday for the license - $1.25.

I didn't work today because I still don't feel well. Dave and I spent the afternoon in my room "shooting the breeze".

I bought 3 gallon of gas Saturday - $0.50. I got a letter from Herb today.

Although I didn't feel well, I got cleaned up and went to church for supper tonight and had a good time. They call it "Social League" and after supper they have some kind of entertainment. It is the third Tuesday of every month.

January 20, 1932 – Wednesday

The weather was cool, cloudy, and the wind was real rough.

I didn't get to fly today, so I stayed home all morning and am beginning to feel a little better.

I went out to the airport this afternoon and worked on that cowling. The other students worked on the other ships we are rebuilding. Harvey went from ship to ship to see that the students were doing the work properly.

Becht and Shucks came by after supper and we loafed the streets. Wrote Mom a letter.

I haven't felt very well the last few days and didn't feel like celebrating, no parties, no date, no nothing.

I am the first and only, so far, licensed student pilot of the Dixie Flying Service.

January 21, 1932 – Thursday

The weather was raining, cloudy, and the wind was real rough. No flying today.

We took the "OX" engine out of the International ship and took it apart. We cleaned the parts with gasoline and examined them far defects.

Dick Oglesby went to Chicago to get another plane so we all loafed around most of the afternoon. I boxed 5 rounds with Max Sittner.

I took my suit to the cleaners and had it cleaned and pressed for $1.00. Also bought 2 rolls of film - $0.50 and picked up my films $0.35.

After supper I went out to Sachleben's and we played euchre and I won 5 out of 8 games. That "home brew" tastes better every time. Some day I am going to get them to tell me how to make it. When I got home I wrote Paul a letter before hitting the "sack".

The Workshop

"OX5" engine out the International that I was overhauling; "Eagle" wing we first covered; and "Velie" crankcase – January 3, 1932.

January 22, 1932. – Friday

The weather was rainy, cloudy, and the wind was rough. No flying today.

I ground the valves and made new gaskets for the "OX" engine after cleaning out the carbon.

As Dick was still out of town we all loafed around the rest of the day. I boxed 5 more rounds with Max, then played "casina" with Bud the rest of the afternoon.

I paid my room and board, $16.50 for 3 weeks. Earl S. sent me a nice letter.

January 23, 1932 – Saturday

The weather was rainy, cloudy, cold, and the wind was very wild. Still no flying.

I spent the morning starting to put the "OX" engine back together after putting new rings on the pistons. It is slow work, as every piece has to be examined thoroughly. It only takes one worn or defective part to cause an engine to breakdown. It doesn't make much difference on my Cheve but it sure would up in the air.

As it is Saturday and Dick is still out of town, everyone took the afternoon off.

I went back to the hanger and worked on my Cheve. I cleaned the engine, adjusted the valves, and found one of the valve tappets was stuck. It took quite a while, with penetrating oil, to get it loose. I had to work on that darn horn again.

Joe S. and I rode around this evening looking for girls to pick up but we didn't have any luck. After dropping Joe off at his home I went back home. Becht and Shucks came by and we went loafing up town.

January 24, 1932 – Sunday

The weather was cold, cloudy, and the wind was very rough. Still no flying.

After a breakfast of bacon and eggs, pancakes and molasses, orange juice, and coffee, I got all "shined" up, put on my best and only newly cleaned and pressed Sunday suit and went to Sunday School with the Sachlebens. I am in Joe's class.

Becht and Shucks came by after dinner and we went out to the airport.

Dick got back. He brought a Stinson-Detroiter ship from Chicago. It is a swell 5-place cabin job. A pilot and co-pilot operate it and has a wheel instead of a "joy stick". It even has brakes. The engine is a NJ5, 9-cylinder Wright Whirlwind. I messed around in the cabin and cockpit hoping that I would get a chance to fly it someday. But if it is like the rest of the planes we get it will have to be rebuilt. It would be quite a change to use a wheel instead of the stick and brakes to stop the ship.

Becht, Shucks, Joe S. and I went for a walk through the Sachleben pasture on the east side of Michigan Road. The east end of the pasture drops off into a steep deep valley where we found a miniature "Clifty Falls". We are going to see it next Sunday.

I got back to the boarding house just in time for supper. You better not be late because there is a very good chance there will be no food left. Generally supper is the leftovers from dinner. I got spruced up and went to church.

Marlin with Stinson

January 25, 1932 - Monday

The weather was cool, clear, and the wind was bumpy. I finally got to fly one hour today.

A tire went flat on the ship as I was about to take off but I had plenty of time to stop. After fixing the tire I took off into a pretty bad crosswind. I sure

145

didn't want to use that north-south runway again. I steered the ship slightly south of west to compensate for the cross wind, but it still almost blew me into the trees.

I practiced the figure eight, spiral, 180 degree, and 360 degree turns. I got bored doing the same old maneuvers all the time so I went up about 4,000 ft to do a couple of stalls.

I took a real good look at the ground below me for the first time. Up to now I had concentrated entirely on perfecting the different maneuvers.

The earth looked like a giant brown quilt; a kind of crazy quilt pattern. Each farm was a different size and shape patch, identified by fences of irregular types, as well as roads. Most farmhouses, barns, and sheds were distinctly different in color, type, size, and shape, making each farm easily recognizable. Now I know how to keep from getting lost in my home territory.

Most of the roads seem to go north and south or east and west, but some seem to go northeast and southwest. A few look like worms twisting and turning in different directions.

The Ohio River looked like a giant snake, whose head in Pittsburgh and tail in Cairo was out of sight. It twisted and turned in every direction imaginable and at times seemed to reverse itself. It crawled in a southwesterly direction from Pittsburgh and formed a giant U with Madison sitting at the base of the U on its north side. It looked calm and peaceful, not menacing, laying there in the sun.

It sure was a beautiful picture but my time was running out so I quickly did a couple of stalls. I get a "kick" out of doing them as there is just a minute or so between regaining control of the ship or completely losing it. That is why it is necessary to have plenty of altitude. The odds are against you if the ship stalls close to the ground as a lot of pilots have found to their sorrow.

A "whip stall" can be very dangerous, and should be avoided if at all possible. It generally results from losing flying speed in a very steep zoom or an uncompleted loop, and starts with the ship falling tail first and then whipping around as the nose drops. The controls should be held steady to avoid any more strain or whip on them than necessary.

I made one three-point landing and two normal landings with the wheels touching the ground slightly ahead of the tailskid. The ship didn't bounce so that was an improvement. I barely undershot the field twice and overshot it once: not bad considering the rough crosswind.

Once coming in for a landing a gust of wind got under the left wing and almost turned the ship over. Twice just as I was about to set the ship down on the runway a gust lifted the ship a couple of feet off the ground. I held the controls steady and the ship settled back down on the runway.

The other students started taking the Stinson apart. I spent the afternoon sanding a "Fleet" wing and it was a very boring job that had to be done carefully.I got a letter from Ma. Becht and I loafed around town after supper.

January 26, 1932 – Tuesday

The weather was cold, rainy, and the wind was awful rough. No flying today.

I sandpapered another "Fleet" wing while the other students took the wings and tail assembly off the Stinson fuselage. Then we dismantled the engine.

Dick is in the hospital with heart trouble.

We were going to initiate Sam after supper but decided to do **Art Prosek** first. That was a mistake. Art is the tallest student, and he is not skinny. He is bald and that makes him look a lot older than the rest of us. Before it was all over we smashed Sam and Linn's bed so we quit. I let Sam sleep in my bed because it was his bed that we broke. Frankly I never liked hazing of any kind, I always seemed to come out with the short end of the stick. Mrs. Barber didn't appreciate this monkey business so we all pitched in to buy another bed. This is going to hurt, we are all as poor as church mice, honest.

Linn, Art, and I had to sleep in one bed. There are three beds in each room and two students have to sleep in each bed. That is, all but me, I sleep alone in that little "cubby hole" at the head of the stairs, the farthest away from the bathroom.

January 27, 1932 – Wednesday

The weather was cool, fair, and the wind was medium rough. The other students had flying lessons so I didn't get a chance to fly.

After taking the old fabric off the wings, fuselage, and tail assembly, we measured the Stinson to find out how much fabric was needed to cover it. We sanded the frame smooth that comes in contact with the fabric. Joe and I removed, cleaned, oiled, and replaced the magneto on his Waco. Then we removed, cleaned, adjusted the gaps, and replaced the spark plugs. No one is allowed to work on his Waco unless he is there to supervise the job.

After supper we all jumped in and on the old International truck and went out to the hanger to school. Harvey reviewed *Lesson No. 27, "Principles of the Engine"*, and then *Lesson No. 28, Carburetion"*. He wants each of us to draw a 4-cylinder airplane engine showing the cylinder, pistons, connecting rods, crankshaft and bearings, valves and tappets, and a camshaft.

We are to draw a carburetor showing the various jets, venturi, choke, float and valve, butterfly valve, and adjusting needles. Also, show the flow of fuel from the gasoline line through the carburetor and into the intake manifold. These drawings are to be given to him at the next class. This should be right up my alley, I took art in high school, and I am always working on my Cheve engine.

I will never forget that class in art I took at Peabody High School. It was an elective course and I was probably an average student. An Art Supply company sponsored a contest that was restricted to high school art students and their teachers. The object of the contest was to design a front cover for their catalog. We turned our designs into the teacher and a long time later we found out that the company had chosen our teacher as the winner. I was curious to see what her design looked like and when I saw it I was surprised. She took credit for my design.

My Mom and Dad always told us kids that if we got in trouble in school we would be in double trouble when we got home. So I was always quiet and timid, but this time I wanted to know how she got credit for my design. It didn't do any good. I got the catalog and kept it for many years.

I was always pretty good at drawing pictures but I didn't know anything about painting them. The teacher read a short story and told us to paint a picture that had something to do with the story. When she saw my painting she got real upset and held it up so everyone could see it and ridiculed it up one side and down the other. I was so embarrassed I wanted to disappear. After the outburst she went down the hall and showed it to several teachers. When she came back she said I didn't have a chance of passing the course. I couldn't understand what all the commotion was about but it seems that all the colors were too bright. I thought it was a nice picture. She sure ruined any thought I might have had to become an artist. I never painted another picture.[19]

---

[19] How strange! In 1998 I walked up Main Street and saw 2 new

I was in the doghouse the rest of the semester and I needed the credit to graduate. Each student, as a final assignment, was to give a speech on art in assembly hall. I chose as my topic "Prehistoric Architecture", and studied it every night. I always hated to get up in front of a group and give a talk. When it was my turn I had them put up a large blackboard so I could illustrate what I was talking about. In that way I wouldn't be constantly facing the audience. My dad always said that if you knew your topic well enough you would do okay.

Of course, before I even opened my mouth I forgot what I was going to say. So I turned to the blackboard and started to draw as I told a joke. As they all started to laugh I looked over at the teacher and she was laughing. That gave me time to get back on track, and as I really knew the subject, I sailed right on through. I couldn't resist telling another joke as I finished my speech.

The teacher said she didn't have the heart to flunk me after I gave that speech as she thought I was really trying to do my best. I got a passing grade.

January 28, 1932 – Thursday

The weather was fair, cool, and the wind was medium rough.

I flew 45 minutes this morning and made some nice smooth, gentle and steep figure eights, as well as a couple of exciting stalls. On the "180 degree turn and land" I made a great 3-point landing, but on the first "360 degree turn and land" I overshot the field and made a wheel landing. On the second and final "360 degree turn and land" I made a nice normal approach, and made a wheel landing that was very close to a 3-point landing.

We closed up shop today. Dick died at the hospital last night from a heart problem. They shipped him to his home in Memphis, Tenn.

We had some arguments, but I think he was a great manager and I am sorry he died. This is going to have an effect on the company. He was the brains, dreamer, and promoter; he ran the whole show. I don't know what is going to happen in the future, or who will replace him. But I am very sure if he had lived, the school would be a big success. I think we liked each other.

After writing a letter to Earl, Paul, and Earl S., I worked on my drawings of the engine and carburetor.

paintings at least 3 ft x 3 ft in a store window, painted by a local artist. The colors were even brighter than mine.

January 29, 1932 – Friday

The weather was cold, rainy and snowy, and the wind was extremely rough. So no one flew today.

We started to build a crate to ship the LeBlond engine to Canada. It was quite a job and it is not quite finished.

All the fellows, and I, have started making ashtrays out of used airplane engine pistons. I made a nice watch fob out of a used J5 engine piston. No one seems to know what to do.

Art Prosek and I took Don Long to Louisville, KY, and he paid all my expenses both ways. It snowed all the way down. This was the first time I was ever in Kentucky. Most of the streets in Louisville are narrow. As he got out of the car Don promised to keep a lookout for a job for me. He is a darn nice fellow. We slid all the way home.

I spent the rest of the evening finishing both drawings and they look pretty good.

January 30, 1932 – Saturday

The weather was cold, cloudy, and the wind was a gale. No one flew today.

We made a crate for the propeller, and continued building the crate for the engine. In between times I worked on my ashtray.

We all went to school at the airport this afternoon. These classes are very informal, and Harvey does the teaching. We gave Harvey our drawings of the carburetor, and the 4-cycle engine. I should do pretty well according to all the other drawings I saw.

Today was *Lesson No. 31 – Lubrication, Fuel System, and Superchargers.*

Friction is the resistance to motion caused by two metallic surfaces sliding over each other, and it results in wear or abrasion on these surfaces. When a lubricant of some kind separates these surfaces, the amount of friction is greatly reduced.

When two surfaces are partly separated by oil or grease and the friction is reduced by the quality of the oiliness of this lubricant, the lubrication is known to be partial or incomplete. When a layer of viscous oil separates the surfaces, so that the two metals are not in contact, we have what is known as perfect or complete lubrication.

Another type of friction known as rolling friction is found in ball or roller bearings. Steel balls and rollers have a certain amount of elasticity and the friction is due to the deformation on their surfaces or the surfaces on which they contact under heavy loads. Lubrication in bearings that have rolling contact is not so important as in the case of journal bearings. The principle reason why a lubricant of some kind is necessary is to prevent rust from occurring on the metal surfaces.

In a plain journal bearing, the journal would contact with the bearing at its lowest point along a narrow line when at rest. When turning, the oil that is introduced between the bearing and journal will be rolled in the direction of rotation until a film is formed to separate these two surfaces. Oil must be introduced at the proper point and under sufficient pressure to insure the formation of this separating film, otherwise perfect lubrication will not be obtained.

In well lubricated bearings, such as are used for the journals and crank-pins of aircraft engine crankshafts, it has been found best to use a lining of white metal commonly known as babbit. The composition of which the greater part is lead or tin, preferably the latter, to which is added small percentages of copper and antimony forming small copper-antimony and copper-tin crystals. These hard crystals that support the journal are dispersed throughout the soft tin. Such a composition offers some of the advantages of a hard bearing surface, but at the same time is sufficiently pliable to adjust itself to the journal and thereby overcome slight inaccuracies in machining.

There is a further advantage in the use of babbit bearings from the point of view of safety. Should there be a failure in the lubricating system, a bearing would quickly overheat. With some materials the bearing would seize and cause a great deal of damage, while babbit would merely become soft or melt at elevated temperatures and therefore not cause any serious damage to the journal or shaft.

The lubrication of aircraft engines is a far more exacting problem than that of an automobile engine and most other types of internal combustion engines, which are not subjected to such high temperature and bearing loads. This means that only lubricating oils of the best quality and having a viscosity should be used.

The oil must be delivered to the bearings under fairly high pressures in order to insure an ample supply for good lubrication and to maintain the temperature of the bearings within safe limits. The oil supply is no longer carried in the engine crankcase. Separate tanks are mounted at a convenient point in the airplane and the oil is piped to the inlet side of the pressure pump that feeds it to the various parts of the engine. The oil is returned to the tank by a scavenging pump after it has passed through the bearings.

The oil should have adhesive qualities known as oiliness that insures clinging to the surface of the bearing. It should also have cohesive qualities known as viscosity that insures that the particles of oil will cling together and resist the tendency to separate under load. The oil must withstand high temperatures and it must remain fluid at low temperatures. This limits the choice to oils made from paraffin base crudes.

Oils made from Pennsylvania crudes are of the highest order. Oils having a viscosity of 100 or above at 210 degrees F. are recommended in most aircraft engines for general use.

In aircraft engines there are three types of oil pumps used; gear, sliding vanes, and piston types.

The Gear Pump is most commonly used in American engines. Two spur gears closely meshing in a housing carries the oil introduced at one side to the opposite side. Usually the pumps are built up in more than one compartment; one delivers the oil to the engine under pressure, while the other one scavenges the excess oil collecting in the engine crankcase.

Oil pumps are usually located at the lower part of the engine and can be removed for inspection and cleaning. Screens are provided for both pressure and scavenging gears. The oil pressure is regulated by a relief valve that is adjusted by means of changing washers behind the spring, and this pressure is usually set around 40 lbs. Per square inch at an engine speed of 1,700 rpm.

Sliding Vane Pumps. This type of pump has the drive shaft slightly eccentric to the bore. This shaft is provided with slots receiving vanes that are usually held apart by small coil springs that force the vanes outward against the walls of the pump body. The vanes slide in and out of the slots as the shaft revolves due to its eccentricity. Oil is taken from the inlet side of the pump and delivered to the pressure side with positive results.

Piston Oil Pumps. The Basse Selve pump consists of two double-acting plungers working vertically in aluminum barrels and rotated by a worm gear and simultaneously reciprocated by a scroll-cam. During every stroke of the two plungers oil is drawn from the reserve tank to one of the inner pump chambers and is then discharged from the other chamber into the main delivery line. At the same time oil is drawn from the engine sump into the reserve tank.

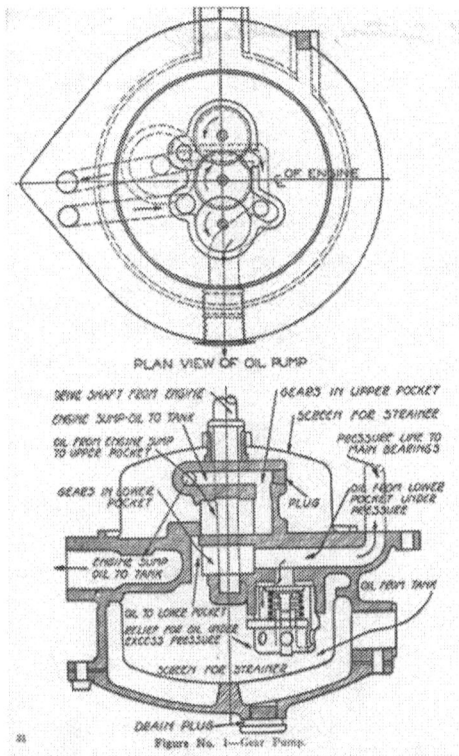

PLAN VIEW OF OIL PUMP

Figure No. 1—Gear Pump.

These are the questions for Lesson No. 31

(1) (a) What is friction? (b) How can friction be reduced?
(2) (a) What are the various degrees of Lubrication? (b) Why are lubricants used?
(3) Describe the lubrication in a plain journal bearing.
(4) (a) What is babbit and where is it used? (b) What are its advantages?
(5) What lubrication systems are used in aircraft engines that differ from the automobile?
(6) What characteristics should aircraft lubricants have?
(7) Name and describe the three different types of oil pumps.
(8) Where are the oil pumps located?

This lesson was too long and covered a lot of information, but I did learn a lot.

I found out why I was having all that trouble with my Cheve engine. Whenever I needed to add a quart of oil between oil changes they gave me a quart of aircraft oil.

What I didn't realize was that aircraft oil had a viscosity of over 100 while automobile engines used oil viscosity of only 30. This meant the oil was over three times too thick to flow freely in my Cheve engine, as it doesn't have an oil pump to force the oil where it is needed.

My Cheve engine has metal scoops attached to the connecting rods to guide the oil to the bearings as well as splash the oil on the cylinder walls, camshaft, etc. so the moving parts of the engine didn't get enough oil to prevent friction from causing wear and breakdown. From now on I'll buy my oil.

Loafed downtown with Ed Becht and Bob Shucks tonight. I met both of them at Sunday school and they are about the only two local fellows who will have anything to do with us. All the students have girl friends so after supper we usually go our own way.

I bought a valet set for 75c, oranges for 20c, peanuts 12c, gloves 25c, watch chain 10c, and a drawing set 10c.

January 31, 1932 – Sunday

The weather was cold, clear, and the wind was medium rough. I flew a grand total of ten minutes today; it took me a lot longer to get ready to fly. I made a 360-degree turn and a very bad landing; undershot the field and made a wheel landing.

I went back to the boarding house, got all cleaned up and went to Sunday school with the Sachlebens.

After Sunday school I drove out to the Sachleben's for dinner. Fred, Joe, and I walked over to the airport, as it is right next to the Sachleben's house.

We watched a Monocoupe, with a 7-cylinder Warner engine, fly in from Cincinnati for a visit. I took two pictures of Fred and Joe with the "coupe" and one of the shop.

The fellow came and took that Aeronca. It was a special built job and he put on quite a show.

We walked back to the house for supper. I enjoyed both meals a lot today; they were quite different from the boarding house meals.

I went to church with the Sachleben's tonight.

Fleet Wing in workshop

Joe and Fred Sachleben with Monocoupe

February 1, 1932 – Monday

The weather was cold, fair, and the wind was smooth. I flew for one hour today and all things considered did very poor work today. I made several poor figure eights, several stalls, 180-degree and 360-degree spirals. I made one fair landing but overshot the field two times, and undershot it one time. The landings were different this time; I landed on the left wheel and the next time I landed on the right wheel. I've never done that before.

Clyde H. Beyer

We finished crating that engine today and we worked on our ashtrays for a while.

I went down to Maddox's garage on Second St. after supper, and loafed with the fellows for a while.

February 2, 1932 – Tuesday

The weather was warm, cloudy, foggy, and the wind was fine. I didn't get to fly today.

We had a little excitement this morning. I had just finished my ashtray when I found an old cigar box filled with old 45 caliber bullets. I thought they would make nice feet for the ashtray but didn't know how. One of the students said he could take the shells apart and we could use the lead points for feet.

He took 8 or 9 apart without any trouble by putting the shell in a vise and chiseling off the case and powder. But the last one exploded taking the end off one of his fingers and ricocheting around the shop before going through the wall. It's a good thing it didn't hit anyone cause it sure left a big hole in the wall.

I drilled a hole in four of them, tapped threads in them, screwed short bolts in, and cut off the heads. I drilled four holes in the ashtray, tapped threads in them and screwed the four feet in.[20]

They decided to have class this afternoon on water pumps and safety. We didn't do any work this morning so I slept till dinnertime.

We went back to the shop and had a stern lecture on safety of all circumstances, including fooling around with bullets. They gave us our drawings we turned in on Saturday. When they calmed down we discussed water-cooled engines and water pumps.

---

[20] Sure made a nice ashtray and I still have it.

Ashtray made from Velie M5 Piston

THE DAYTON SCHOOL OF AVIATION

The rate of air flow is another determining factor, for if we increase the velocity of air passing over a high surface, we can decrease in almost the same proportion the time in which this heat will be dissipated. This applies to both air-cooled cylinders and radiators for water-cooled engines. The rate of cooling a water-cooled cylinder will be increased

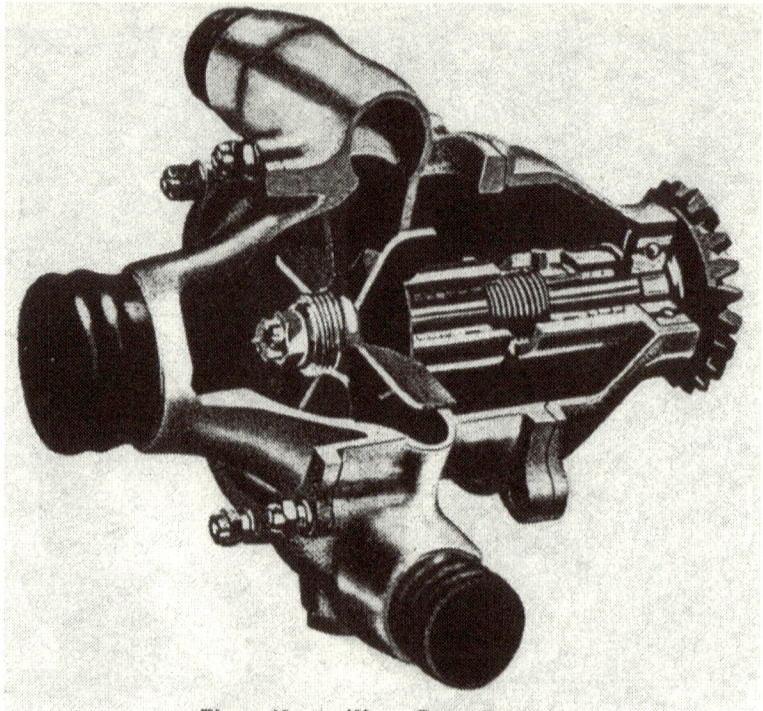

Figure No. 1—Water Pump Details.

by a more rapid circulation of water up to a certain critical velocity. This is the point where turbulent flow begins, and the water apparently scours all the surfaces.

Perhaps the most important factor affecting the rate of cooling is the difference in temperature between the cooling

4

THE DAYTON SCHOOL OF AVIATION

medium and the surface to be cooled, and it especially important in water-cooled engines in designing an efficient radiator.

Water Cooling Systems: An aircraft engine when cooled by water must be provided with the following elements suitably piped together. There must be jackets about the cylinders, a reserve or surge tank above the engine, a circulation pump, and a radiator having a core or cooling element with the proper header and collection tanks. Very often a radiator shutter is provided, thus making it possible to regulate the rate of cooling to meet the wide variations in air temperature.

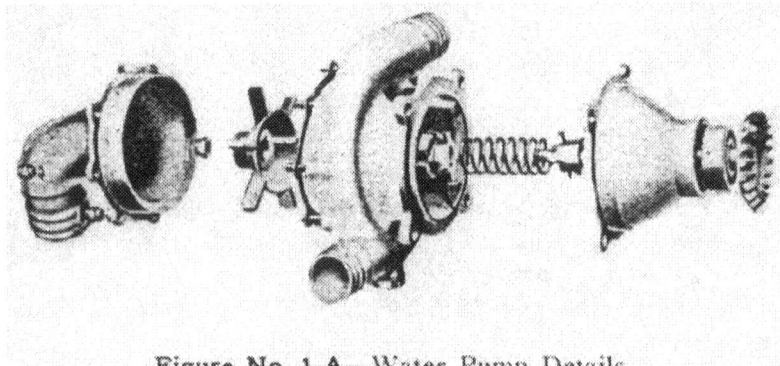

Figure No. 1-A—Water Pump Details.

In the ordinary open system, the pump forces the water through the jackets against a normal head, and while passing through the jackets it becomes heated. The water passes through a reserve or surge tank and then to the header and radiator core. The top of the radiator is vented to the air so the core must pass the full delivery from the pump without offering any positive resistance, otherwise the water would overflow and be lost at this point. The water flows through the radiator core by gravity assisted by the suction at the inlet side of the pump. The water temperature is re-circulated ....

5

Lesson No. 30, "Cooling of Aircraft Engines".

Aircraft engine cylinders are cooled either by air or water. By water - the heat is given to the water circulating through the jackets surrounding the cylinders. In the water-cooled engine the heat is finally given to the air circulated through a radiator.

The water pump is usually of the centrifugal type, not unlike the pumps used in automobiles, except that it must be more efficient and have greater capacity. More care is exercised in designing the impeller and the shape of the inlet and discharge passages. The pumps are usually driven at a speed higher than that of the crankshaft.

The pump forces the water through the jackets and becomes heated. The water passes into a reserve tank, and then to the radiator. The water flows through the radiator and it is then again circulated through the cylinder jackets by the pumps. All water-cooled systems must be provided with a drain at the lowest point.

We took turns taking a water pump apart and putting it back together. Pumps are not very complicated but great care must be used in assembling them.

Marlin Christianson, he's from Minnesota, and I are having a checker contest. He won 3 games and I won 3 games. I used to play checkers with Earl Seitz dad and he taught me a lot about the game. I bet I lost about 25 games before I won one.

I wrote Fred a letter, and a two-sentence letter to Rosalie. Then I walked down to the garage and shot the breeze with Jim Oakley and Francis Ferguson who used to be my rival. [21] We might go to Louisville Saturday night.

February 3, 1932 – Wednesday

The weather warm, cloudy, rainy, and the wind was pretty smooth.

The only work I did today was to sweep the floor. Since Dick died everyone does their own thing. There is no work schedule, no assignments, no regular classes, and we don't know who is going to fly and when. Up to now we all worked on rebuilding the ships or overhauling the engines on a regular basis. Each of us was assigned a specific job so that we would get a thorough education. I don't know what is going to happen to this place if someone doesn't take charge. Joe seems to be interested only in

---

[21] We have been friends ever since, until he died a couple of years ago.

teaching the students to fly and that's all he does. Harvey, the mechanic, works at whatever he thinks is necessary to keep the ships flying.

I am making Dad an ashtray like mine just to keep busy. Two days ago, a new student, Bill Wolter from Madison, Ohio, showed up. I guess he didn't like what he saw because he didn't stay. If Dick were still here I bet the new student would have stayed.

I lost my card case yesterday but found it today. After studying water pumps for a while Becht came by and we loafed around town. Herb sent me a letter.

February 4, 1932 – Thursday

The weather was cool, cloudy, and the wind was very strong. I didn't get to fly today.

We finally got back to work on the "Fleet" today. We put new fabric on the wings and after sewing the fabric to the frame we applied the "dope". This took us all day except for one hour to eat dinner.

Ma sent me a letter with my Private Pilots License in it, so I wrote her a letter saying I got the license. Went down to the garage this evening and "shot the breeze" with the fellows.

February 5, 1932 – Friday

The weather was cold, fair, and the wind was pretty rough. I didn't get to fly today.

Bud and George Dykstra were taking off in the NC122K when the engine quit. They made a forced landing safely in a cow pasture. Some fellows have all the luck. That would be a great experience. I can just imagine how deafening quiet it must have been with the wind whistling by and the "prop" standing motionless as a gravestone.

We jumped in the truck and drove over to where they had landed. The ship wasn't damaged, but the engine was ruined. We went back to the hanger, got a bunch of tools and took the wings off so we could tow the ship back to the hanger. The owner of the farm was friendly and he helped us get the ship off his pasture and onto the county road.

The master rod bearing had failed and completely ruined everything in the engine except sparkplugs and the valves.

I spent the rest of the day helping sew the fabric for the other wing of the "Fleet".

Tonight we went to school at the hanger and continued *Lesson No. 31* started last Saturday regarding *Oil Systems*.

After school I went over to Joe Sachleben's for a while. We had a little "homebrew" and "beer wine" and it sure tasted swell.

February 6, 1932 – Saturday

The weather was cool, fine, and the wind was nice and smooth. I flew for 55 minutes and did rotten. I guess it is because I am not used to nice flying weather.

I climbed up to 3,000 ft and flew over Madison and the Ohio River for the first time. After practicing figure eights, 180 degrees, and 360 degrees, and just before going down for the last time, I circled Barber's Boarding House three times. I undershot two times and made three poor landings.

Joe took Dave up and gave him spins, loops, and wingovers. When they landed Joe said it would be my turn next time.

Since it was Saturday there was no work in the afternoon so I worked on my Cheve. I tightened the brakes and tried to take the shimmy out of the front end. Paul sent me a letter.

There is always a big crowd in Madison on Saturday night so Art and I walked around town looking for something to do. I bought a pack of gum – 10c.

February 7, 1932 – Sunday

The weather was cool, fair, and the wind was pretty smooth. I didn't get to fly today so I slept in.

Joe and I went to a church dedication at Osgood, Indiana, this afternoon and sure did enjoy it.[22] It is about 30 miles north of Madison on Michigan Road, but the church is out in the country and is our preacher's home church. He preaches there in the morning and then comes to Madison and preaches at our church at night. I went to church tonight and Art Prosek went with me.

After church, Dave Gentrup, his girl friend Malva Lohrig and I went for a ride. I ended a very swell day by taking the girls home and putting my Sunday suit under the mattress.

February 8, 1932 – Monday

The weather was cool, fair, and the wind was a little rough. I didn't get to fly today.

---

[22] My wife and I were married in that church on March 16, 1938

We cleaned the hanger and rearranged all the planes. We cleaned and rearranged everything in the shop. Then we all went back to working on the "Fleet" wings.

Becht and I went to the Earl Lockhart trial this evening. He was convicted of shooting some men and was sentenced to 1 to 10 years in prison.

February 9, 1932 – Tuesday

The weather was warm, fine, and the wind was pretty rough.

I flew for 55 minutes today, practicing figure eights, 180 degrees and 360 degree turns. All normal landings except they were wheel landings. That's a bad habit; I seem to have forgotten how to make a three-point landing. For the first time in a long time I didn't overshoot and undershoot the field. I just have to fly more often if I am going to pass my L.C. test.

I started to build Joe a cabinet but was transferred to washing the "Prep". To be honest I don't know a thing about building a cabinet but I never get a decent job anymore. It's the same old baloney; the others need the experience that I already have. I sure don't need anymore experience washing the "Prep". We played football and won 60 to 30.

We went to school tonight at the hanger, and Bud spent the evening asking us questions about aircraft engine oil systems. He told us to draw a typical aircraft engine oil system and turn it in at the next class. After class I went over to the Sachleben's and talked to them for a while.

February 10, 1932 – Wednesday

The weather was warm, clear, and the wind was very rough. I didn't get to fly today.

I cleaned a whole bunch of parts for Bud while he was giving the other students flying lessons. There was nothing else to do the rest of the day so I studied the oil system that I am going to draw, probably tomorrow. We are spending a lot of time studying oil systems because it is very critical to the operation of an engine. Without a constant and sufficient amount of oil distributed throughout the entire system an engine cannot run. They think the master rod bearing on the "Prep" failed due to the lack of sufficient amount of oil getting to it.

Earl and Harry sent me an interesting letter and I wrote Ralph a letter.

After supper the whole outfit decided to go to the movies and watch "Halloween". We all marched down Main Street by twos and you should have seen the people look. One little feller said, "Here comes that Dixie Flying Service outfit". It was a swell movie.

"The Complete Outfit" – May 19,1932
Kneeling from left: Clyde Beyer; Rodney Pert; Pete Nemetsky; George Hayden Back Row L-R: John Yellitz; Frank J. Peternell; Howard Brennan; Walter Etchinson; Andy G. Nufer Jr.; Arthur Prosek; Linwood Wright Jr.; Harvey Kattelmann; Daniel Booth; Dave Swett

February 11, 1932 – Thursday

The weather was warm, cloudy, and the wind was awful rough. I didn't get to fly today.

I spent the day washing the "Eagle", "Prep", and "Wright Jr." Then I polished the "Waco" streamline flying tie rods, and the streamline flying wires with fine sandpaper. That was sure a tedious job.

I finally finished my other ashtray. Those worn out pistons make nice ashtrays, but are a lot of work.

I made the drawing of an aircraft engine oil system that has to be turned at the next class.

Ma sent me a letter so I was able to send $16.50 to Mrs. Barber for 3 weeks of room, board, and laundry. I wrote Ma, Paul, and Herb.

I bought 5 gallons of gas last Saturday – 75c. I took some castor oil and it made me sick.

February 12, 1932 – Friday

The weather was warm, clear, and the wind was strong, but steady.

I flew for 50 minutes, the figure eights were fair, and the spirals and 180 degrees were pretty good. My take-offs are great but the landings could be better. I undershot the field once and made one two wheel landing.

Spent some time straightening up the woodpile. I cleaned a large cardboard box of nuts, bolts, and washers and put them in separate boxes this afternoon.

We all went out to the hanger tonight to class and handed in our drawings of an aircraft engine oil system. We took a test on *"Cooling of Aircraft Engines"* followed by a test on *"Lubrication and Fuel Systems"*.

February 13, 1932 – Saturday

The weather was cool, clear, and the wind was pretty strong. I didn't get to fly today.

We finally got the "Eagle" ready for inspection by the Department of Commerce Inspector. We double-checked the wing trim and angle of incidence. Then thoroughly checked the ship from the "prop" to the tailskid. We finished by washing and waxing her.

There was a 1917 "standard" wing with all the stuff we brought from Milan sometime ago. I got the job of taking the fabric off and taking the frame apart. I got to keep one of the ribs and am going to make a neat tie rack out of it.

No one works Saturday afternoon so we all loafed around the airport. Becht stopped by after supper so we loafed around town all evening. The streets sure are crowded on Saturday night.

February 14, 1932 – Sunday

The weather was cool, clear, and the wind was fairly quiet. I flew for 35 minutes today.

After a great takeoff and plenty of altitude I made a couple of nice stalls. Then climbing back up to a safe altitude I made my first power dive by pushing the stick and throttle forward, keeping the rudder pedals in neutral and the wings level. The ship was flying so fast toward the ground it seemed to jump up at me. When the controls got stiff I throttled the engine and gradually pulled the stick back to flatten out the dive. As the dive continues it increases the strain on the controls when pulling out of it. It was swell.

A power dive is a descent in which the air speed is grater than the maximum speed in horizontal flight. When the controls become stiff it is time to flatten out the dive, but this should be done gradually with the engine throttled, and never with a jerk of the stick.

As I still had some flying time I decided to try the wingover maneuver. Not knowing how it would turn out I went up to about 3,000 ft. I pushed the stick and throttle forward keeping the rudder petals neutral and the wings level. When I felt the ship had enough speed I pulled the stick back steadily in a very steep climb or "zoom". Before reaching the stalling point I moved the stick to the left and pushed the left petal forward

As soon as the nose came around and the ship had completed the 180-degree turn, I pulled the throttle back and brought the stick and rudder pedal back to neutral. I was now in a position for a dive. This is a great way to make a fast turn and is a lot of fun. But it will take a lot of practice to do it right, although it did turn out pretty good.

They told me this maneuver was invented during World War I when a German pilot trying to escape from an American plane who was on his tail, went into a steep power dive. He pulled the stick back and "zoomed" into a steep climb, then quickly banked left in a 180 degree turn, and dove in the opposite direction. This gave him a chance to attack the other plane. This maneuver has a name but I don't know how to spell it.

Wing Over.This maneuver is nothing more than dive, a zoom, and a 180-degree turn. The turn being made in the zoom with "gun", and the engine throttled just as soon as the nose comes around and the ship has completed the 180-degree turn, usually ended in a position for dive. The stick goes forward for the dive and back steady (never jerk controls) for zoom, then left stick and rudder for the turn, and when around, stick and rudder back to neutral.

Zoom. This is when the stick is pulled back for a climb after horizontal flight with full throttle on, the ship climbs at a greater angle than that which is normal and, this steeper climb is called a zoom. A pilot must watch that the stick is not brought back too suddenly at the start, and that the zoom is not held until a stall results.

Lesson 52 – Flying

THE DAYTON SCHOOL OF AVIATION

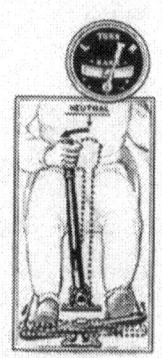

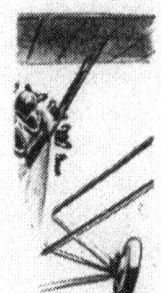

Figure No. 2
Motion of stick for take-off
and climb.

Figure No. 4
Controls for a gentle right turn.
Ripples on lake are good
indications of wind direction.

(Courtesy Standard Oil Co., of New York)

8

167

THE DAYTON SCHOOL OF AVIATION

made sharp and the second turn gradual so as to be sure of the field. After the second turn, is made so that the ship is in proper direction to come in to land, a straight glide should be made for the field. If the glide is short, never try to stretch it and run the danger of losing the flying speed and stalling but open the throttle (forward), gun the engine, and be sure of the landing with sufficient flying speed. If it is apparent that the glide is going to be too long, the experienced pilot will slip his ship, thereby losing altitude without gaining speed. This maneuver, however, should not be tried until you have had considerable instruction in slipping at a safe altitude and in landing from a slip.

Figure No. 5—The Side Slip an Altitude Loser.

A 180° Turn and Land. This maneuver is the same as the 360° turn except that it is started in opposite direction that is going away from the field at 1000 ft. altitude with the wind. The gun is cut and a glide is started and a gentle 180° turn is made to come back on the leeward side of the field for a normal or three point landing – these last two maneuvers are to show that a pilot can bring the ship down and land at a specific point without the use of the engine.

Side Slip. This maneuver is the very best way to lose altitude without gaining speed and is used by the majority of pilots when landing if altitude must be lost. It is performed by banking in the direction of the slip and then correcting with proper amount of rudder in the opposite direction.

9

THE DAYTON SCHOOL OF AVIATION

in order to hold the ship in the spin. This will cause the ship to spin in a stalled condition and is held in the spin until the stick is moved forward to drop the nose and the rudder and stick are brought back to neutral or beyond as correction is need to stop the spin. As soon as the nose drops sufficiently to create lift, and the ship stops spinning then the stick is brought back to neutral for normal flight

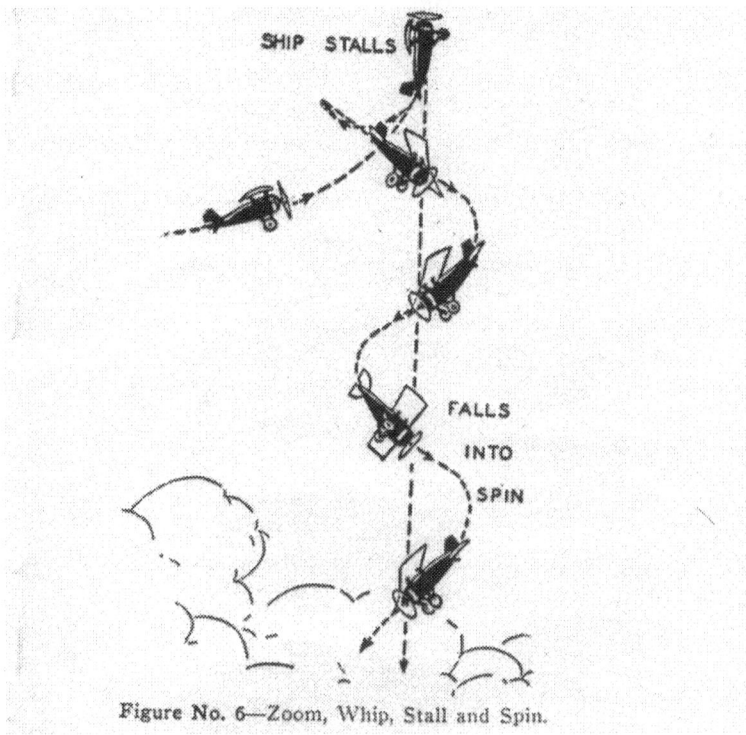

Figure No. 6—Zoom, Whip, Stall and Spin.

and the ship is leveled off. The pilot should perfect his maneuvers so that the direction of recovery is at all times assured. This takes practice and makes it possible to start maneuvers in the desired direction as well as to recover and start out onto an even keel in the desired direction – unless a pilot practices and perfects these points, his stunts will have a sloppy appearance to the observer on the ground.

11

THE DAYTON SCHOOL OF AVIATION

Wing Over. This maneuver is nothing more than a dive, a zoom, and a 180o turn. The turn being made in the zoom with gun and cut or engine throttles just as soon as the nose comes around and ship has completed the 180o turn, usually ended a position for dive. The stick goes forward for the dive and back steady (Never jerk controls) for zoom, then left stick and rudder for the turn and when around stick and rudder back to neutral.

Dive. As you already know, a dive is a descent in which the air speed is greater than the maximum speed in horizontal flight. The wires will increase their hum and the controls become still which is warning to flatten out the dive

Figure No. 7—Zoom and Dive.

but this should be done gradually with engine throttled or gun cut and never with a jerk of the stick.

Zoom. When the stick is pulled back for a climb after horizontal flight with full throttle or gun full on, the ship climbs at a greater angle than that which is normal and this steeper climb is called a zoom. A pilot must watch in this maneuver that the stick is not brought back too suddenly at the start and that the zoom is not held until a stall results.

Loop. This maneuver is started in a dive and followed by a zoom with gun on until ship has reached an inverted position, when nose goes over the top the gun is cut and stick eased up a little so that recovery is possible on the downward or last half of the loop. The rudder should be moved throughout

12

THE DAYTON SCHOOL OF AVIATION

the entire loop so as to keep the nose of the ship straight and the loop a good one. The ordinary loop is a simple maneuver and if properly executed the pilot will remain in his ship the entire loop. However the outside loop is just the opposite for the pilot is on the outside and the centrifugal force tends to throw him out of the ship. This should never be tried on an ordinary ship for the wing loads are reversed and the landing wires take the strain. Besides power, speed, and a strong ship are the only possibilities of completing or accomplishing this maneuver.

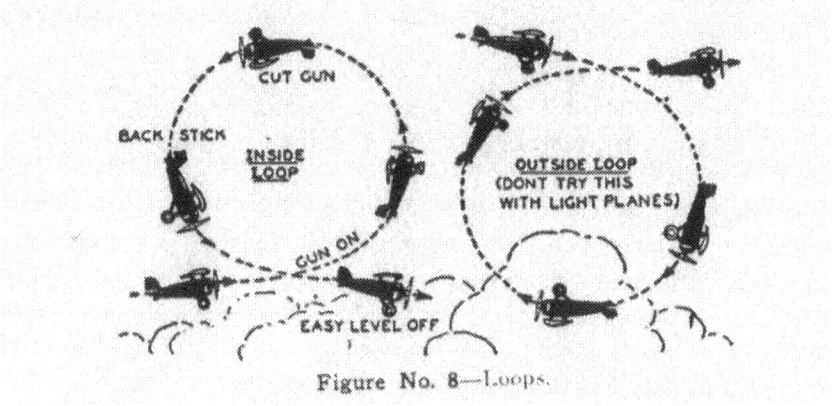

Figure No. 8—Loops.

Whip Stall. This usually results from losing speed in a very steep zoom or an uncompleted loop and starts with the ship falling tail first or a tail slip and then whipping as the nose drops. The controls should be held steady to avoid any more strain or whip on them than necessary.

Roll. This maneuver is to roll the ship 360o about the longitudinal axis and remain in horizontal flight. It is accomplished by horizontal flight with full gun and then given stick and rudder to the side of the desired roll, with stick back a bit. This starts the roll and when more than half around the stick is eased up and put to the other side, which helps the wings around the last half of the roll. The engine is throttled slightly if nose tends to drop. When roll is about completed stick and rudder are used to correct for level flight.

13

It sure was swell to make the power dives, zooms, and wingovers rather than the monotonous figure eights, 360 and 180 degrees. But, as Joe says, when you master those three maneuvers, plus the takeoffs and landings, you can fly, and all the other maneuvers will naturally fall in place. The 360-degree landing was nice for a change.

Joe must have been in a good humor. He took me up on his time, and taught me how to do wingovers, loops, and spins. The loop was sure great. He put the ship in a power dive, throttle on, pulled the stick back in a zoom and as the nose went up above the horizon all we could see was the sky. He held the stick back, moving the rudder to keep the nose straight, and as the engine started to labor the nose went over the top and we were upside down sitting firmly in our seats.

He pulled the throttle back and eased up on the stick, as the horizon came in view we were in the downward half of the loop. It was a perfect loop; otherwise we could have left our seats while upside down, or the ship could have fallen into a whipstall.

It is important to do the loop properly so the gas, oil, or water stays in place, by centrifugal force, and keeps the engine running normally. That's why a ship can't fly upside down in a straight line very far before the engine stops because the flow of gas, oil, or oil is disrupted.

It is the same principle as whirling a bucket full of water over your head and not spilling a drop.

Loop. This maneuver is started in a dive and followed by a zoom with the throttle wide open until the ship has reached an inverted position. When the nose goes over the top the throttle is cut and the stick is eased up a little so that recovery is possible on the downward or last half of the loop. The rudder should be moved throughout the entire loop so as to keep the nose of the ship straight. If properly executed the pilot will remain in his seat the entire loop.

The Outside Loop is just the opposite and should never be tried with an ordinary ship. It takes power, speed, and a strong ship to do the outside loop.

As Joe finished the loop in level flight he pulled the stick back and we went back up to about 4,000 ft. This is what I had been waiting for, a tailspin.

He pulled the stick back and throttled the engine. When the controls got sloppy (didn't respond) and the ship was about to stall, he pushed the left rudder petal all the way forward. At the same time he held the stick back and moved it to the left. The ship slowly turned to the left, and the nose dropped toward the ground with the tail toward the sky. As we began to turn faster, the earth looked like a giant record running and the longer we held the ship in the spin, the faster it turned. We were losing a lot of altitude in a hurry after 3 complete turns.

So Joe pushed the stick forward to create lift, then brought the rudder pedal and stick back beyond neutral to correct and stop the spin. He brought the rudder petal and stick back to neutral for normal flight and we leveled off. Wow! I begged Joe to spin the ship to the right so we went back up and he went through the same maneuvers spinning to the right with the same results.

Boy, it sure was swell! This is right down my alley. That's what I call flying. I can't wait to try it. I know I'll get to practice it, when he gives the okay because it is part of the Limited Commercial Pilots flying test. You have to spin three times to the left, and then spin 3 times to right, beginning and ending in the same direction. It is not as dangerous as a lot people think as long as you have plenty of altitude to recover, and it does take a lot.

Tail Spin – When a ship is placed in a stalled position where the wings do not create the necessary lift, and the controls become sloppy, it is then at the correct position to start a spin. If a spin is to be avoided, the nose must be dropped to secure flying speed and controllability.

If a spin is desired then the rudder should be pushed all the way forward in the direction of the desired spin, and the stick back and to the side in order to hold the ship in the spin. This will cause the ship to spin in a stalled condition, and is held in the spin until the stick is moved forward to drop the nose, and the rudder and stick are brought back to neutral or beyond as correction is need to stop the spin.

As soon as the nose drops sufficiently to create lift and the ship stops spinning, then the stick is brought back to neutral for normal flight and the ship is leveled off. Spins should be practiced at high altitudes as it takes quite a bit of lost altitude to come out of a spin, depending on the number turns. Spins should be started in a desired direction and come out of the spin on an even keel in the same direction.

Becht and Shucks came out to the airport after dinner. We spent the rest of the afternoon hiking with Joe and Carl. It helped to bring me back down from the clouds. I stayed at the Sachleben's for supper. It has been a great day.

I went to Sunday school this morning

February 15, 1932 – Monday

The weather was warm, cloudy, and the wind was pretty strong. I flew for 35 minutes today, that's two days in a row and very unusual lately.

I practiced 180-degree turn and land, 360-degree turn and land, and spirals. I overshot and undershot the field once and all were wheel landings.

Frank Peternell soloed this morning and did a pretty nice job.

Joe Shumate, the Department of Commerce inspector, came in this morning to inspect and test hop the Eagle we rebuilt. He had been here a couple of times checking the progress we were making. The framework, etc, has to be inspected before the fabric, or covering, is put on. After inspecting the ship thoroughly he let Joe take it up for a test hop. Inspectors never go up first. When he came down and reported that the ship performed pretty nice, the inspector took the Eagle up and gave it a good workout. He licensed the ship and said it was ready to fly. He is a swell chap.Max, George, and I made a list of all the equipment, parts, tools, supplies, and planes; everything except the office equipment and supplies. After finishing that job we went up on the roof of the hanger and played cards.

A Monocoupe, with a 7-cylinder Warner engine, License No. NC8951, flew in yesterday for a visit. It was a darn nice "crate" and I took a picture of it.

A Curtiss-Robin with an OX5 engine stopped in for a visit yesterday. I wrote Ralph a letter and loafed around town with Becht tonight.

Pop Sachleben

February 16, 1932 – Tuesday

The weather was cool, cloudy, rainy, and the wind was pretty strong. No flying today.

We finished taking stock that we started yesterday and then cleaned up the workshop.

We went back to school this afternoon at the airport. The tests we took last Friday, *"Cooling of Aircraft Engines" and "Lubrication and Fuel Systems"* were discussed. It seems as though "I didn't do so good". The students who graduated from the Dayton School of Aviation Correspondence course did pretty well.

I made a swell necktie rack out of the rib I took off the 1917 "Standard" wing. I bought 5 gal. Gas 25c.

I wrote Earl and Harry a letter then spent the rest of the evening reading a couple of airplane stories.

February 17, 1932 – Wednesday

The weather was cool, cloudy, and the wind was extremely violent. No flying today.

I cleaned and inspected a Dixie Magneto; it is ok and ready to be installed. I cleaned and inspected a whole mess of OX water pumps; they are all in good shape.

I drained the gas out of the "Prep" and took the carburetor off the engine. After taking the carburetor apart, I cleaned and inspected the float and valve, all the jets, and the throttle valve. There was no sign of wear or damage, so I put the carburetor back together and installed it on the engine.

I brought a Waco 10, OX license number, a couple of aviation magazines, and my necktie rack home. The boarders thought the tie rack was great.

I went up the Hanging Rock hill, barely, to the Wakefield junkyard and bought a used 4-cylinder block for $3.00 for my Cheve. The cylinder walls in my engine are so badly scratched and scored, it uses a lot of oil and has very little power.

Then I went down to Bob Jones' garage and bought a set of 4 pistons and pins $1.50, and a set of piston rings $1.55. All dirt cheap, but I can't pay him until I get money from home which is ok with him. Guess I'll start working on the Cheve tomorrow.

I drove out to Sachleben's and spent the evening playing cards (Chinese Rum). Had a great time arguing, and enjoyed the "homebrew" and "beer wine".

February 18, 1932 – Thursday

The weather was clear, warm, and the wind was mild. No flying today.

During pre-flight inspection we found that the crankcase in the "110" engine was cracked. That puts both 'Preps' out of business.

They have been repairing the "122" engine that broke down while Bud and George were taking off on Feb. 5, 1932. Practically everything had to be ordered so it was taking a long time to get the engine ready to run.

Pete Nemetsky

I took **Pete Nemetsky**, a new student who came in yesterday from Illinois, to Seymour in my car. After he took his physical examination we made the 41-mile trip home in 58 minutes.

I cleaned the engine block I bought yesterday at the Wakefield junkyard and put it on the engine stand.

I spent the evening trying to catch up on my homework. I got a letter from Ma today.

February 19, 1932 – Friday

The weather was clear and cool. The wind was mild. No flying today as both training ships' engines are being repaired.

We finally got the "122 Prep" engine repaired so they decided to put it in the "110 Prep".

After installing the engine in the ship, which was quite a job, we chocked the wheels and tied the ship down. They started the engine and let it run all day to "break it in".

I started to take the engine out of my Cheve and it sure was a tough job. I took the hood, battery, radiator, starter, generator, carburetor, distributor, intake and exhaust manifold, cylinder head, vacuum pump, and hoses out of the car.

Joe Vail wants someone at the hanger twenty-four hours a day to protect the ships, etc. During the day the students, instructors, and office workers are there. Each student will take his turn and spend one night at the hanger from 7 PM to 7 AM. Each student will also take his turn and stay after work until the student who is staying all night, comes back from supper.

Today was my turn to stay until **Max Sittner** came back from supper. By the time I finished eating the late supper Mrs. Barber had waiting for me, it was too late to attend class at the airport, so I wrote Ma a letter.

February 20, 1932 – Saturday

The weather was clear, cool, and the wind was fairly rough.

I flew for 35 minutes today practicing spirals, 180-degree and 360-degree turn and land. I am beginning to do these maneuvers automatically without too much concentration, which is a good sign. I undershot the

field once and overshot it one time, and can't seem to make three point landings anymore, as they were all wheel landings.

It is strange that I have never seen a ship from the Crozier airport flying, so while doing a spiral I accidentally made it bigger than usual and flew over the airport, but didn't see anything going on. It looked "dead as a doornail". Maybe some of the fellows down at the garage can tell me what is going on over there.

Linn Wright and I walked around town after supper.

February 21, 1932 – Sunday

The weather was cold and it rained all day, so there was no flying. I had to sit around all day on account of the rain.

Dad, Earl, and Ralph came for a short visit today. They drove down in Dad's new "Marmon". It sure is a beauty and is built in Indiana.

We went out to the airport and looked at all our ships. I showed them how to operate the controls and explained what happened, and why. Then we went in the workshop and I showed them how we were rebuilding a wing and overhauling several engines. On the way out of the shop they stopped to examine a couple of ships we were just starting to rebuild. They asked about a million questions.

Dad and Earl have a room down at the Madison Hotel, southwest corner of Mulberry and Second Streets. Mom and I stayed there the first night we were in Madison.

I slept in Andy Barrett's bed so Ralph could sleep in mine.

February 22, 1932 – Monday

The weather was clear, cold, and the wind was rough.

I flew for 45 minutes today, practicing spirals. Then I took Ralph up for a 15-minute ride. The ceiling was only 800 ft so I gradually flew up into the clouds. We were in a little world all our own, we couldn't see anything except each other, and no one could see us. As soon as we lost sight of the ground I leveled off and slowly pushed the stick forward until we caught a glimpse of the ground. I sure didn't want to get lost in the clouds while my brother was with me. He seemed to enjoy it so we skipped in and out of the clouds a couple more times, and then headed for the runway.

We bounced a little while landing and ran through a big hole filled with water, literally splashing water and mud all over us. After cleaning up the "Prep" I took Earl up for a 15-minute ride. We flew in and out of the clouds a few times then flew over Madison.

I treated Earl a little better than I did Ralph. As we approached the field I realized I was going to "overshoot" the runway. After crossing over the electric high lines I sideslipped the ship in and made a fair three point landing, no bounce, and we didn't even get splashed. He didn't know what it was all about when I said I sideslipped in but thought it was different.

I didn't want to take Dad up because I didn't want anything to happen to him if something went wrong. My time was up so I had a good excuse. Before leaving for home right after the plane rides they gave me some fruit, cookies, and a cake.

A stunt pilot and his custom built, high-powered Aeronca flew in today for a visit. He took off almost straight up for 50 ft and then promptly "barrel-rolled". He did every maneuver possible except one outside loop, and was never more than 150 ft in the air. I sort of like these new Aeroncas.

"Barrel Roll" - To roll the ship 360 degrees about the longitudinal axis and remain in horizontal flight. It is accomplished by horizontal flight with the throttle wide open and then giving the stick and rudder to the side of the desired roll with the stick back a bit. This starts the roll and when more than half way around the stick is eased up and put to the other side, which helps the wings around the last half of the roll. The engine is throttled slightly if the nose starts to drop. When the roll is about completed the stick and rudder are used to correct for level flight and direction. The roll should be completed in the same direction as it is started.

Marlin Christianson soloed today.

I worked on the Cheve all day. I took the cylinder block out of the car, cleaned it, and removed the crankshaft, camshaft, timing gear, piston rods, and pistons. I took the pistons off the piston rods and installed the pistons, pins, and rings I bought from Bob Jones. Then I put the crankshaft, camshaft, timing gear, pistons, and rods into the engine block I bought from the junkyard. After fitting the main and piston rod bearings I put on the oil pan.

February 23, 1932 – Tuesday

The weather was cool, clear, and the wind was fairly mild. I didn't get to fly today.

Chester T. McKinley soloed today after 9 hours of dual instruction.
Bud Guild, the other instructor, must have been feeling good today because he taught all the students how to spin today.
I cleaned all the carbon off the Cheve cylinder head and ground the valves and seats. It is ready to be installed after I put the engine block in the car.
We had school after supper at Barber's Boarding House.

The lesson was on the following questions:

1. (a) What is friction? (b) How can friction be reduced?
2. (a) What are the various degrees of lubrication? (b) Why are lubricants used?

3. Describe the lubrication in a plain journal bearing
4. (a) What is babbit and where is it used? (b) What are its advantages?
5. What lubrication system is used in aircraft engines that differ from the automobile?
6. What characteristics should aircraft lubricants have?
7. Name and describe the three types of oil pumps
8. Where are oil pumps located?
9. Name and describe the various type fuel systems.
10. How do super-charged engines affect the fuel system?
11. What ways are used to increase the engine power at high altitude?
12. Describe superchargers and state the advantages and disadvantages of each type.

I gave the "gang" my cookies during class.

February 24, 1932 – Wednesday

The weather was fine, clear, and the wind was fairly mild. I didn't get to fly today.

I put the engine block in the Cheve today and installed the cylinder head, starter, generator, water pump, distributor, intake and exhaust manifold, carburetor, vacuum pump, radiator, battery, spark plugs, wiring, and hoses. After adjusting the valve tappets and putting on the valve cover I added the oil, and put water and alcohol in the radiator.

Then I turned the ignition switch on, called out "contact", and stepped on the starter button. To my surprise the engine started right off. Boy, she sounded perfect and I couldn't want it to run any smoother so I left it to run about an hour to "break it in".

Joe Sachleben and I went to church tonight.

February 25, 1932 – Thursday

The weather was warm, clear, and the wind was pretty mild.

I got grounded instead of flying today. Bud Guild tried to "gyp" me out of my turn to fly so I told him where to get off.

I told him the only time I got to fly was when the weather was too bad for the other students to fly. And they were getting more flying time because the school had to pay their room and board, so they wanted the students to get their licenses as soon as possible. I have to pay my own room and board as I am under a different contract. Thinking about it, I am better off learning to fly in bad weather; those guys will be "fair weather pilots". Two new students came in today.

I let my car engine run for 2 hours today to "break it in". Then Marlin Christianson and I took a ride out to Paris Crossing.

I got a letter from my sister Pearl and one from my Sunday school friend Herb Schwartz. We all played cards after supper.

February 26, 1932 – Friday

The weather was fair, clear, and the wind was pretty strong. I flew for 50 minutes today practicing gentle and steep figures eights, stalls, and 180 and 360-degree landings and did ok, but they were all wheel landings. I can't seem to get rid of that habit and back to making those three-point landings again. Joe says that I am trying too hard to be perfect, that the tailskid was only about one inch off the ground when the plane touched down. I sure wish they would let me practice spins. Boy, if Bud and Joe ever leave the airport while I am flying I am going to do some spinning! They say it can be hard on the ship and can be dangerous, but it is part of the Limited Commercial Pilots License flying test. So when am I suppose to practice it? The old "saw" says "Practice Makes Perfect".

Kenneth B. Russell, Sanford, Maine, and **David Swett**, White River Junction, Vermont, are two new students who came in yesterday.

I spent the rest of the day filling in all the ruts and holes in the runway.

Mom sent me a letter with $5.00 today. Ed Becht came by after supper and we loafed around town.

February 27, 1932 – Saturday

The weather was cool, clear, and the wind was strong and bumpy.

I flew for an hour today practicing spirals, stalls, vertical power turns, and 180 degree and 360-degree landings. I overshot the runway twice and undershot it once. I made 3 three-point landings today and only one wheel landing.

That discussion (?) I had with Bud on Thursday must have done some good as this is the second day in a row I got to fly.

I spent the rest of the morning filling in all the ruts and holes in the runway, in between landings and takeoffs.

An Army flyer from Sufridge Field flew in today in a Boeing Pursuit ship. It is powered with a 9-cylinder Pratt-Whitney engine. His name is Mr. Eckert[23], a very highly ranked officer, and a native of Madison. He is a local hero, and a swell guy. Boy! Oh Boy! It sure is a keen ship. MY ambition will be realized when I fly one of them.

A neat American Eagle ship powered with a Kinner 5 engine also dropped in today from Columbus, Ohio.

Art Prosek soloed today. After supper I went out to the airport and spent the evening "shooting the breeze" with Linn Wright.

Art Prosek
soloing

February 28, 1932 – Sunday

The weather was cool, cloudy, and the wind was rough.

I flew for 50 minutes today and did all the maneuvers fairly well. But I am still losing too much altitude during the vertical power turns, not as much as I have been but still too much. I overshot the field one time and made one rotten landing; that is, a wheel landing, so I am doing better than usual.

After going to Sunday school, I spent the day at the airport helping Joe Vail. We cranked up "Peg", his Waco biplane, filled it full of gas and gave it the pre-flight inspection.

---

[23] Lieutenant General William Eckert: Born in Madison in 1909 and residing on Michigan Road. Eckert managed the first war-time defense production system and was a top-ranked aide of the Secretary of the Air Force and Air Force Controller. He was also named Commissioner of Baseball from 1965 to 1968 before dying in 1971.

He flew quite a few passengers, two at a time in the front cockpit at $2.00 each[24]. If there were small kids he would squeeze 3 in the cockpit. Usually he flew them over Madison, the Ohio River, Cragmont State Hospital, and Clifty Falls State Park. But if there were a lot of people waiting, he would cut the rides short. As the Waco is an open cockpit ship we loaned each passenger a helmet and goggles. Joe sure can fly; he takes off so smooth and lands "three pointers" most of the time. When he opens the throttle the engine lets out a reassuring roar that is sweet music to my ears. The plane takes more strength, weight, and skill to crank, as it is bigger than the "Prep" engine.

Mr. Eckart, the Army pilot, made a number of power dives before leaving for his home base. Boy, what that ship can't do, can't be done. That engine's roar makes Joe's Waco sound like a kitten's purr.

A Waco F ship, with a Kinner K5 engine, dropped in on us today. Not a bad job!

After the last passenger left we parked the ship in the hanger and closed down the airport for the day.

I didn't go to church this evening, as the preacher was sick. Joe Sachleben and I went for a ride on some new, to me, country roads in my Cheve.

February 29, 1932 – Monday

The weather was clear, warm, and the wind was fairly rough.
I flew for one hour and thirty minutes today.

All my maneuvers, spirals, figure eights, sideslips, stalls, dives, zooms, wingovers, whipstalls, and vertical power dives were ok. The vertical power turn are the hardest, turning to the left is quite a different maneuver than turning to right due to the engine torque. The engine torque affects all turning maneuvers.

I overshot the field one time and undershot it one time. I made pretty darn fair three-point landings for a change.

Joe Vail divided us up into groups today. There are four groups of students and each group is to take care of the hanger for a week. My group has to keep the hanger clean and neat this week. We have to see that the ships are parked properly and ready to fly. So I spent the rest of the day straightening out the hanger.

---

[24] $2.00 is a day's wages for the average Madison worker.

183

I stayed home after supper and read several lessons in my Dayton School of Aviation Correspondence course book. Dave Gentrup came by and we spent the rest of the evening "woofing" ("shooting the breeze").

March 1, 1932 – Tuesday

The weather was cloudy, cool, and the wind was pretty rough.

I flew for 55 minutes today and spent most of the time practicing the vertical power turns. This is the only maneuver that is giving me any trouble and it is part of the Limited Commercial Pilot's flying test. I overshot the field once, but set the ship down in pretty darn nice three point landings again today. I just hope it gets to be a habit.

E.D. (Bud) Guild, the flying instructor from Cincinnati, Ohio, was laid off today. There isn't enough flying time to keep both instructors busy, and as Joe Vail is part owner of the school, Bud was the one that had to go. Since he and I had that "discussion" about flying time last Tuesday we have become friends and I hated to see him leave. He was a very good pilot and instructor. He had the talent to teach the why and how, so you could understand and do, what you were being taught.

Linwood Wright, Jr. soloed today.

We are getting the "Eagle" and the "Fleet" in pretty good shape and they will be ready to rig pretty soon. My group was in charge of the hanger today. As there was nothing else to do today I tightened the rod bearings on my Cheve. I hope I don't have a piston slap but it sure sounds like it.

We went to school at the airport after supper and Joe Vail was the instructor. Does he know his stuff? I'll say.

He discussed the Private Pilot written examination and flight test that is going to be held here soon. I have my Private Pilot license so I daydreamed and dozed off and on during that part of the lesson. But when I heard him mention that the Limited Commercial Flight test was going to be held at Bowman Field in "Looeyville", Ky, I snapped out of it.

The written test is to be held here at the airport and if we pass it they will tell us what day and time to be in "Looeyville". Joe said, in very strong language, that we better study and memorize as much of the following as we can if we hope to pass the written exam. There will be three of us taking the tests.

The Secretary of Commerce shall by regulation provide for the granting of registration to aircraft eligible for registration.

Aircraft means any contrivance now known or hereafter invented, used, or designed for navigation of a flight in the air, except parachute or other contrivance designed for such navigation, but used primarily as safety equipment.

An aircraft to be entitled to license and registry, must be airworthy and equipped in accordance with the requirements of the Secretary of Commerce and owned by:
- *A. A citizen of the United States*
- *B. A partnership of the U.S.*
- *C. A corporation of the U.S.*
- *D. The government of the U.S., State, Territory, or possession, or a political subdivision thereof.*

Aircraft licenses will be issued for a period of not exceeding one year and will identify the airplane, specify the authorized type of the engine and the authorized gross weight. The aircraft license must be carried in the aircraft whenever it is in service and must be conspicuously posted where it may be readily seen by passengers or inspectors.

The owner or operator of every licensed aircraft shall keep a navigation and engine log book and shall quarterly transmit to the Secretary of Commerce a navigational summary report, in duplicate, showing the number of hours and the approximate number of miles the aircraft has been flown during the quarter, the duration of the use of each engine, the engine installation and repairs, and the plane structure and rigging changes and repairs.
The logbooks shall be carried in the aircraft at all times when such aircraft is away from its home airport.

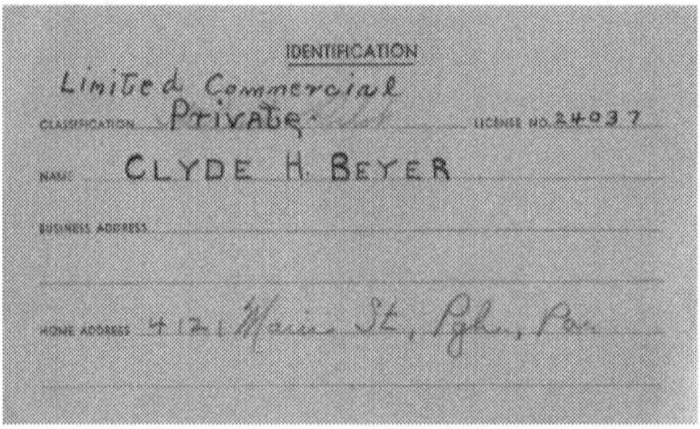

IDENTIFICATION

Limited Commercial

CLASSIFICATION Private LICENSE NO. 24037

NAME CLYDE H. BEYER

BUSINESS ADDRESS

HOME ADDRESS 4121 Main St, Pgh., Pa.

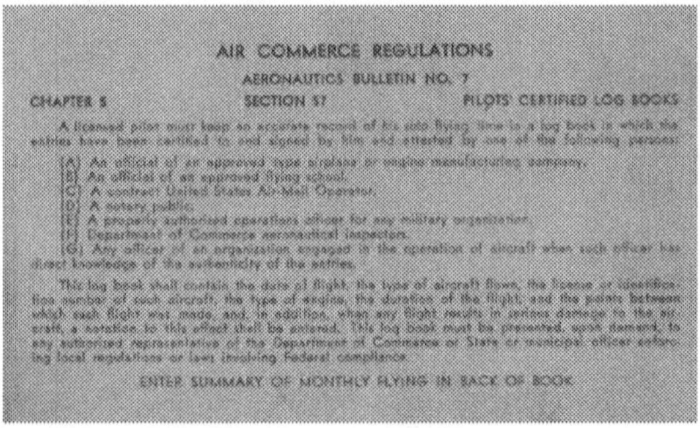

AIR COMMERCE REGULATIONS

AERONAUTICS BULLETIN NO. 7

CHAPTER 5 SECTION 57 PILOTS' CERTIFIED LOG BOOKS

A licensed pilot must keep an accurate record of his solo flying time in a log book in which the entries have been certified to and signed by him and attested by one of the following persons:

(A) An official of an approved type airplane or engine manufacturing company.
(B) An official of an approved flying school.
(C) A contract United States Air-Mail Operator.
(D) A notary public.
(E) A properly authorized operations officer for any military organization.
(F) Department of Commerce aeronautical inspector.
(G) Any officer of an organization engaged in the operation of aircraft when such officer has direct knowledge of the authenticity of the entries.

This log book shall contain the date of flight, the type of aircraft flown, the license or identification number of such aircraft, the type of engine, the duration of the flight, and the points between which each flight was made, and, in addition, when any flight results in serious damage to the aircraft, a notation to this effect shall be entered. This log book must be presented, upon demand, to any authorized representative of the Department of Commerce or State or municipal officer enforcing local regulations or laws involving Federal compliance.

ENTER SUMMARY OF MONTHLY FLYING IN BACK OF BOOK

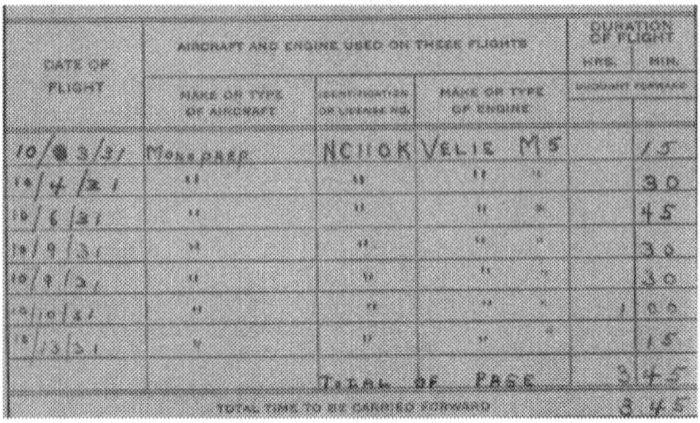

DATE OF FLIGHT	AIRCRAFT AND ENGINE USED ON THESE FLIGHTS: MAKE OR TYPE OF AIRCRAFT	IDENTIFICATION OR LICENSE NO.	MAKE OR TYPE OF ENGINE	DURATION OF FLIGHT: HRS.	MIN.
				BROUGHT FORWARD	
10/3/31	Monoprep	NC110K	Velie M5		15
10/4/31	"	"	" "		30
10/6/31	"	"	" "		45
10/9/31	"	"	" "		30
10/9/31	"	"	" "		30
10/10/31	"	"	" "	1	00
10/13/31	"	"	" "		15
		Total of	Page	3	45
TOTAL TIME TO BE CARRIED FORWARD				3	45

FROM	TO	REMARKS
LOCAL	4 FLIGHTS	Solo - 100 & 460' LANDINGS
"	4 "	" " " "
"	2 "	" - Foggy.
"	4 "	" - 150 & 263' LANDINGS
"	1 "	Check Time
"	2 "	Solo - 100 & 360' LANDINGS
"	1 "	Check Time

I CERTIFY THE ABOVE FLIGHTS WERE MADE.

ATTESTED BY

DATE OF FLIGHT	AIRCRAFT AND ENGINE USED ON THESE FLIGHTS			DURATION OF FLIGHT	
	MAKE OR TYPE OF AIRCRAFT	IDENTIFICATION OR LICENSE No.	MAKE OR TYPE OF ENGINE	HRS.	MIN.
				6	50
12/5/31	MONOPLANE	NC110K	VELIE M5		30
12/8/31	"	"	"		30
12/13/31	"	"	"		10
12/15/31	"	"	"		40
12/16/31	"	"	NO. COUNTER "		40
12/16/31	"	"	"		30
12/17/41	"	"	N.T. COUNTER		30
	TOTAL OF PAGE			2	20
	TOTAL TIME TO BE CARRIED FORWARD			9	10

A licensed aircraft shall bear an identification mark consisting of the license number of the aircraft preceded by –

1. The Roman capital "S" for aircraft used solely for governmental purposes.
2. The Roman capital "C" etc. for all other licensed aircraft.
3. The letter 'N" must precede the license symbol and numerals on licensed aircraft engaged in foreign air commerce and may precede it on other licensed aircraft.

The identification marks shall be located as follows:

On the lower surface of the lower left wing and the upper surface of the upper right wing, the top of the letters or figures to be toward the leading edge, the height of the letters and figures need not be more but shall not be less than 30 inches.

In the case of a monoplane the mark shall be displayed on the lower surface of the left wing and the upper surface of the right wing.

The marks shall also appear on both sides of the rudder, of size as large as the surface will permit, leaving a margin of at least two inches.[25]

The Secretary of Commerce shall by regulation provide for the periodic examination and rating of airmen serving in connection with aircraft of the United States as to their qualifications for such service.

The term "airmen" means any individual (including the person in command and any pilot, mechanic, or member of the crew) who engages in the navigation of aircraft while underway, and any individual who is in charge of the inspection, overhauling, or repairing of aircraft.

It shall be unlawful to serve as an airman in connection with any aircraft registered as an aircraft of the United States without an airman certificate.

Any person who violates any provision of this shall be subject to a civil penalty of $500.00.

Licensed pilots are classed as commercial or private pilots. Commercial pilots are licensed as transport, limited commercial, or industrial pilots. Private pilots are designated as private pilots or as student pilots. A person may hold a plurality of licenses, such as pilot's and mechanic's license.

Transport pilots may pilot any type of licensed aircraft, but not unlicensed aircraft, carrying persons or property for hire. He shall have all the privileges of navigating aircraft conferred upon other classes of pilots, which shall include the right to instruct students.

Limited Commercial pilots shall have all of the privileges conferred and be subject to all of the restrictions imposed on transport pilots, except they shall not pilot aircraft carrying persons for hire outside of the areas mentioned in their licenses, nor shall they, for hire, instruct students in the operation of aircraft in flight.

Industrial pilots may pilot any type of licensed aircraft not carrying persons for hire, and shall not, for hire, instruct students in flight.

Private pilots may pilot licensed aircraft but shall not carry persons or property for hire. Private pilots shall not, for hire, instruct students in flight.

As it was getting late and Joe was just halfway through the lesson, he decided to quit for the night. I don't think I'll ever make my L.C. Phew!! And this is just half of what I'll have to know. I'm not worried about the flying, but this?

---

[25]   The license numbers of the two "Prep" training ships are NC110 K and NC122 K respectively.

March 2, 1932 – Wednesday

The weather was cool, cloudy, and the wind was still strong.

I flew for one hour and 5 minutes practicing spirals, sideslips, climbing turns, and recovering from stalls, and took off and landed in a crosswind. Everything went ok and I don't even think about keeping the wings level, or the nose on the horizon anymore. I got rid of the "One Wing Low" nickname quite some time ago.

I overshot the field once and undershot it once. I sure better not do that during the test. I am still setting the ship down in pretty nice three-point landings, but my luck can't last forever. I guess I'll make up for it tomorrow.

There was nothing to do at the airport so I had Bob Jones tighten the main bearings on the Cheve; they were pretty loose. I paid him $1.00, and the $1.50 I owed him for the pistons. I paid Fred Sachleben the 20¢ I owed him, bought 2 gal of gas 35¢, and a pack of cigarettes 10¢.

My oldest brother, Paul, sent me a letter, so did Harry Pyatt and Earl Seitz.

After spending the rest of the day studying the lesson we had last night I went out after supper and flirted with a couple of girls to clear my mind.

March 3, 1932 – Thursday

The weather was cool, cloudy, drizzly, and the wind was fairly strong.

I flew for 50 minutes, practicing gentle and steep figure eights, climbing turns, 180-degree and 360-degree turn and lands, stalls, and spirals. Not bad. I am still setting the ship down in pretty darn fair 3-point landings. What's going to happen? I overshot and undershot the field once.

Andy Nufer broke the "prop" on the "Prep" yesterday. He landed with the tail too high and the propeller hit the ground before the wheels did.

Lawrence Burklow broke the landing gear and propeller on the other "Prep" today. He landed so high in the air that when the ship hit the ground the landing gear broke down and the "prop" hit the dirt.

Joe was mighty unhappy and I learned a few new cuss words. I have been lucky so far but the "law of averages" generally catches up with you. At least we will have something to do in repairing it.

Sam Poletunow soloed today.

I spent the rest of the day studying for the written exam. My group cleaned the hanger and workshop, parked the ships and locked up the hanger for the day.

We went to school at the hanger and Joe was the instructor. He was still in a real bad mood. No "monkey-shines" tonight. He discussed the Limited Commercial test that he started Tuesday evening.

An application for a pilot's license must be filed, under oath, with the Secretary of Commerce. An applicant must appear for a physical exam before a physician designated by the Secretary of Commerce and pass such exam. An applicant must be of good moral character. The minimum age requirements are 16 yrs for Private pilots and 18 yrs for Industrial, Limited Commercial, and Transport pilots. A Private pilot may be a citizen of any country. An Industrial, Limited Commercial, or Transport pilot must be a citizen of the United States

An applicant must have at least the following flying experience:

Transport pilot – 200 hrs of solo flying of which at least 5 hrs must have been within the last 60 days.

Limited commercial pilots – The solo flying required of Industrial pilots.

Industrial pilots – 50 hrs of solo flying of which at least 5 hrs must have been within the last 60 days.

Private pilots – 10 hrs of solo flying of which at least 2 hrs must have been within the last 60 days.

Pilot's Physical Qualifications:

Good past history; sound pulmonary, cardiovascular, gastrointestinal, central nervous, and genital urinary systems; freedom from material structural defects or limitations; freedom from disease of the ductless glands; normal central, peripheral, and color vision; normal judgment of distance; only slight defects of ocular muscle balance; freedom from ocular disease; absence of obstructive or diseased condition of the ears, nose, and throat; no abnormalities of equilibrium that would interfere with flying.

Pilot's Examination and Test

Transport pilots – Examination of the air traffic rules and those portions of the Air Commerce Regulations pertaining to pilot's privileges and limitations and to the inspection and operation of aircraft.

Practical and theoretical examination: in elementary engine and plane mechanics and rigging; and a theoretical examination in the fundamentals of meteorology and air navigation.

Practical Flight Test

In addition to normal takeoffs and landings the following maneuvers will be required:

From 1500 ft, with engine throttled, make a 360-degree turn and land in normal landing attitude, by wheels touching ground in front of and within 200 ft of a line designated by examiner for the Department of Commerce.

From 1000 ft, with engine throttled, make a 180-degree turn and land in normal landing attitude, by wheels touching ground in front of and within 200 ft of a line designated by examiner for the Department of Commerce.

A series of 5 gentle and 3 steep figure eight (8) turns from 800 to 1000 ft, respectively. Spiral in one direction from 2000 ft, with engine throttled, and land in normal landing attitude, by wheels touching ground in front of and within 200 ft of a line designated by examiner for the Department of Commerce.

The examiner now climbs aboard.

Flying in emergency maneuvers, doing spirals, sideslips, climbing turns, and recovering from stalls and spins, crosswind landings and takeoffs.

Stalls must be started and recovered in exactly the same direction after three complete turns.

Spins must be started and recovered in exactly the same direction after three complete turns.

Fly over a triangular or rectangular course at least 100 miles, landing at place of takeoff within at least 5 hours. This flight shall also include 2 obligatory landings, not at point of departure, where craft must come to rest.

Limited Commercial Pilots – The same examination and tests as are proscribed for transport pilots except the examination on elementary meteorology and navigation, and the cross-country flight.

Industrial Pilots – The same examination and tests as proscribed for transport pilots except the examination on elementary meteorology and navigation.

The practical flight tests proscribed for transport pilots except the distance for the cross-country flight shall be 60 miles.

Applicants for the pilot's licenses who have failed the prescribed theoretical or practical tests may apply for reexamination at any time after 90 days.

Transport and Limited Commercial pilot's licenses shall remain in force for 6 months, and Industrial and Private pilot's licenses one year from date of issuance.

Lesson 54 – Air Commerce Rules

THE DAYTON SCHOOL OF AVIATION

(B) Carelessness or inattention to duty.

(C) Any demonstration of incompetence in the repair or overhaul of aircraft.

(D) Being under the influence, or using, or having personal possession of intoxicating liquor, cocaine, or other habit-forming drugs while on duty.

(E) Refusal to exhibit license upon proper demand.

(F) Making any false statement in application for license or in any reports required to be submitted by these applications.

(G) Doing any act in connection with aircraft which is contrary to the public safety or interest or detrimental to the morale of pilots or mechanics.

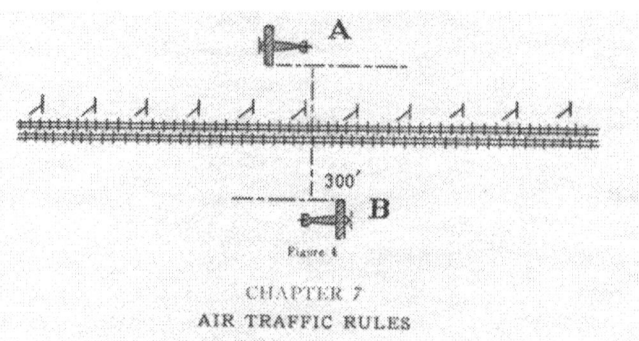

CHAPTER 7
AIR TRAFFIC RULES

Sec. 70. Law.

"The Secretary of Commerce shall by regulation establish air traffic rules for the navigation, protection, and identification of aircraft, including rules as to safe altitudes of flight and rules for the prevention of collisions between vessels and aircraft." (Air commerce act of 1926, sec. 3 (e).)

Sec. 71. Unlawful Acts.

"It is unlawful *** to navigate any aircraft otherwise than inconformity with the air traffic rules." (Sec. 11 (a) (5).)

Sec. 72. Penalty.

"Any person who violates any provision of subdivisions (a) of this section *** shall be subject to a civil penalty of $500." (Sec 11 (b).)

Sect 73. Application of the Law.

In order to protect and prevent undue burdens upon interstate and

3

THE DAYTON SCHOOL OF AVIATION

foreign air commerce the air traffic rules are to apply, whether the aircraft is engaged in commerce or noncommercial, or in foreign, interstate, or intrastate navigation in the United States, and whether or not the aircraft is registered or is navigating in a civil airway." (Statement of managers accompanying conference report, air commerce act of 1926).

Sec. 74. Flying Rules.

(A)Right-Side Traffic. Aircraft flying in established civil airways, when it is safe and practicable, shall keep to the right side of such airways. (Fig. 4.)

(B)Giving-Way Order. Craft shall give way to each other in the following order:

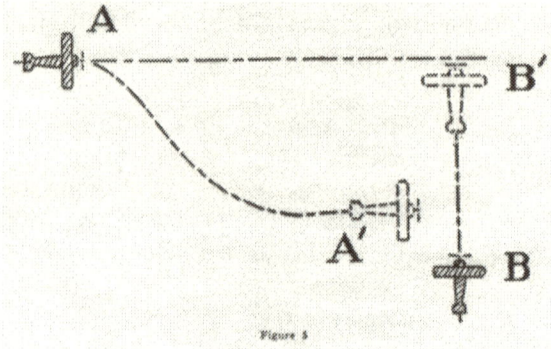

Figure 5

(1)Airplanes
(2)Airships
(3)Balloons, fixed or free

An airship not under control is classed as a free balloon. Aircraft required to give way shall keep a safe distance, having regard to the circumstances of the case. Three hundred feet will be considered a minimum safe distance.

(C) Giving-Way Duties. If the circumstances permit, the craft, which is required to give way, shall avoid crossing ahead of the other. The other craft may maintain its course and speed, but no engine-driven craft may pursue its course if it would come within 300 feet of another craft. 300 feet being the minimum distance within which aircraft, other than military aircraft of the United States engaged in military maneuvers and commercial aircraft engaged in local industrial operations, may come within proximity of each other in flight.

33

THE DAYTON SCHOOL OF AVIATION

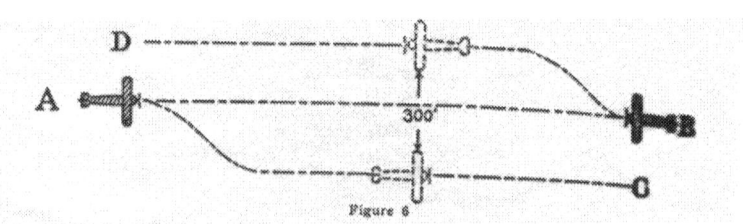

Figure 6

(D) **Crossing**. When two engine-driven aircraft are on crossing courses the aircraft which has the other on its right side shall keep out of the way. (Fig. 5)

(E) **Approaching**. When two engine-driven aircraft are approaching head-on or approximately so, and there is risk of collision, each shall alter its course to the right, so that each may pass on the left side of the other. This rule does not apply to cases where aircraft will, if each keeps on its respective course, pass more than 300 feet from each other. (Fig. 6)

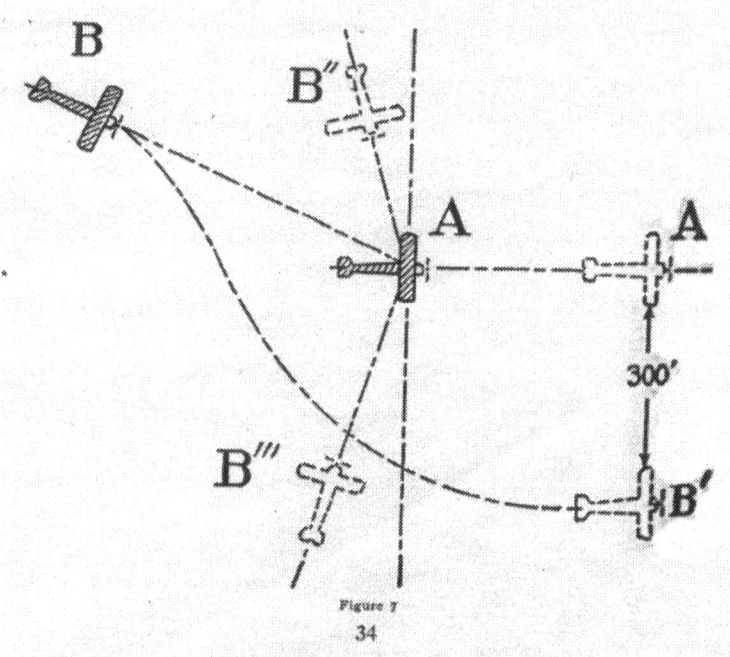

Figure 7
34

THE DAYTON SCHOOL OF AVIATION

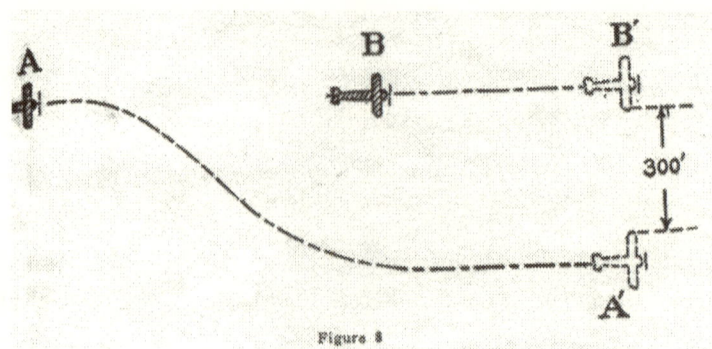

Figure 8

(F) Overtaking. (1) Definition. An overtaking aircraft is one approaching another directly behind or within 70° of that position, and no subsequent alteration of the bearing between the two shall make the overtaking aircraft a crossing aircraft within the meaning of these rules or relieve it of the duty of keeping clear of the overtaken craft until it is finally past and clear. (Fig. 7.)

(2) Presumption. In case of doubt as to whether it is forward or abaft such position, it should assume that it is an overtaking aircraft and keep out of the way.

(3) Altering course. The overtaking aircraft shall keep out of the way of the overtaken aircraft by altering its own course to the right, and not in the vertical plane. (Fig. 8.)

(G) Height Over Congested and Other Areas. Exclusive of takeoff from or landing on an established landing field, airport, or on property designated for that purpose by the owner, and except as otherwise permitted by section 79, aircraft shall not be flown ----

(1) Over the congested parts of the cities, towns, settlements, except at a height sufficient to permit a reasonably safe emergency landing, which in no case shall be less than 1,000 feet. (Fig. 9.)

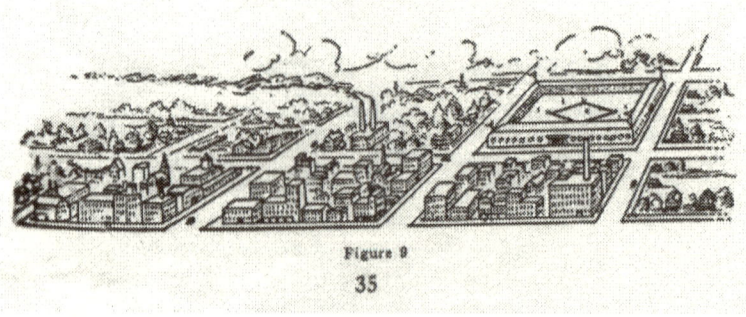

Figure 9

35

THE DAYTON SCHOOL OF AVIATION

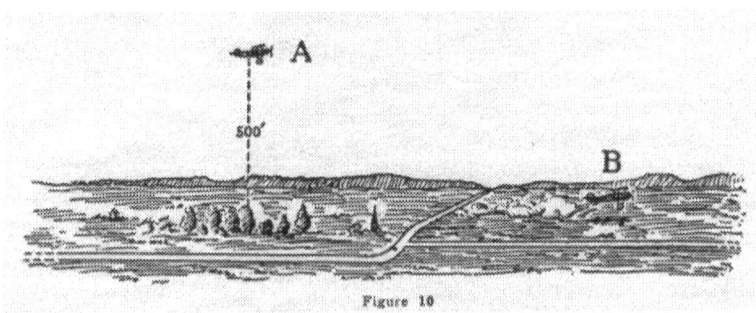

Figure 10

(2) Elsewhere at height less than 500 feet, except where indispensable to an industrial flying operation. (Fig. 10.)

(H) Height Over Assembly of Persons. No flight under 1,000 feet in height shall be made over any open-air assembly of persons except with the consent of the Secretary of Commerce. Such consent will be granted only for limited operations.

(I) Acrobatic Flying (1) Acrobatic flying means intentional maneuvers not necessary to air navigation.
(2)No person shall acrobatically fly an aircraft -------
(a) Over a congested area of any city, town or settlement. (Fig. 11.)

NO ACROBATICS IN THIS PROHIBITED SPACE

Figure 11

36

THE DAYTON SCHOOL OF AVIATION

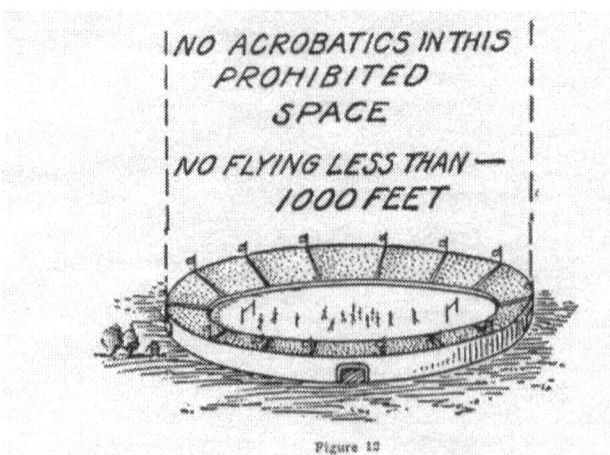

Figure 12

(b) Over any open-air assembly of persons or below 2,000 feet in height over any established civil airway, or at any height over any established airport, or landing field, or within 1,000 feet horizontally thereof. (Figs. 12, 13, and 14)

(c) Any acrobatic maneuvers performed over any other place shall be concluded at a height greater than 1,500 feet. (Fig. 15.)

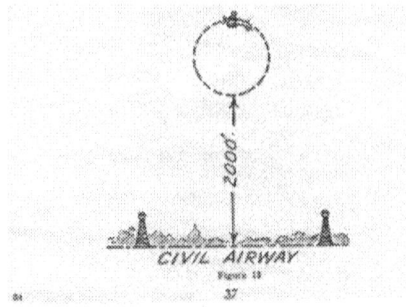

Figure 13

37

37

Licenses will be renewed for like periods where the prescribed physical condition of the holder is shown by the same method as when the original license was issued, except that a Transport or Limited Commercial pilot must prove that he has had at least 10 hrs of solo flying within the last 60 days; Industrial pilots at least 25 hrs within the last year; and Private pilots at least 10 hrs within the last year.

The pilot's license shall be kept in his personal possession when he is piloting aircraft, and must keep an accurate record of his flying time in a logbook.

Pilot's licenses will be suspended or revoked for violating any provision of the Air Commerce Act of 1926, or violating air traffic rules. Or doing any act in connection with aircraft that is contrary to the public safety or interest, or detrimental to the morale of pilots or mechanics.

Aircraft shall not be flown over the congested parts of the cities, towns, or settlements, except at a height sufficient to permit a reasonably safe emergency landing, which in no case shall be less than 1000 ft, or over any open air assembly of persons. Aircraft shall not be flown elsewhere at heights less than 500 ft.

There are special rules regarding acrobatic flying.

Geez! Now I know I'm not going to make it. Joe said: "this is what you have to know and what the test will be about. You better remember, no pass, no fly. The school's responsibility is to teach you guys to fly, and prepare you for the examination for your license. As soon as you begin the examination, whether you pass or fail, your course has been completed". No horsing around tonight; he picked up his papers and left.

March 4, 1932 – Friday

The weather was cool, cloudy, and the wind was fairly strong.

Check time today. Joe went up with me for 30 minutes, and I practiced gentle and steep figure 8's, climbing turns, stalls, and vertical power turns. He must have been satisfied because he took me up higher and we practiced spinning the ship to right and to the left. That is one maneuver I really like, and now I'll get a chance to practice it. We came down and hedgehopped across a few farms before I made a fair 3-point landing.

Joe wasn't very talkative so I concentrated on doing the best I could. He never yelled at me so I guess I did pretty well.

He climbed out of the ship saying, "take it away" so I took off. I never have any trouble taking off anymore; it's just like driving my Cheve down Main Street.

After reaching a safe altitude, I tried to spin the ship to the left. It was not as easy as it looked.

To pass the spin test, the ship is to start the spin in a desired direction, make 3 complete turns, and come out of the spin on an even keel in the same direction.

I use the Ohio River as my desired direction so I can see it out of the corner of my eye as the ship comes around, which gives me time to start the ship in the spin just as the nose is pointed directly at the river. It also gives me time to start to pull out of the spin so that the ship is on an even keel with the nose pointing directly at the river.

I throttled the engine, pulled the "joy stick" back and just as the ship began to stall I moved the stick to the left of neutral and pushed the left rudder petal forward. The nose dropped below the horizon and the ship started to spin to the left, but by this time the nose was past my desired direction. I held the stick back and rudder petal forward to keep the ship in the spin for 3 turns. During the third turn I moved the stick forward and to the right, at the same time pushing the right rudder petal forward to stop the spinning. Moving the stick and rudder petal to neutral I moved the throttle forward and came out of the spin on an even keel at cruising speed. But the nose was past my desired direction. In fact, the river was almost out of sight.

I went back up and spun the ship to the right. This time the ship started spinning before the nose was facing the river. After 3 turns the ship came out of the spin on an even keel before reaching my desired direction, that is, the nose pointing directly at the river.

I practiced spinning the ship first to the left and then to the right 5 times, but was never able to start or stop the spin with the nose pointing directly at the river.

To start and stop the spin in a desired direction is difficult because the speed and direction of the wind is never the same. It is going to take a lot of practice. Hooray! I sure like the spinning sensation, and the chance to try to bring the ship out of a dangerous and life-threatening situation. Spins are not part of flying, except stunt flying, but are the result of losing flying speed and control, and are very dangerous.

As my 55-minute solo time was up I spiraled down and made a fair 3-point landing. Overshot and undershot the line once.

My group washed and waxed Joe's Waco, then spent the rest of the day studying. There is not enough repair work, or rebuilding, to keep everyone busy. We are not getting any damaged or wrecked ships in since Dick Oglesby died. Even Harvey, our mechanic, has to "goof-off" once in a while. Things have been going downhill ever since, and I am beginning to think that there might not be a job waiting for me when I get my L.C. license.

We parked the ships, locked the hanger, and everyone went down town to supper. I stayed at the hanger till they came back to attend class. One of them will spend the night. I had a late supper, and as it was too late to go back out to school, I wrote Mom, Pearl, and Paul a letter. Spent the rest of the evening studying. There sure is a lot to learn. I know the theory, and can repair and rebuild gasoline engines. It is the rules, regulations, laws, and theory of the why and how airplanes fly, as well as the construction, repair, rebuild, and maintain the airframe that is my weakness.

I paid Andy Barrett $1.70 for Cheve parts.

March 5, 1932 – Saturday

The weather was cold, snow, and the wind was very rough. So, no flying today.

We, my group and I, cleaned the hanger, workshop, and office this morning. After dinner Pete Nemetsky and I went down Main Street to the picture show and saw "Women of the Big House". It was good.

After supper I wrote a letter to Earl Seitz, Harry Pyatt, Red Sandstrom, and Herb Schwartz. Then I studied the last lesson we had and I am having trouble remembering all that stuff.

About 9:30 pm, Jim Oakley, Francis Ferguson and I drove down to "Looeyville" to go to a picture show. We really slid all the way down as the roads were covered with snow. When we got there at 12:00 am we found all the picture shows were closed. After sliding all around town trying to find a show that was open we decided to go home. We didn't have any trouble till we got halfway down Hanover Hill and had a flat tire. I had to fix it without any lights, in all that cold wind, about 5 degrees, at 3:30 am. Boy was it cold! Otherwise we had a good time.

March 6, 1932 – Sunday

The weather was very cold, and the wind was gusty. No flying today.

We all hung around the house all day, as it was too cold outside. Spent the day eating, sleeping, shooting the breeze, and studying for the L.C. examination. It took me all day to get thawed out, and catch up on my sleep from that trip to "Looeyville". As the airport wasn't open today, my group didn't have to do any maintenance on the hanger. Now it will be at least 3 weeks before it is our turn again. That sure is a big relief. It is Sam Poletunow's turn to stay at the hanger all night. After supper I went out and "shot the breeze" with him for a couple of hours. It gets awful boring staying out there all alone from 6 pm to 6 am.

On the way out to the hanger I stopped and bought some Tuxedo tobacco 12c to make cigarettes, chewing gum 10c, and eats 25c.

March 7, 1932 – Monday

The weather was cold, and the wind was gusty. No flying today.

Nothing going on out at the airport so I took Mrs. Barber and her daughter Inez, who helps her mother run the boarding house, to "Looeyville" so they could do some shopping. They paid for the gas and gave me a week's free room and board. I parked the car on 3rd Street and walked around town till it was time to meet them at the car to come home. I had a good time. The car quit running on the way home and I had to take the carburetor apart to clean out the water. There must be water in the gas tank. I am going to buy gas at a different gas station.

Went to the picture show tonight and saw "Cheaters at Play", it was pretty good. Went to bed early.

March 8, 1932 – Tuesday

The weather was cold, snow, and the wind was extremely rough. The airport was closed all day, so there was no flying. Everyone stayed home.

I drove the Cheve down Second Street to the Maddox garage. They let me drive up on the second floor so I could drain the water and gas out of the gas tank. There was quite a bit of water in the tank. That station must be selling water as gas. Bought 2 gallons of gas at the garage and went back to the boarding house. The Cheve is just about "broken in".

I studied the "Air Commerce Rules and Regulations" again, and I am beginning to know what it is all about. The supper bell rang so I made a mad dash to the dining room. "Heaven help the hind most".

We all went to school at the hanger after supper. The lesson was about "Rigging" the ship.

The pilot's life, the speed and climb of the ship, its controls and general efficiency in flight, and its duration as a useful machine all depends on the rigger. While the engine may fail the pilot may still glide safely to earth, but if the airplane fails, then all is lost unless a parachute is worm.

Flight is secured by driving through the air, a cambered surfaces inclined to the direction of motion. Such inclination is called the "angle of incidence".

In this way the cambered surfaces secure a lift from the air, and when the speed through the air is sufficient, the lift will become greater than the weight of the airplane, which must then rise. The resistance of the air to the passage of the airplane is known as drift, and is overcome by the propeller thrust, which moves the airplane through the air and so overcomes the drift.

There are 4 forces acting on an airplane. The lift, which is opposed to the weight, and thrust that is opposed to the drift. The lift is useful – the drift is the reversal of useful. The proportion of lift to drift is lift-drift ratio. This is of paramount importance for upon it depends the efficiency of the airplane.

If the angle of incidence is increased over the angle specified in the rigging instructions, then both the lift and the drift are increased also – but the drift is increased in greater proportion than the lift. If the angle if incidence is decreased, then the lift and the drift are decreased, but the lift decreases in greater proportion than does the drift.

The whole weight (center of gravity) is balanced upon or slightly forward of the center of lift. If the weight (center of gravity) is too far forward, then the ship is nose heavy, and if the weight (center of gravity) is too far behind the center of lift, then the ship is tail heavy.

The stability of the airplane is the tendency of the plane to remain upon an even keel and to keep its course.

Directional stability - If this did not exist it would be continually trying to turn to the right or to the left, and the pilot would not be able to control it. For the plane to have directional stability it is necessary for it to have in effect more keel surface behind its vertical axis than there is in front of it. Directional stability is sometimes known as "weathercock stability". Directional stability will be badly affected if there is more drift, drag, or resistance on one side of the plane than there is on the other side.

Lateral stability – Stability with reference to disturbances involving rolling, yawing, or sideslipping. The only possible thing that can make the machine fly one wing down is that there is more lift on one side than on the other. This may be due to the following reasons: the angle of incidence may be wrong or the surfaces are distorted.

Longitudinal stability – If the fore and aft balance is not perfectly right then the machine will try to fly nose down or tail down. This may be due to the following reasons: The stagger may be wrong or the angle of incidence of the main planes is not right.

Owing to the propeller torque the airplane has a tendency to turn over sideways in the opposite direction to which the propeller revolves. It is offset by decreasing the angle of incidence on the side tending to rise. It is far better however to give half such increase on one side, at the same time making a similar decrease on the other side.

When the angle of incidence is increased permanently near or toward the wing tip that is called "Wash-In".

When the angle of incidence is decreased permanently near or toward the wing tip that is called "Wash-Out"

These are the questions to Lesson No. 20

1. (a) What are the forces acting on an airplane?
   (b) How do they affect the airplane?
2. (a) What is directional stability?
   (b) How does rigging affect directional stability?
3. (a) What is lateral stability?
   (b) How does rigging affect lateral stability?
4. (a) What is longitudinal stability?
   (b) How does rigging affect longitudinal stability?
5. Define "wash-in" and "Wash-out"
6. What is the common practice to offset propeller torque?

Lesson 20 - Rigging

THE DAYTON SCHOOL OF AVIATION

be possible to change the angle of incidence by adjusting merely the incidence wires, the result of such practice is to throw other wires into undue tension, and that will cause the framework to become distorted.

DIHEDRAL ANGLE – One method in determining the dihedral angle of an airplane is to place the "set-up" gage at the intersection of a wing panel and the center section as shown in Fig. No. 10. The short member is leveled with the

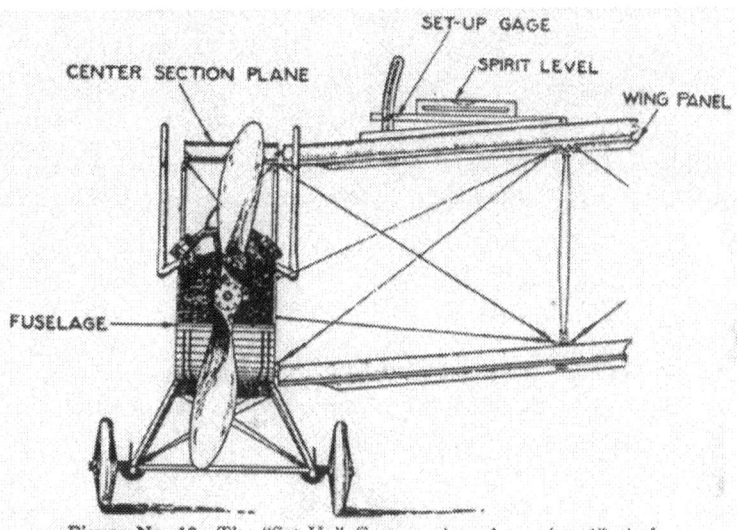

Figure No. 10—The "Set-Up" Gage used to determine dihedral.

spirit level and the degrees of dihedral are shown on the protractor. Thus, to wire a wing panel for a known dihedral, the end of the wing is either raised or lowered until the upper member of the gage is level.

Another method of securing dihedral angle, which is the upward inclination of the wings from the center section towards their tips, is as follows, and this method will at the same time, give you the angle of incidence.

18

THE DAYTON SCHOOL OF AVIATION

The strings, drawn very tight, must be taken over both the front and rear spars of the top plane. There must run between points on the spars just inside the outer struts. The set measurement is then from the strings down to four points on the front and rear spars of the center section plane.
(Fig. No. 11)

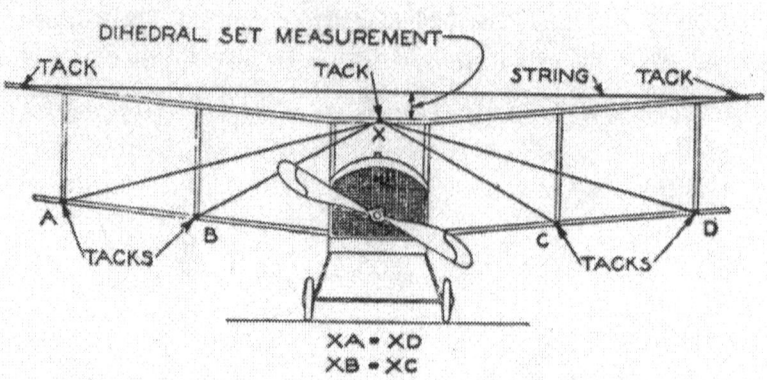

Figure No. 11—The string method used to determine dihedral and angle of incidence.

These points should be just inside the four center section struts; that is to say, as far as possible from the center of the center section. **Do not attempt** to take the set measurement near the center of the center section.

The string should be as tight as possible and, if it can be down, the best way to arrange it is as shown in Figure No. 11, i.e. by fastening the down string to the spars by means of tacks. This will give a tight and motionless string.

However careful the string adjustment is made there is almost sure to be some slight error, and it is necessary to take certain check measurements as follows:

Each bay must be diagonally measured, and such diagonal measurements must be the same on each side of the airplane, as shown in Figs. No. 11 and 12. As a rule these diagonal measurements are taken from the bottom socket

19

Joe picked up a bunch of papers, started out the door, and said, "To be continued". Geez! Sounds like the Saturday afternoon matinee movie serials.

March 9, 1932 – Wednesday

The weather was cold, snow, and the wind was extremely wild. No flying today.

There was nothing to do all day. So I bought an autograph book and spent the morning at the hanger having the fellows autograph it.

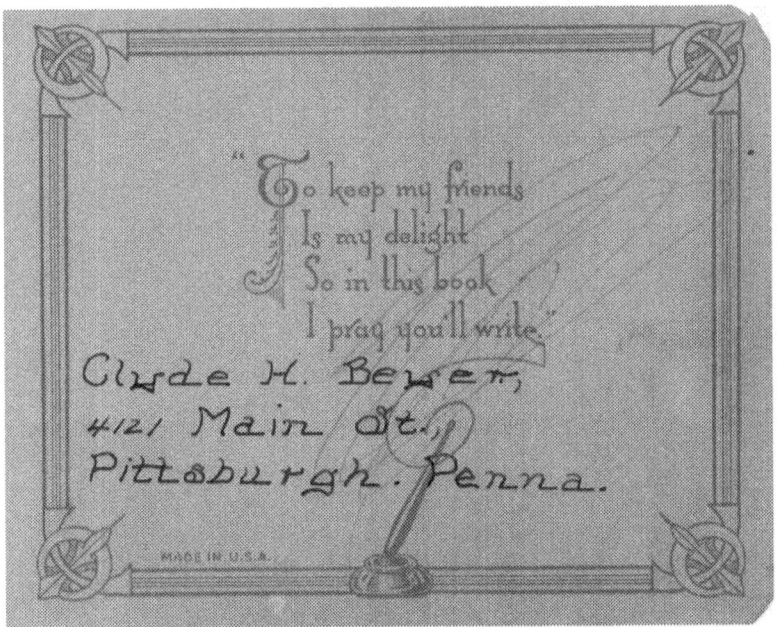

Madison, Ind.
March. 9, 1932

Dear Clyde
Needles and Pins
Needles and Pins.
When you get Married
Your trouble begins.
Happy Landings
Max C. F. Siltner
Hudson
Kansas

Madison, Indiana.
March 9, 1932.

Greetings from another student
from Maine,
Some day he too will fly a plane.
Happy landings!
Kenneth B. Russell
49 Emery St.
Sanford
Maine.

Madison, Ind.
March 9, 1932

Dear clyde:

When ever your car start
knocking bring it to the hangar
and I'll help you fix it.

Arthur Rosert
Box 2,
Hurley, Wis.

Madison, Ind.
March 9, 1932

Dear clyde
When you grow older and your hair is Gray
Please clyde remember our dope so Gay
That we spent at the Dixie Service
Yours for Three Points
Marlin Christianson
Box 22
Clarks Grove,
Minn.

417 Walnut St.
Madison Ind
Mar. 4, 1932.

Dear Clyde.
    To the boy who is never
late.
    He gets up at six and
wakes up at eight.
                    Happy Landing
                        Jack.

Madison, Ind.
Mar. 9, 1932.

Dear Clyde,
    When you are gone
And far away
Just remember
That you still have one
                friend
Formerly of R. F. S.
        Happy Landings
                    Sam

Sam Poletnow
    Box 121
        Bairdford Penna

Dear Clide

I hope we will met in or future some day and talk over or days at the Dixie flying Service

Yours truly

Lawrence Burshaw
920 n. Center St
Park Ridge
Ill.

Madison, Indiana
March 9, 1982

The little Dutch Boy
With the long curly hair
Whenever you need him
He's always there.
Hooray for Clyde!

C. L. (Lin) Hright, Jr.
Churchland, Va.

Madison, Ind.
Mar. 9 — 32

Dear Clyde,

When you get a big job in aviation think of your old friend from Vermont once in a while.

David Swett
Box 212
White R. Jct.
Vermont

Madison, Ind.
Mar. 9 – 32

Dear Clyde,

When you get a big job in aviation think of your old friend from Vermont once in a while.

David Swett
Box 212
White R. Jct.
Vermont.

315 Orchard St
Zeigler Ill
Mar 9, 1932

Dear Clyde:

From a student from Illini who with the rest of the boys is learning to fly.

your Pal
Pete Nemetsky

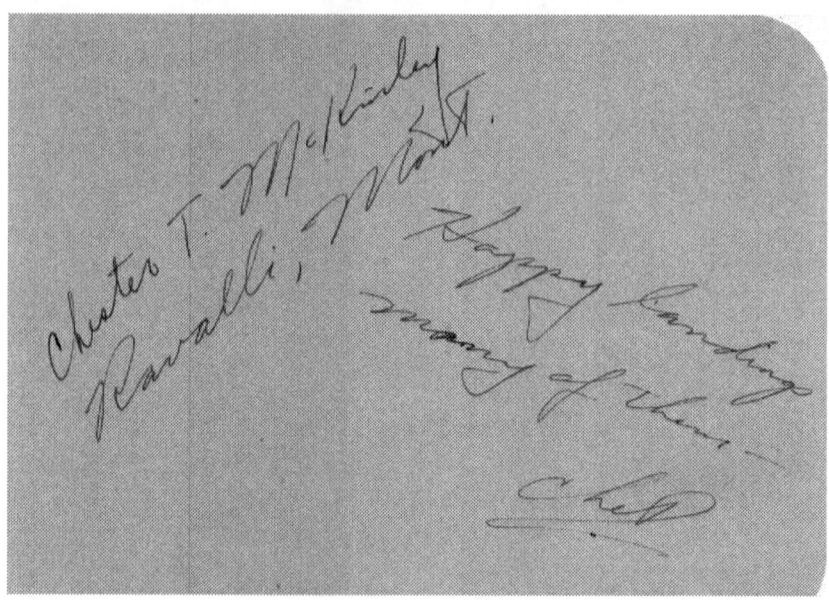

315 Orchard St
Zeigler Ill
Mar 9, 1932

Dear Clyde:

From a student from Illini
who with the rest of the boys
is learning to fly.
your Pal
Pete Nemetchey

Chester T. McKinley
Ravalli, Mont.

Happy landing
many of them
Chet

Madison, Ind.
Mar. 9, 1932.

Dear Clyde,

When you are a Transport
Pilot,
Make lots of happy three
Pointers,
For your old pal,

Frank J. Peternell
Meadowlands, Pa.
Box 185.

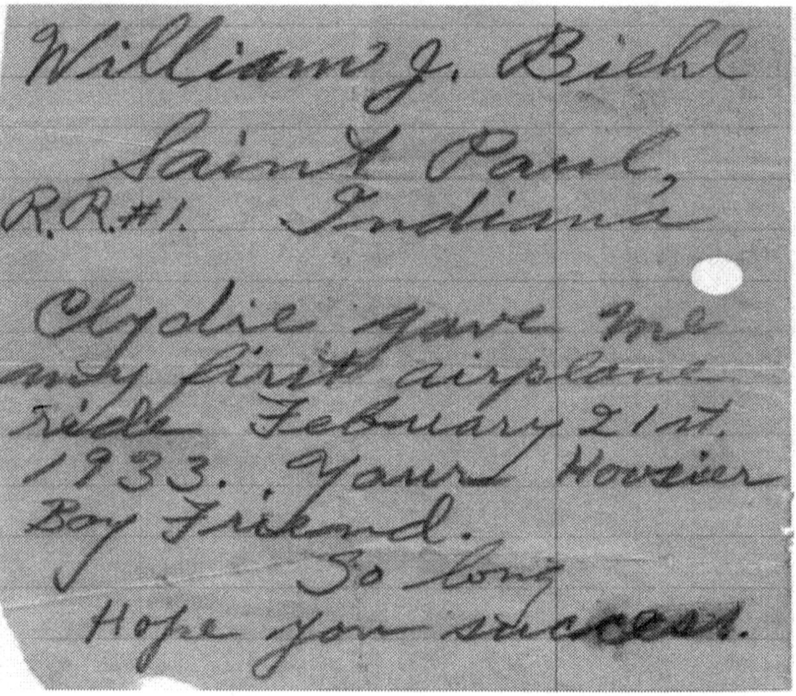

William J. Biehl
Saint Paul,
R.R.#1. Indiana

Clyde gave me
my first airplane
ride February 21st.
1933. Your Hoosier
Boy Friend.
So long
Hope you success.

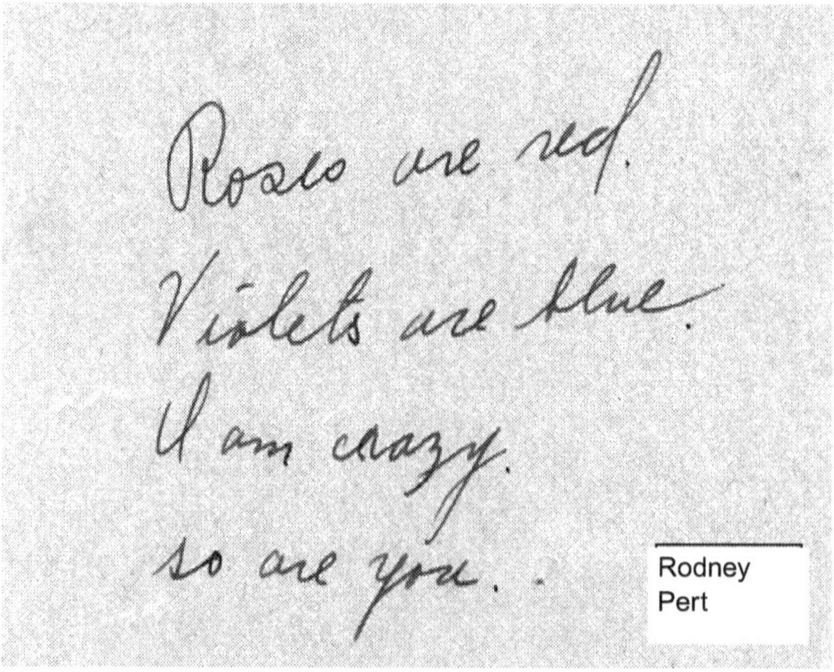

Roses are red.
Violets are blue.
U am crazy.
so are you. .

Rodney
Pert

3/10/32.

Madison, Ind.

Dear.????:—

When you leave this joint
and have a swell job never
lose your temper and argue
with your superiors.

Your all powerful
Friend.

Winamac Ind   "Andy"
Rt. 3.   A. G. Musser Jr.

John Hoover
548 Montgomery Ave.
Chambersburg, Penna.

Just a word to remember me by,
and wish you many happy landings,
(like the first one we made this afternoon)

Dear Clyde

   In all your flying career may you have many frog landings sometimes when I get a ship we sure will raise hell.

     Your old Pal
     David Gentry
     North Madison
     Indiana
     Box 153

Madison Indiana

Remember our solo

3/10/32

Dear Clyde

   Here's hopeing we get an ride together before we leave. Happy Landings and lots of luck.

     Bob
     Box 362
     Galax Va.

tnod eb os
denrub suoiruc.

Madison, Ind.
April 5, 1932.

Dear Clyde:

I'm always standing on
burning desks,
Some people think me
flighty.
I'm always in trouble
up to my neck:
"Geepers Cripes A'mighty."

Howard Brennun
307 Bowman St
Wilkes-Barre,
Pa.

Madison, Ind
March 16, 32

Dear Clyde
While the sun is setting red,
And the sky is still blue.
Hope Dave will fly a T.A.T
and you will, too,

Your Pal
Joe Sachluben

May 18, 1932

Here's wishing you
the best of luck.
And I hope the landing
gear will hold you up.
             Your Friend
             Bud Lawrence
             338 N. Defiance St
             St. Marys, Ohio

May 19, 1932

Dear Clyde;
     Greetings from a friend
of the Dixie Flying Service
days.
             Daniel Booth
             Tower Ind.

May 19 - 1932

Dear Clyde;
Don't forget your little friend from Madison Ind. Wishing you lots of success.
Yours till the Statue of Liberty falls flat.
"Alvin"
Cleone Grimes
N. Walnut St.
Madison, Ind.

(Remember Tarzan"
What a ——.)

May 19, 1932.

Dear Clyde
Here's wishing you much success.

George Hayden
Bangor, Mich.

221

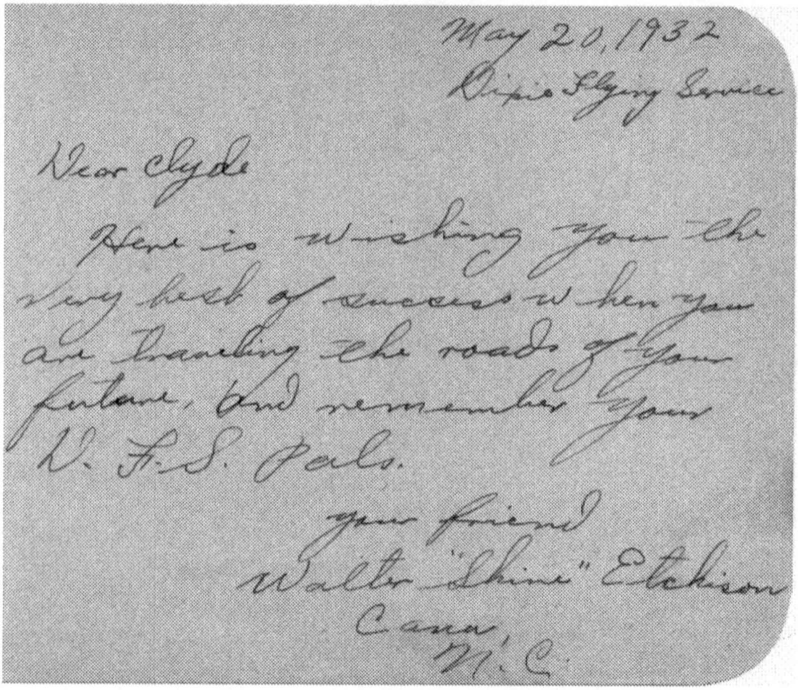

May 20, 1932
Dixie Flying Service

Dear Clyde

Here is wishing you the very best of success when you are traveling the roads of your future, and remember your D. F. S. pals.

your friend
Walter "Shine" Etchison
Cana,
N. C.

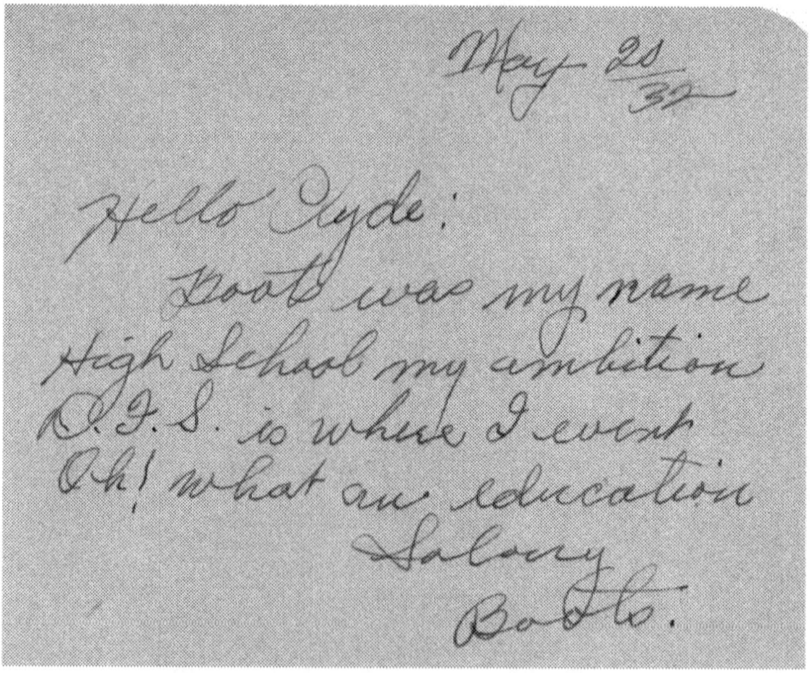

May 20
32

Hello Clyde:

Boots was my name
High School my ambition
D. F. S. is where I went
Oh! what an education
Salary
Boots.

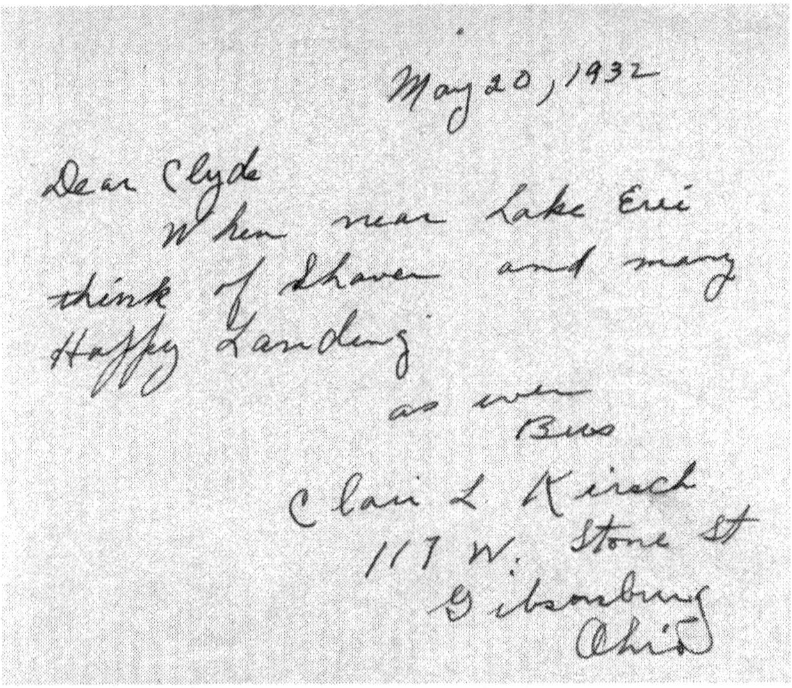

May 20, 1932

Dear Clyde
    When near Lake Erie
think of Shawen and many
Happy Landing.
                as ever
                    Bus

        Clair L. Kirch
        117 W. Stone St
            Gibsonburg
                Ohio

Madison, Ind.,
August 13, 1932.

Dear Clyde:

I haven't much to say
but I wish you lots of luck & success.
Remember your pals of the D.F.S.

Your pal

Dick Boepple

717 N. Walnut St.,
Madison, Ind.

Aug. 5, 1932.

Dear Clyde,

Friends are like wool lace
Must I tell you why?
To find one good! You must
a hundred try.

Howard Price
Darlington, Md.

When in doubt
Or all goes wrong
Get that nose down
And sail right on.

Yours

Harvey Kattelmann
1841 Rockford Place
Cincinnati Ohio

I loaned Jack Yellitz 25c. I spent the rest of the day studying "Rigging", and after supper went right back to studying. It is going to take a lot of work but if I can learn all of this I know it will make me a better pilot. I'll know what the ship is capable of doing, and why it does what it does or can't do.

The local "toughs" still don't like us "foreigners". It is not a good idea to go west of West Street alone, so most of the time several of us go together. It is tricky dating a girl in the west end so most of the students double date. Several times when I dated a west Madison girl it was always as a double date with Dave Gentrup and his girl friend. He was born and raised in Madison and his family is well known.

Most of the people I know and run around with live east of West Street so I don't have much of a problem. My church, the movie house, the garage I loaf at, all are in safe territory. It is similar to Pittsburgh, where there were a couple of areas beyond my neighborhood that wasn't safe, which I found out, to my sorrow, on more than one occasion.

March 10, 1932 – Thursday

The weather was cold, clear, and the wind was wild. No flying today.

There was nothing to do so I bought film, 25c, for my camera and took pictures of the students. It wasn't easy, as I had to hunt them down.

My Cheve froze up and I hope it didn't hurt anything. I think the alcohol must have evaporated. Clyde Weber helped me thaw the engine and fill it with alcohol and water. He treated me to the alcohol, and Max Sittner bought me 2 gallons of gas.

We are the poorest bunch of students I ever heard of. At times it is hard to borrow a lousy dollar from the whole bunch. Everyone borrows from everyone, and always pay it back when the money comes from home. It works pretty well as we generally receive our money at different times. Just about the time everyone is flat broke one of us will get money in the mail.

There is no way we can look for part-time work as our contract with the school says that when we are not taking flying lessons, going to school, or studying, we have spend the rest of each day working for the school. This is part of the tuition fee. Lately there has not been enough work for everyone so we get to loaf now and then. Even if we were allowed to look for part-time work it would not do any good because there are simply not any jobs, part-time or otherwise. Also, I'm told, they hire only local people, relatives, friends, and neighbors they grew up with.

After supper I went out to Joe Sachleben's and we played cards. There was an awful lot of arguing as usual, especially between Fred, Carl, and Joe. "Ma" Sachleben never argues and no one better argue with "Pop" Sachleben. I never argue, as card games are not important to me. I just play for fun. I finally talked "Pop" into giving me a full tin cupful of "home brew" and another sample of his "beer wine". Tastes good.

March 11, 1932 – Friday

The weather was cool, clear, and the wind was fairly steady. I flew for 35 minutes today, and as this was the first time since last Friday I spent the entire time on landings and takeoffs.

I overshot the line one time and made four fairly decent landings. If I could only get the wheels and tailskid closer than 3" to the ground when I set the ship down. Now that would be perfect, and I am getting there.

There was nothing to do so I studied "Rigging" for a while then put new ignition wires on my Cheve. After cleaning the carburetor I loafed around the hanger until it was time to clean the hanger and put the ships in their proper places.

We locked up the hanger for the night and went down to "Barber's" for supper.

After supper I went down to Maddox garage and shot the breeze with the fellers. While there I bought Permatex 50c, hose 25c, and tools 50c. I got a letter from Ralph today.

March 12, 1932 – Saturday

The weather was cold, cloudy, and the wind was pretty rough. I flew for 30 minutes today, taking off great as usual, and climbed to about 3000 ft and practiced my favorite maneuver, tail spins.

It is also a part of L.C. flight test so that gives me a good excuse to practice it. However, it is not all fun because you have to begin and finish the spin in exactly the same direction for the test. It would be more fun if you could begin and finish the spin in any direction but I guess the test is to prove you have complete control of the ship at all times. The spins were ok. A new student, from Gibsonburg, Ohio, came in today. His handle is **Clair L. "Bus" Kirsch**. So I'll get to give him his first plane ride, which doesn't count toward my flight time, and probably take him to Seymour for his physical exam.

I helped clean the workshop, hanger, and rearranged the ships. I am holding my breath, I think it is about time to wash and wax all the ships. A

227

stray Sheppard pup showed up at the hanger today and we adopted him. Quite a keen dog, called "Jack".

Clair L. "Bus" Kirsch

As we were getting ready to close the airport for the day, Joe told us he had to go out of town for about a week on business. If anyone wanted to go home for a visit it would be ok, but be sure to be back by Sunday, March 20th.

After supper most of the fellers decided to go home for a week, so I decided to go home too.

I filled up the gas tank, bought 2 quarts of oil to take with me, cranked up the Cheve and headed north on Highway 29 at 6:30pm. At Versailles I turned right on Highway 50 drove east through Cincinnati, Ohio where I filled the gas tank again as there are very few stations open all night. At the intersection I turned right on Highway 22 and drove NE at a steady 40mph. I went through Wilmington, Ohio, Washington Court House, Ohio, Circleville, and at Lancaster I filled the gas tank.

As I drove through the tiny town of White Cottage, Ohio every single house and building that I could see was painted white. Maybe that's why they chose to name the town White Cottage. When I go back I am going to try to find a building that is not painted white.

Driving through the small towns and deserted streets, with most of the houses and all of the businesses dark, only the streetlights were burning, and the few traffic lights blinking. I wondered about the people in those homes, and what they did for a living. Were they mechanics, clerks, plumbers, carpenters, laborers, or what? What did they do besides work, and what did they do for fun? Every town I went through I tried to imagine what it would be like to live there, and if there were very many pretty girls and a lover's lane.

Going through the ritzy part of town I wondered if a doctor, lawyer, dentist, storeowner, banker, or maybe even a pilot lived in that house. I never got tired of dreaming about those strangers, and they were very interesting thoughts that kept me from getting sleepy.

The larger towns were not much different but the big cities were more impersonal.

I breezed through Zanesville, Ohio, and as I drove through Cambridge, Ohio the sky began to lighten up. By the time I reached Steubenville, Ohio it was almost daylight. People began to come out of the houses, cars and trucks drove up and down the streets, and the town came to life. As I drove through town I glanced down the alleys and wondered about the backside of this town, I guess because I spent most of my grade school years in the alleys around our neighborhood. Alleys still fascinate me.

I left Steubenville and those thoughts behind and I was anxious to get home. I crossed the Ohio River, drove through Weirton, WV, filled the gas tank, and got home in Pittsburgh at 7:00 AM. The trip home was breeze and the Cheve ran great. I had the highways to myself and didn't see a dozen cars on the open highway after midnight. There were very few people or cars in the towns I went through. I saw two cops.

The trip took 12 hours and the Cheve used 25 gallons of gas and two quarts of oil,

March 13, 1932 – Sunday

The weather was cloudy, cold, and the wind was pretty mild.

I parked the Cheve in the alley, walked up the outside stairs and knocked on the kitchen door, which was above the bakery shop. Dad opened the door and I said, "Hi Pop" and shook his hand. He never believed in wasting any words, all he said was "Hello Skipper". They sure were surprised to see me.

They were just starting to eat breakfast so Ma threw another plate on the table for me. While Mom, Dad, Earl, Ralph, and I were eating a great breakfast, I told them why I came home. I started to snatch and grab at the food. Before I remembered that I wasn't at the boarding house, Mom reminded me that I was at home. I sure was hungry; I didn't have anything to eat since supper yesterday.

Dad always called me "Skipper", I guess because ever since I could stand up I never bothered to walk, that took too much time. I ran up and down the stairs and through the house all the time. It drove Mom nuts. I never learned to walk naturally and always felt uncomfortable when I did. I ran to Peabody High School, which was about 2 ½ miles from home in the morning and saved the 10c streetcar fare. After school I ran home and saved another 10c. It paid off when joined the track team and specialized in the mile and cross-country.[26]

I was about to go upstairs and hit the sack, I'd been up since 7:00am Saturday, when Mom told me I better hurry up and get ready to go to Sunday school and church with them at the First Evangelical Church of Pittsburgh, organized in 1843.

The Sunday school seemed real glad to see me. The teacher didn't get to do much teaching as they all wanted to know what I had been up to, and wanted to hear about my flying experiences. What a change. Before I went away they wouldn't have noticed me if I dropped dead. So when I told they about the school I did a bit of exaggerating. I had been "sweet" on the preacher's daughter but she never even gave me the time of day. She probably did me a big favor.[27]

As we left the Sunday school room to go in the church sanctuary Don Bicht invited me to eat dinner at their house on Wednesday. Don and his brother Jack (who is quite an artist) are members of my Sunday school class. They live in a very ritzy part of town.

Our church has a lot of rich members. The Haller family owns a big bakery company with a fleet of trucks. Walter Haller has been superintendent of the Sunday school for many years and is really a nice guy. The Rall family owns the Pittsburgh Pipe and Steel Corp.[28]

You simply could not get a job unless you were a good friend of someone important, or married into the family. Paul eventually married into the Niebaum family and they got him jobs in a number of Westinghouse

---

[26]   I went at a "trot" everywhere I worked. But when I went to work for the government at Jefferson Proving Ground I got a reprimand, and told not to do it anymore because it was a safety hazard.

[27]   Many years later Paul told me she had been married so many times she must have thought it was going out of style.

[28]   Charles Rall gave Paul a job, chief of security, in his factory, and Paul said it took some getting used to being bossed by such a young President

factories. The Charles Mayer family owned a big construction company, and there were several rich doctors and lawyers.

The church I go to in Madison is quite different. The members are mostly farmers and working people. I don't know of any really rich members, maybe a few "well-to-do".

I got my first job after I graduated from high school because my Dad was a member of the Masonic Order and also the President of the Retail Bakers Association of Pennsylvania.

We hurried home after church and had a great dinner. I snatched up the Sunday comic strips, laid down on the davenport, then fell asleep. We had the usual leftover Sunday supper, and then Pop and I walked out Penn Avenue to East Liberty, crossed over Baum Blvd. (it is called gasoline alley because most new car dealers are there) and after looking at all the new cars we circled back on Baum Blvd. and headed home. Dad sure loved cars. He always smoked a "stogie" (a cheap cigar) while out walking. Mom ain't suppose to know this but I'll bet she does. No smoking, no drinking, no "cussin" is allowed in the house, and the only card games allowed is something like "Old Maid".

After writing all this in my journal it was snack time, then sack time.

I went to sleep dreaming about Eleanor, the other girl in my Sunday school class. She sure is a "knockout", wow! But I never had the guts to ask her to go out with me, although I did get up the nerve to trade bibles with her. The first thing I noticed about her is that she walked like a boy in a hurry. She made straight A's in high school, graduated a semester before I did, and went on to college at Bryn Mawr (a high-class girl's college) on a scholarship. Her dad is a wealthy doctor.[29]

March 14, 1932 – Monday

The weather was cloudy, warm, and the wind was breezy.

After a late breakfast I helped Dad and Charles in the bakery shop. I fried the donuts, put icing on the rolls, and squirted custard into éclairs, then put chocolate icing on top of them.

I watched Dad put bread, etc. to be baked in the oven with a wood peel. It is flat wood paddle with a 6 ft or so handle. He put the rim-filled pies, like egg custard and coconut custard in the oven without spilling a drop. I always expected him to, and waited for him to cuss in German, but he always disappointed me.

---

[29] I've wondered since how her life turned out.

The oven was the size of an ordinary room, and our back porch, where Mom hung the laundry to dry, sat on top of it. It was lined with special bricks, and the slightly curved ceiling was only about 3 ft above the floor of the oven. At one time it had been heated by burning coal and wood in a pit on the right side. Dad had large tubes, about 3" in diameter, installed and heated it with natural gas.

One time there was a problem with the ceiling bricks. It took a couple of days to let it cool down enough so Dad could slide in on his back and take care of it. Paul and Charles stood right by the cast iron door, which was about 3 ft wide and 2 ½ ft tall, in case they had to pull him out in a hurry as there was only limited circulation of air. When they pulled him out, his clothes were all wet with sweat and they told him he was half-baked. They didn't mean he didn't have all his marbles.

I went to see Tony Fellow, 4106 Penn Avenue, this afternoon. His family owns a laundry and dry cleaning store. His back yard is just 3 yards down the alley from ours. We found Red Sandstrom, 4621 Friendship Avenue, and several other fellers, including Bus Frank, 4412 Penn Avenue, (his family owns a barbershop and his backyard is across the alley from ours) that we used to play ball with and spent the afternoon "shooting the breeze".

His folks asked me to have supper with them and we had a great Italian meal, lots of spaghetti and everything that went with it, fresh Italian bread, and a large pitcher of wine in the center of the table, nothing else to drink. Tony's dad made sure I didn't have too much wine. Tony took me down the cellar and showed me all the big wooden barrels of wine his dad made. There sure were a lot of them stacked to the ceiling. He said wine was the only drink they had at meals. Tony's dad would go down to the freight yards and if they were auctioning off a fright car of grapes that were beginning to spoil he would bid on it and get the grapes real cheap. When I lived at home I ate at Tony's house quite a few times.

As I left Tony gave me two quarts of wine to take back with me. He said they called it "Dago Red". It sure is powerful stuff and I not going to open it till I get back to Madison.

I spent the rest of the evening studying "Rigging" and it is getting easier. If I could practice rigging on a ship as I study, it would be easier and stay with me. Practice is the key.

March 15, 1932 – Tuesday

The weather was cloudy, warm, and the wind was calm.

I helped down in the shop all morning. I got a letter from Joe Sachleben. After dinner Earl Seitz took me out to his shack in the country at RFD 1, Beatty Road, Turtle Creek, Pa. where he taught me how to shoot a revolver and rifle. We shot a 38-caliber revolver at a target, then a 45-caliber revolver that wants to kick back when you pull the trigger. The 22-caliber pistol was the easiest to shoot. Finally we practiced shooting at the target with a hair trigger rifle. A couple of times I barely touched the trigger and the rifle went off before I had aimed it at the target. It is sensitive. We soon ran out of ammunition so we pitched several games of horseshoes. Earl won two games and I won one.

Earl took me to his folk's house for supper and it was great. After supper, Earl, his dad, and I went into his den and we took turns playing checkers all evening. I lost more games than I won.

March 16, 1932 – Wednesday

The weather was cloudy, warm, and the wind was gusty.

I got home late last night so I got up late this morning. I spent the morning fooling around the shop with Charles. You didn't fool around with Pop while he was working, which was most of the time.

After dinner I got all "shined up", put on my Sunday best and drove out to Don Bicht's house at 7500 Rosemary Street. It is a spiffy house in a ritzy neighborhood. My poor Cheve stuck out like a sore thumb sitting in front of the house. He showed me through the house, and then we went out in the yard and shot the breeze. He said his mother had been an invalid for a long time, and her sister was their housekeeper. I told him what it was like at the school and how I loved to fly.Jack and Mr. Bicht showed up at the same time. He works in office downtown. I think he is draftsman. We all went in the house and had a fine "dinner" in their dining room, which is a beauty. My table manners might not have been the best, but at least I didn't spill anything on the fine tablecloth, and I enjoyed it.

I thanked them for the "dinner", jumped in the "sorry" Cheve and headed home. On the way home I started to think about my model biplane I had built which was sitting on the piano in the parlor.

I took the biplane, sat it on the table beside me and as I studied "rigging", I used it to help me understand the principle and how important it was to rig the ship properly. A biplane is more difficult to rig than a monoplane because the upper wing must be rigged perfectly in relation to the lower wing, which must be perfectly rigged in relation to the fuselage. But a monoplane wing only has to be rigged perfectly in relation to the fuselage.

Of course the tail surfaces and landing gear has to rigged perfectly in relation to the fuselage.

If the ship is not rigged properly it either won't fly at all, or will fly poorly making the pilot work constantly to compensate for the rigging errors. Snack time, sack time.

March 17, 1932 – Thursday

The weather was cloudy, warm, and the wind was real rough.

Ralph stayed home from school today and helped me work on the Cheve. We took the cylinder head off and cleaned the carbon off. We drained the oil and took the oil pan off and cleaned it. Then we took the pistons and rods out of the cylinder block and cleaned the carbon off the pistons. We put new inner rings on the pistons and put the pistons and rods back in the cylinder block. This should keep the engine from burning so much oil. We tightened the bearings, put on the oil pan and cylinder head, and filled the engine with oil. We cranked up the engine and it sounded great.

We spent the rest of the evening loafing around. I am ready to go back to Madison. Snack time, sack time. I went to sleep dreaming about flying those things.

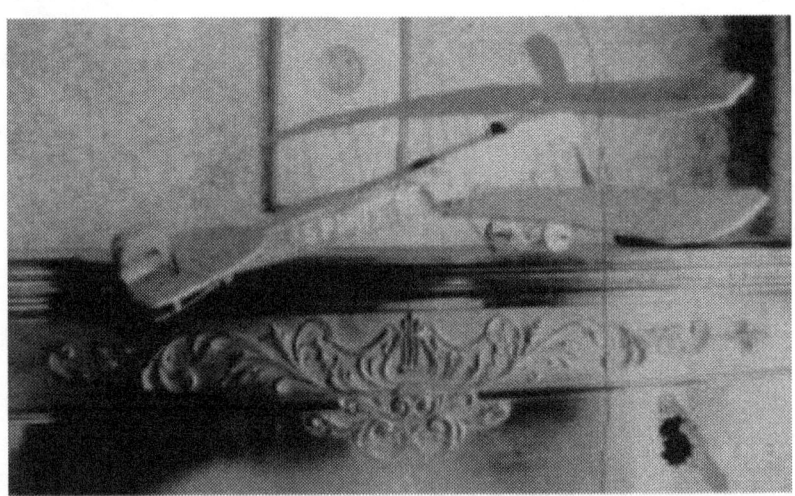

March 18, 1932 – Friday

The weather was clear, cool, and the wind was strong.

I had a real bad day. Someone stole my fancy radiator cap off the Cheve and tore the neck off the radiator to get it. I spent the whole day hunting a radiator neck and cap, and finally found a Studebaker neck and cap at a junkyard.

I took the radiator out of the Cheve and had the neck welded on at the junkyard. I put the radiator and cap back in the Cheve and filled it with alcohol and water. It is ready to go back to Madison and so am I. I sure hated to lose that fancy cap.

After supper I glanced through the rigging lesson and I am ready for the second lesson on it. Fred Moss came by and after shooting the breeze, we went to a midnight show. It was good. Fred lives at 727 N. Mathilda Street.

March 19, 1932 – Saturday

The weather was clear, cool, and the wind was rough.

I had a late breakfast. Mom bought me a nice pair of shoes and I bought a new pipe.

I checked the Cheve, filled her up with gas, bought 2 quarts of oil to take with me; now she is ready to head back to Madison.

I read in an interesting magazine article promoting an airplane factory in Lincoln, Nebraska. It showed several pictures of a plane taking off and landing on the main street. That seemed like a great idea to promote the plane. I'm going to keep the article; it might come in handy.

After dinner I went around the neighborhood and told the fellers I was leaving for Madison after supper. Red Sandstrom introduced me to a young Irish girl, about 17 years old I guess, who had just moved in the neighborhood a couple of weeks ago. She was very nice, very pretty, and sure is stacked up. Hoop-ala-la, I wish I had met her last Monday, and was tempted to stay another day.

Red left, Maureen and I chitchatted awhile. She is Irish all right. I put on the dog, telling her what a pilot I was but I don't think she was impressed. As it was near suppertime I asked her about pitching a little woo.

She wanted to know what I meant, but I had a hard time putting it into words. Finally I said, "You know, lolly-gagging, holding hands, hugging real close, kissing, and stuff like that". She laughed and said, at home, they call that schmultzying. Well I didn't get any schmultzying, but she seemed to be a very nice girl.

Mom fixed a big supper. I loaded the Cheve and put the 2 quarts of wine under the seat. Pop gave me a lot of cigars, and Ralph bought me 5 packs of 1876 pipe tobacco. I hate not having any money of my own. Mom gave me a big box of cookies and cakes.

I cranked up the Cheve and left Pittsburgh at 5:30pm, Pittsburgh time. I filled the gas tank in Steubenville, Ohio. As I drove through the tiny town of White Cottage I slowed down and tried to find a building not painted white. If there is any I didn't see it.

I filled the gas tank in Lancaster, Ohio and added a quart of oil. The Cheve ran great, chugged along at a steady 40 mph (top speed is 50 mph). It didn't sound like the 16-cylinder Marmon I test drove, which sounded more like an electric sewing machine motor than a gasoline engine. But the 4 cylinders fired in a steady sequence sounding like some chicks cackling in a hen house.

I filled the gas tank in Cincinnati, Ohio, and added a quart of oil.

March 20, 1932 – Sunday

The weather was cloudy, drizzly, warm, and the wind was wild.

I got into Madison at 5:30am. The trip took 12 hours, 15 gallons of gas, and 2 quarts of oil. No trouble at all. The new inner rings didn't seem to make any difference.

I had a big breakfast, dashed up the steps into my cubbyhole, fell in the bed and went to sleep thinking about that cute Irish girl.

They woke me up in time for supper, then I unloaded the Cheve and put my Sunday-go-to-meeting suit between the cardboard under the mattress. I spent the rest of the evening in the living room swapping lies with the fellers. They all got back today. I am dying to get back in the cockpit.

March 21, 1932 – Monday

The weather was cloudy, cold, and the wind was rough.

I had an early breakfast and went out to the airport. I didn't get to fly today. The pup "Jack" sure is growing.

Joe Shumate, the Inspector for the Department of Commerce flew in this morning to give the private pilot's test to Andy G. Nufer, Lawrence Burklow, Rodney Pert, Frank, J. Peternell, Max C.F. Sittner, and Chester T. McKinley. They all passed the written test but the Inspector decided the wind was too rough to take the flying test.

They all went home tonight. I don't know if they have to come back and take the flying test or if they are going to get their licenses without taking the test. Frank will have to as he is taking the Limited Commercial Course.

I spent the day helping the fellers clean up the hanger and the workshop. I inspected both "Preps".

After supper we all went to the hanger and Harvey taught us the second lesson on rigging a ship.

In order to rig a machine it is necessary to have a correct idea of the work every wire and every part of the airplane is doing. The work the part is doing is known as stress. If, owing to undue stress, the material becomes distorted, then such distortion is known as strain.

All the spars and struts must be perfectly straight. Struts and spars must be symmetrical, undamaged, and must be properly fitted into their sockets or fittings. Interplane struts have to keep their planes apart so that the latter are in their correct attitude.

The greatest care must be exercised in properly rigging the aileron, rudder, and elevator, for the pilot entirely depends upon them in managing the ship. The ailerons and elevators should be rigged so that when the machine is in flight they are in a fair true line with the surfaces in front, and to which they are hinged.

If the surfaces to which they are hinged is not a lifting surface, then rig the control surface to be in a fairly true line with the surface in front. If the control surface is hinged to the back of a lifting surface, then it is necessary for it to be rigged a little below what it would be if it were in a fair true line with the surface in front.

As a general rule you will be safe in rigging it down so that the trailing edge of the control surface is half-to-three quarters of an inch below where it would be if it were in a fairly true line with the surface in front.

When adjusting the control surfaces the pilot's controls must be in their neutral position. They should be blocked into position with wood braces.

Before rigging the machine it is necessary to place it in what is known as "flying position". This best secured by blocking up the machine so that the engine foundation is perfectly horizontal both longitudinal and laterally, or in a position specified in the rigging instructions.

A simple "setup" gauge can be used to determine the angle of incidence, dihedral and stagger. It consists of two wooden members hinged at one end and spread at the ends by a brass protractor and wing screw. The position of the spirit level differs with the use of the gauge.

The dihedral angle is the upward inclination of the wings from the center section towards their tips.

Stagger is the distance the top plane is in advance of the bottom plane when the machine is in flying position.

The last thing to go onto the machine is usually the propeller and you must be very careful to see that this is fitted on true and straight.

The exact angle of incidence of the tail plane is given in the rigging diagram. Be careful to see that the spars are horizontal when the machine is in flying position.

The rudder, ailerons, and elevator must be not be distorted in any way.

The landing gear must be very carefully aligned. When rigging the landing gear, the airplane must be blocked up in its flying position and high enough so that the wheels are off the ground. When in this position the axle or a line between the fittings must be horizontal.

These are questions to Lesson No. 21

1. *What is your idea of stresses and strains?*
2. *How would you rig the aileron and elevator on a lifting and non-lifting surface?*
3. *(a) What is the first step taken to rig an airplane? (b) How is it done?*
4. *(a) What is a "setup" gauge and how is it used in rigging? (b) What other methods can be used besides that already named?*

Lesson 21 – Airplane Construction – Assembly & Rigging

THE DAYTON SCHOOL OF AVIATION

TRUING UP THE MAIN PLANES – The dihedral is 1° for both the upper and lower main planes. Adjust the tension on the front flying and landing cables between bays until the planes are true. (Fig. No. 7). Check by set-up gage or string method. Stretch a cord from the bottom of the front kingpost socket on the upper right main plane. The vertical distance from this stretched line to the top of upper center section at a rib should be 2 ¾ inches. The stagger should be 16 inches throughout, a continuation of the

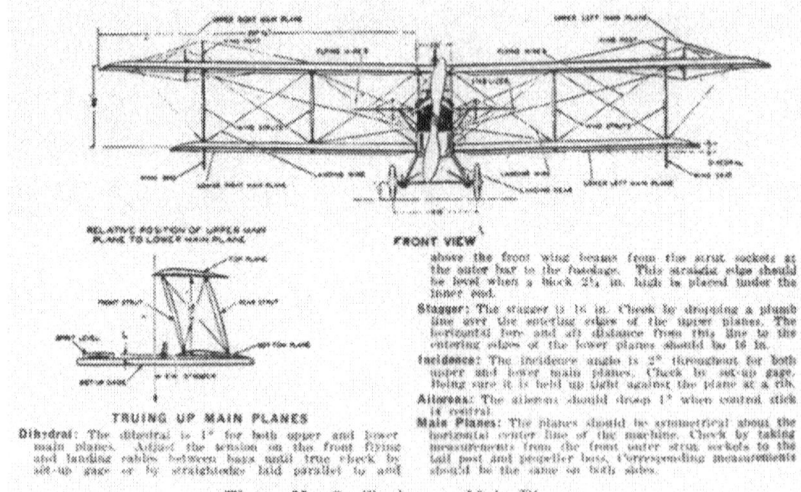

Figure No. 7—Truing up Main Planes.

center section plane. Adjust the incidence cables between each set of front and rear interplane struts and check by measuring the horizontal distance fore and aft from a plumb line dropped over the entering edge of the lower main planes.

The angle of incidence is 2° throughout for both the upper and lower main planes, arriving at the proper angle by

10

240

THE DAYTON SCHOOL OF AVIATION

adjusting the incidence cables and the rear flying and landing cables. Check by set-up gage, being sure it is held up tight against the plane at a rib.

The main plane should be symmetrical about the longitudal center line of the machine. Check by measuring the distance from the front outer strut No. 1 to the tail post at the rear of the fuselage, and the distance from this strut to the drag cable plates on the sides of the nose of the fuselage. Corresponding measurements from the right strut No. 4 back and forward

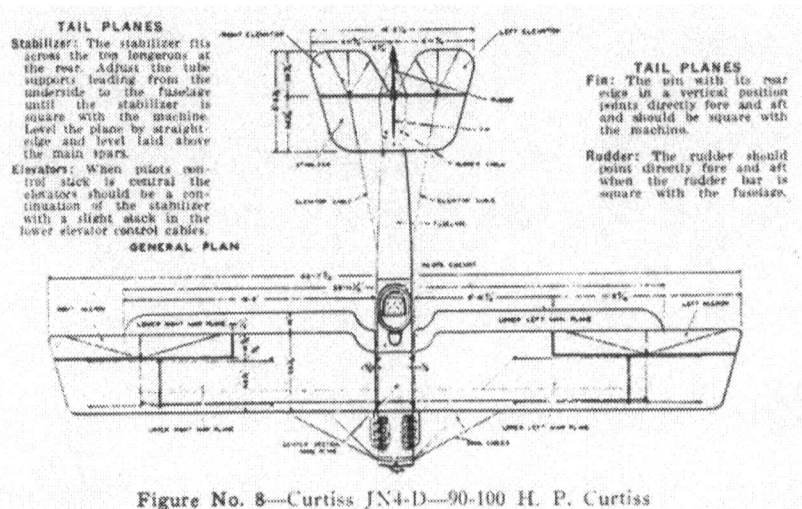

Figure No. 8—Curtiss JN4-D—90-100 H. P. Curtiss

should be the same as on the left, or within 1/8 of an inch. After the planes have been trued up, attach the upper and lower nose drag cables to the fittings at upper and lower strut sockets Nos. 2 and 3. All cotter pins are bent out, and turnbuckles for flying cables can be safely wired together.

ATTACHING TAIL PLANES AND TRUING UP –

The stabilizer is bolted down on top of the rear end of the fuselage by "U" bolts and held rigidly by the steel tube sup-

11

Clyde H. Beyer

THE DAYTON SCHOOL OF AVIATION

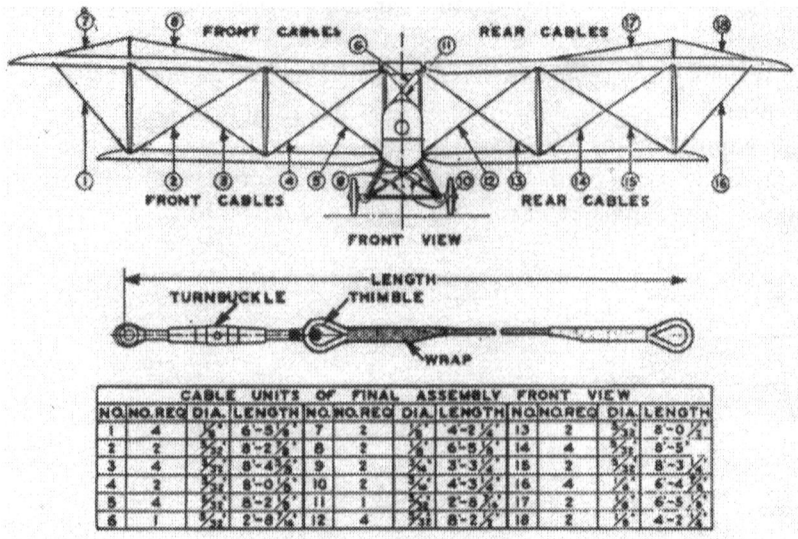

CABLE UNITS OF FINAL ASSEMBLY FRONT VIEW											
NO	NO.REQ	DIA	LENGTH	NO	NO.REQ	DIA	LENGTH	NO	NO.REQ	DIA	LENGTH
1	4	1/8	6-5/8	7	2	1/8	4-2 1/4	13	2	5/32	8-0 1/2
2	2	5/32	8-2 3/8	8	2	1/8	6-5 1/2	14	4	5/32	8-5
3	4	5/32	8-4 3/8	9	2	1/8	3-3 1/2	15	2	5/32	8-3 1/4
4	2	5/32	8-0 1/2	10	2	1/8	4-3 1/4	16	4	1/4	6-4 3/8
5	4	5/32	8-2 3/8	11	1	5/32	2-9 3/4	17	2	1/4	6-5 3/8
6	1	5/32	2-8 3/4	12	4	5/32	8-2 1/4	18	2	1/4	4-2 3/4

Figure No. 11—Cable Lengths Curtiss JN4-D Assembled Wing Panels

Check main planes for dihedral, planes and center section for stagger, and the incidence angle of main planes and center section.

Look over all interplane struts and see that they are tightly bolted in place, streamlined in the direction of flight.

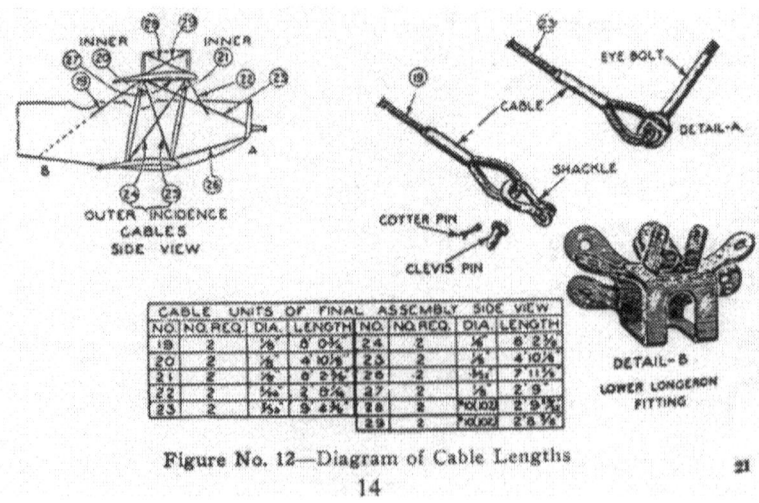

CABLE UNITS OF FINAL ASSEMBLY SIDE VIEW							
NO	NO.REQ	DIA	LENGTH	NO	NO.REQ	DIA	LENGTH
19	2	1/8	8-0 1/2	24	2	1/4	8-2 1/4
20	2	1/8	4-10 1/2	25	2	1/4	4-10 1/4
21	2	1/8	8-2 3/4	26	2	5/32	7-11 3/4
22	2	5/32	3-0 3/4	27	2	1/8	2-9
23	2	5/32	9-4 3/4	28	2	No.10x2	2-9 3/4
				29	2	No.10x2	2-9 3/4

Figure No. 12—Diagram of Cable Lengths

14

21

After dictating the lesson, Harvey took us out in the hanger and went through the whole lesson on Joe Waco. He performed and explained each step as he went along. It is always easier for me to understand the whys and wherefores when I am actually doing the work, and I remember it better. Now when I study this high fallutn lesson I'll know what they are talking about. Each type of airplane will have its own set of rigging instructions.

It was real late when Harvey finished and locked up the hanger. It was my night to stay and act as night watchman so as soon as they were all gone I hit the sack (cot).

Fit as a fiddle and ready for love. My "bestest" love is that little 5-cylinder gal called "Prep" sitting out in the hanger. "Jack" had better be hanger broken.

I am worried about the school and the job I am supposed to get. Everything is at a standstill. No new students and no wrecked ships to practice on and repair for sale. Everything has been going downhill since Oglesby died.

March 22, 1932 – Tuesday

The weather was cold, cloudy, and the wind was pretty rough.

When the fellers showed up I went down to Barber's for a late breakfast, and then went back out to the airport.

I flew the NC 122K for 35 minutes today. Practiced 180° turns, 360° turns, and spirals. Right on the money. Overshot the line twice and undershot it once. Made four near perfect landings. I got to be able to put the ship on or near the line every time and soon. It won't be long before I take the L.C. test. Vertical power turns is the only maneuver giving me trouble and not very much at that.

Art Prosek filled the crate with gas and took off down the runway like a drunken sailor. But he did get it in the air. I don't usually watch the others fly, I got other things to do, but when Art took up much of the field taking off I wanted to see his landing. I thought I would die laughing. He touched down with a thud and looked like he was playing hopscotch, bouncing all over the field. He has a lot to learn. We sure gave the landing gear and tailskid a good inspection.

We washed and waxed the "Fleet". Joe wants the newest students to do what little work there is to do on the ships. That gives me a lot of free time. We wash, wax, maintain and inspect the ships more often than necessary.

We all went out to the hanger after supper to school. Instead of Joe teaching us, we held a good old-fashioned singing school, and boy I sure can't sing.

The people at church have been bugging me to sing in the choir. Joe and Carl Sachleben do, and Carl is a great bass singer. I tried to tell them I couldn't carry a tune in a bucket. Finally I did join the choir, and the first Sunday the choir director kept giving me dirty looks every time I sang. They never asked me to sing again.[30]

I had a chance to talk to Joe alone and told him about the article I read in a magazine while I was in Pittsburgh. The article said that an airplane factory in Lincoln, Nebraska promoted their place by landing and taking off on Main Street.[31] I asked Joe if he thought we could do something like that to attract more students. I wondered if someone could go around and buy up junk and wrecked ships to rebuild like Dick Oglesby did. I also asked him what he thought about flying under the bridge. I'll volunteer. He said he would think about it but flying under the bridge was against the law.

March 23, 1932 – Wednesday

The weather was cool, cloudy, and the wind was furious. No flying today.

A new feller, **Howard Brennan**, 307 Bowman Street, Wilkes Barre, Pa. came in today and I took him to the doctors at Seymour for his physical exam.

I need a four-seater as it is kinda crowded double-dating in my Coupe. When I ain't using it there are students always wanting to borrow it although they gripe. I could make money renting it out every night but the fellers are as poor as I am so all they have to do is keep the gas tank full.

I went down to the junkyard and they had a Cheve Sedan just like my Coupe, same year and everything. As the rest of their car was in real bad shape, I made a deal with them that I would switch the bodies and give them $10.00.

---

[30] The choir director, Gladys Dietrich Bersch, and I are still very good friends.

[31] See picture at end of the book

Art Prosek and the new student Howard Brennan helped me tow the Cheve Sedan up the ramp onto the second floor of Maddox garage. It was real tricky as the ramp was not only steep but it was L shape. If anything went wrong and the car went out of control down the ramp it would crash right through the office.

Mr. P. Maddox always lets me use the second floor as the school rented it until we built the hanger. His son Jim is suppose to run the garage and doesn't like the idea that I use it, but Mr. Maddox told him to let me alone.

We parked the two Cheves side by side near the hoist. I took the hood, windshield, instrument panel, and all the wires and braces off my Cheve. We jacked up both cars so I could slide under them with a creeper and then we went back to Barber's for supper.

I went back to the garage, slid under my car, and began to take off the bolts that held the body to the frame. It was a real hard job; the bolts and nuts were rusty, and some were rounded off causing the wrench to slip. I finally gave up for the night and went back to the boarding house and hit the sack. As I left, Jim Maddox griped that I was using too much of their electricity. I threatened to tell his dad.

March 24, 1932 – Thursday

The weather was cool, cloudy, and the wind was furious, so there was no flying.

I went out to the airport with the fellers and cleaned all the ships windshields and cockpits. I also performed the daily inspection on the ships so they will be ready to fly when the weather is fit. We all went back to town at noon.

I stayed in town and went down to the garage and worked on the Cheve. After supper I went back to the garage and finished taking the rest of the bolts out. The body is ready to be lifted off the frame but I will need some help.

I took the hood, instrument panel, windshield, and all the wires and braces off the sedan, and then quit for the night. I am so stiff and sore all over, and my hands, arms and elbows are all skinned up. It was a chore to walk back to the boarding house.

I am not used to working lying on my back for 2-3 hours at a time. I didn't need to be rocked to sleep tonight, and instead of hitting the sack, I gently eased myself in. Someone will have to pry me out in the morning. Geez.I got a letter from Ma today.

March 25, 1932 – Friday

The weather was warm, clear, and the wind was pretty mild.

I gave Joe Vail the Easter egg cake I brought from home and he said he liked it a lot. Just trying to butter him up a little.

I flew for 1 hour and 15 minutes today and overshot the line 4 times and undershot it once. The calm air probably fooled me, as the wind has always been rough. Rotten work. The touchdowns were real smooth. I practiced figure 8's, 180° and 360° turns, spirals, and stalls right quick.

During the maneuvers I let the ship gradually drift east until we were out of sight of the airport, and several miles up the river. We turned, headed west over the river, and went down to about 100 feet to see how dangerous it would be to fly under the bridge.

Standing on the riverbank the floor of the bridge looked like it was way above the water, and the pillars supporting the bridge were far apart. But as I flew toward the bridge the space that I would have to fly through looked awfully small. I pulled up, turned right climbing over the hill towards the airport.

I decided that flying under the bridge with our low powered ship would be too risky. Wind gusts could push the plane into the pillars, the floor of the bridge or the water, and it is against the law. I just don't have enough experience to try something like this yet. I'm coo-coo, but not crazy.

I made a soft landing and parked the ship. Art said that Joe wanted to see me in the office. I thought he was going to bawl me out for overshooting the line so much. But it was worse. I am grounded again. Someone told him I tried to fly under the bridge. I tried to tell him that I absolutely was not about to try to fly under the bridge, that I was not even near the bridge, that I just wanted to see what it looked like, flying close to the water. He was down right mad and said that if the inspector heard about it the school would be reprimanded.

I don't think giving him that Easter egg cake buttered him up enough.

I piddled around the hanger, and then filled some ditches in the runway with gravel; trying to stay out of Joe's sight until he cooled off. I watched Bob Parker takeoff and he sure is getting reckless. He waited too long to liftoff; but took off anyway and the wheels scrapped through the top of the trees at the west end of the runway. I was sure the ship would flip but it kept right on climbing.

I stayed in town after dinner, walked down to the garage, worked on the Chevy sedan and got the body all ready to be lifted off the frame. Clyde Weber helped me hoist both bodies off their frames and I found the bottom of the door posts on both bodies were rotten. They will have to be replaced and I am beginning to think I bit off more than I can chew. I can't quit now; I don't have any money to hire someone to finish it for me.

Clyde Weber and I went to Paris Crossing after supper to see his redhead girl friend, and it was real late when we got back to Madison.

March 26, 1932 – Saturday

The weather was clear, warm, and the wind was mild.

I was grounded so I stayed downtown and worked on the Cheve. I took the rotten doorposts off to use as guides in making the new ones. I walked across the street and had Naill's Lumberyard cut me four pieces of hardwood. I spent the whole day shaping and fitting them to match the rotten posts, and put on both bodies. I was surprised they fit, as I never had any woodworking experience. I told the fellers who had been borrowing my car they had better help me if they wanted to use it.

After supper Jack Yellitz and Frank Peternell helped me hoist both bodies back on their frames and we bolted them down tight. We put the doors on and took the jacks out from under the cars. Next we put both windshields on and called it a day.

March 27, 1932 – Sunday

The weather was warm, fair, and the wind was rough.

After a great breakfast, I shaved, pulled my best and only suit from under the mattress, got all shined up, and went to Sunday school and church with the Sachlebens. I still get embarrassed at the way the boys tease the teacher, Clara Dieterich. Sometimes it gets pretty bold and I feel like crawling in a hole. I guess the farm boys don't think anything about it as they are around farm animals all the time. We would never dream of talking like that in my Sunday school class back home. And we are called "city slickers"?

I spent the afternoon down at the garage putting the instrument panel, all the wires, braces, hoses tubing, floor board and hood on both cars. I got both cars all done and fired up my Cheve sedan and drove it up to the boarding house just in time for supper. The engine runs kinda rough so I'll have to check it out. It was sorta strange driving a sedan but it looks great. I will have to get my title changed from a coupe to a sedan. Art Prosek and Howard Brennan are going to help me tow the other Cheve back to the junkyard.

I spent the evening reading my Dayton School of Aviation Correspondence Course Book and was glad to hit the sack.

March 28, 1932 – Monday

The weather was warm, fair, and the wind was very rough.I am still in the doghouse so no flying today.

While filling in the pot holes and ruts in the runway, I watched some of the other fellers land and takeoff. If that's the best they can do I must be pretty good.

Linn Wright, the fattest student and weighs the most, tried to take off south on the north-south runway, as there was a real strong crosswind. It was a dumb idea. The runway is way too short and was soft as it is not used much. The ship didn't roll as freely so when it reached the end of the runway the wheels were barely off the ground. He flew right through the barbwire fence and kept right on going. I sure expected him to crash. After he landed on the east-west runway and parked the ship, he and Joe went in the hanger.

I stayed plumb away cause I knew what was gonna happen and I didn't want to hear it. But I did sneak over to the ship and look it over. There was a big gash in the propeller and it will have to be replaced. Several places on the landing gear and engine cowling will have to be touched up with paint. He sure was lucky and he just joined my club, the "Grounded Group".

After dinner, and after Joe had cooled off a little bit, I asked him if would be okay if Art and Howard helped me tow the other Cheve back to the junkyard.

It was real tricky getting that car down the ramp without crashing through the office, but towing it back to the junkyard was a breeze. I went back to the airport and took the Cheve distributor and carburetor all apart, cleaned them, put them back together, and the Cheve runs much smoother.

I spent the evening studying "Air Commerce Rules and Regulations."

March 29, 1932 – Tuesday

The weather was cool, cloudy, and the wind was awfully rough. No flying. I guess I'm grounded permanently.

Several students want to become A & E mechanics so they get to do most of the repair & maintenance work.

I spent the day loafing at the airport, filling the ruts and chuckholes in the runway with gravel. I then checked the oil and filled several ships with gas. I spent the evening bumming around town.

March 30, 1932 – Wednesday

The weather was cool, cloudy, and the wind was awful rough. No one flew today.

I loafed at the airport all morning. The other group had the hanger and ships in great shape. There was nothing else to do so I waxed the Cheve. In about a week it will be my group's turn to maintain the hanger and ships. We've spent the afternoon playing ball. I wrote mom a letter and spent the rest the evening reading my Dayton School of Aviation correspondence course book.

March 31, 1932 – Thursday

The weather was clear, cool, and the wind was fairly mild.

I loafed around the airport all day helping the fellars get both "Preps" ready to fly and watched several of them take off and land. Watching them helps my confidence and my ability to fly.

This afternoon Joe cranked up his Waco "Peg" and took off and headed straight to Clifty Falls Park. He loves to fly over there and chase the buzzards. This time he got more than he wanted. He caught up with one and it splattered all over him, the windshield, cockpit, and the rest of the ship. When he landed he was mad and laughing up a storm at the same time. I offered to help him clean the ship. I should have been listening instead of talking. It was a real mess. We finally got it all cleaned up in time to go downtown for supper.

I spent the evening "goofing off" and telling lies with the fellars in the living room; that is, with the ones that weren't out "smooching" with their girls. It is really interesting talking with boys my age who are from different states and from different backgrounds. We were very different except for one thing; we all wanted to become pilots, and we all love girls.

I wrote Bert a letter.

April 1, 1932 – Friday

The weather was clear, cool, and the wind was breezy. I flew for 55 minutes today and it was great.

I began to think my flying days were over. I was a little rusty but got back in the rhythm. If helping Joe V. clean up his Waco got me out of the "doghouse", I'm willing to do it everyday.

I flew the Monoprep, NC122K, Velie 5 engine, today. I practiced 180°'s, 360°'s, and spirals and did pretty darn good for a change so I snuck up and spun the ship several times to the left and to the right. I didn't come out of the spins exactly where I wanted to but it was close. I don't think "close" counts on the test. It is a good thing I don't get motion sickness as I really wound the "crate" up tight and the ground spun around like a high-speed record player. Joe gets mad if he catches me, he says the trainer wasn't built to withstand the strain of too many turns because the ship falls and spins faster with each revolution. Tis True.

It's been a wild day being as it is "April Fool's Day". Everyone had a different joke to pull. But it was no joke when we were told to be at the hanger after supper for school.

Things were slow at the airport so I watched several students do everything but crash. They got away with murder. I guess they believe that old saying "If you can walk away it was a good landing". Those "Preps" sure take a beating.

We are back on *"Airplane Construction", Lesson No. 17, Control Systems*, with Harvey instructing.

In the early days of flying quite a few ships were designed with the control surfaces in front and rear of the main supporting surface, and the engine in the rear (pusher type), just the opposite of the present day design (tractor type).

It was finally agreed by the early designers that the control surfaces belonged at the rear of the ship. The reason was the fact that both the longitudinal and the directional controls were made more sensitive because of additional air pressure from the propeller which otherwise would be lacking with the control surfaces ahead of the engine. With this development of the controls, it is possible to have control of a ship even though the engine is "dead".

There were 3 types of controls used by the early designers. The "double stick" as used by the Wright Bros, the "shoulder yoke and wheel", as used by Curtis, and the wheel designed by Deperdussin, and known as the "Dep" control. This "Dep" control is now used on all large cabin planes of today.

We must have control of the ship about each of the three axes.
1. *The elevators, which control the pitching of the ship, or its rotation around the lateral axis.*
2. *The rudder, which controls the yawing of the ship, or its rotation around the vertical axis.*
3. *The ailerons, which controls the rolling of the ship, or its rotation around the longitudinal axis.*

This control of the ship, so that it will stay in its proper flying position, is accomplished by the operation of these three control surfaces.

The elevators are airfoil sections hinged to the trailing edge of the horizontal stabilizer. When the elevators are raised the tail is forced down by the increase of pressure above the surface, thereby causing the nose to rise; when the elevators are lowered the reverse takes place and the nose is dropped.

The rudder is an airfoil section hinged to the rudderpost of the fuselage at the trailing edge of the vertical stabilizer. When the rudder is moved to the right the increased pressure on that side cause the tail to move to the left, and consequently the nose turns to the right. If the rudder is moved to the left the tail goes to the right and the nose turns left.

The ailerons are hinged sections of the wing located at the trailing edge and near the wing tip. The ailerons are connected so that when the right aileron is raised the left aileron is lowered. In this position the right wing will drop, due to the air pressure on the raised aileron, while the left wing will rise because of the pressure underneath the wing, due to the lowered aileron. This will cause the ship to roll over, or bank, to the right.

Control Mechanisms and Connections. Materials, rods, tubes, stranded cables, chain or hard drawn steel wire may be used for transmitting control movement. Where control members pass around pulleys or sprockets only stranded cables or chains shall be used. There are two controls – the rudder bar or pedals, and the control stick – and are connected to the various control surfaces by cables, tubes, or rods, as to move them singly or all together at the pilot's ill.

The rudder bar consists of a wooden or metal bar, or tube, or pedals, attached at its center by a moveable pivot to a brace fastened to the floor of the ship. This permits a forward or backward motion of the bar around the pivot as the pressure of the foot is applied. The cable system is used because it is the lightest and simplest.

The control stick, which controls the ailerons and elevators, is supported by means of a connection, which acts as a universal joint, so that the top of the stick may be moved in any direction. The cables are so fixed between the elevators and the stick that a forward motion of the stick causes the elevators to move downward and the ship to nose downward and correspondingly, the ship will move upward if the stick is pulled back.

THE DAYTON SCHOOL OF AVIATION

Figure No. 5—"Cables and pulleys" type of control system.

THE DAYTON SCHOOL OF AVIATION

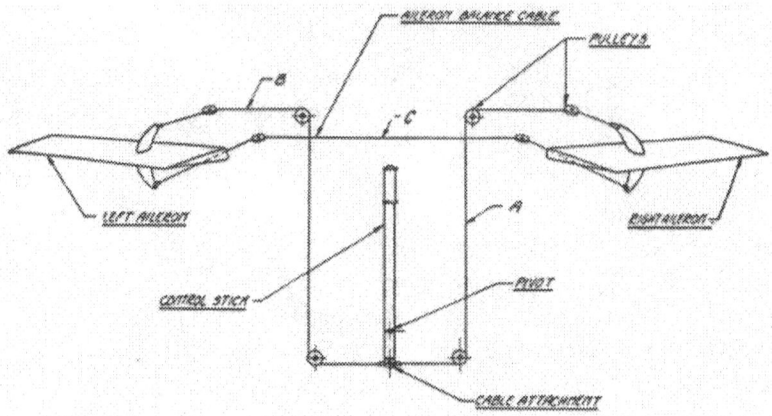

Figure No. 6—A Cable and pulley type of Aileron control system.

The cable and pulley type of aileron control system is shown in Fig. No. 6. Moving the control stick to the right moves the cable attachment to the left, since the stick moves about the pivot. This puts a positive tension on cable A, and a negative tension on cable B – the right aileron is pulled up, and in moving up pulls on the aileron balance wire C, pulling the left aileron down. This system as well as all others should be so designed for linkages that the cable lengths will be the same for all positions of the stick and ailerons.

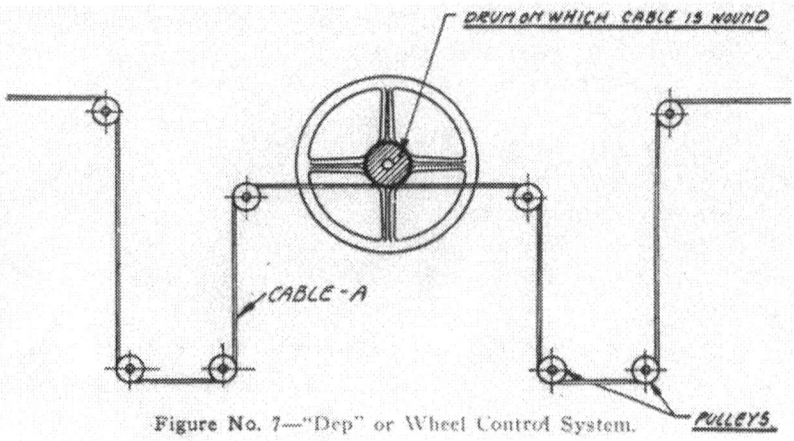

Figure No. 7—"Dep" or Wheel Control System.

13

The cables that run from the ailerons are so attached to the stick that a movement of the stick to the right causes the right aileron to rise and the left to lower, so that the ship rolls in the same direction that the stick is moved.

In large ships it is difficult to give enough movement of the ailerons by means of the control stick so the "Dep" system is used. This system is simply a large wheel on a control stick. To the wheel is attached a drum for control cables to pass several times around. Due to the tremendous leverage exerted by a "Dep" control, the pilot has no difficulty in making the ship respond by moving the control surfaces.

A stabilizer adjusting mechanism is installed in order that the pilot can place his plane on an even keel when in flight by shifting the center of gravity to its proper position by moving the horizontal stabilizer, which is fixed at the leading or trailing edge, and the opposite edge is moveable. The hand lever is hinged, usually at the left of the pilot, and is so built that it is possible to move after disengaging the ratchets, somewhat on the idea of the emergency brake of the automobile.

Questions

1. *Describe the early type of control systems, and tell why they were changed.*
2. *Is it possible to maneuver without the aid of the engine?*
3. *Name the three types of early controls.*
4. *What materials are permitted to be used in control mechanisms and connections?*
5. *Name the five types of modern controls.*
6. *What systems are used with the modern type of controls?*
7. *Describe the rudder controls.*
8. *Describe the aileron controls.*
9. *Describe the elevator controls.*
10. *Describe the stabilizer controls.*

I am taking this lesson to bed with me. It is getting more complicated all the time, and I still think I would be better off if I had taken the correspondence course first. I didn't even know it existed. I wrote Pearl a letter.

April 2, 1932 – Saturday

The weather was clear, cool, and the wind was gusty. I didn't get to fly today.

They kept the NC122K busy as they tried to give as many students as possible some flying time. The NC110K was out of service for inspection. It took me all morning giving that "crate" a thorough inspection from top to bottom, and front to rear. The ship was in good shape, I just had to replace a few bolts and nuts and tighten quite a few others.

After filling the ship with gas and changing the oil, the NC110K was put back in service.

While inspecting the "Prep" I noticed, several times, Joe as he got out of the trainer with a student his face would be deep red, almost purple, and about ready to explode. I have been through that a time or two.

As I brought last night's lesson with me I spent the rest of the day studying. I think everyone is beginning to get a little jittery; it will soon be time to take the Private Pilot and Limited Commercial Pilot examination. The written exam is what has me worried. I wrote Mom a letter.

April 3, 1932 – Sunday

The weather was cloudy, cool, and the wind was calm. It was a nice day for flying, but I didn't get to fly.

The school is shut down on Sundays. Joe spends the day taking sightseers on a short plane ride in his Waco over Madison, the Ohio River, Cragmont, and Clifty Falls Park. Two passengers sit in the front seat.

After breakfast we all took off in different directions. I went to Sunday school with Joe and Carl S. The fellers took up where they left off last Sunday teasing the teacher and the girls. We couldn't get away with that at my church. I met two nice girls in Sunday school, Jessie Eaglin and Annie Eaglin. They are related but not sisters.

After dinner, Frank Peternell, Jack Yellitz, and I walked down Main Street toward the west end looking for pretty girls. We got as far as the R.R. Bridge and ran into the "west end toughs". They started throwing rocks at us and as we were outnumbered, we "high-tailed"

It out of there toward home. I didn't learn to run for nothing and I led the way back up Main Street. The locals sure don't like us.

When we got to Mulberry St., we saw a crowd of people standing around talking and looking at Bear's Jewelry Store. We were told that a flying student was up on the 2nd floor where we had been refinishing airplane sections before the hanger was built. It seems he had locked himself in and threatened to commit suicide because his girl wouldn't marry him. They got the girl to talk to him but she wouldn't agree to marry him. He finally came out and he was a mess, he must have been crying all day. I sure was surprised to see that it was K.B.R., he always seemed to be a "laid back", quiet, easy going kind of feller. I don't know how this is going to turn out, but it sure created a lot of excitement for a Sunday. Joe said the school didn't want any publicity so he took Kenny and they left in a hurry.

The supper was awfully quiet. Everyone concentrated on eating supper and keeping their thoughts to themselves, which was very unusual. Kenny never showed up so I guess he is still with Joe. I can't imagine committing suicide and missing flying forever because a girl wouldn't marry me. As of now I'll take flying, although I'd rather have both. I got a letter from Bert.

April 4, 1932 – Monday

The weather was cloudy, cool, drizzly, and the wind was calm. I didn't get to fly today.

When we got to the airport after breakfast, neither Joe nor Kenny showed up. My crew is to take care of the hanger this week so we dragged all the ships out on the apron and cleaned out the hanger. We even washed all the windows, etc. Then we cleaned the workshop and the windows, etc. We cleaned and oiled all the equipment. We then cleaned and rearranged all the parts and bins as well as all the tools and boxes.

While we were taking care of the hanger Harvey had the rest of the students fill all the ships with gas, and check the oil level. Then to keep everyone busy he helped them inspect the ships so they would be ready to fly when Joe showed up. By the time we put the ships back in the hanger it was dinnertime and still no Joe or Kenny. We went down to the boarding house for dinner and Kenny wasn't there.

When we got back to the airport Joe was waiting for us. He asked us to meet with him in the hanger to discuss the Kenny situation and any other problem. He told us Kenny had left the school and was too ashamed to face us. The school asked him to leave as they felt this suicide incident showed he wasn't temperamentally fit to be a commercial pilot. He also said the school gave him a portion of his tuition. We all hated to see Kenny leave.

Joe then gave us a real pep talk. He told us not to get too serious with the girls, to remember they were local girls and would probably marry local boys. That the girls were attracted to, and curious about us, cause we are strangers in town and gonna be airplane pilots. He said he has had a lot of experience and is still single although he is a local boy and a licensed pilot. We busted out laughing when he told us he had several close calls but always managed to slip away. We all felt better by the time we locked up the hanger and went downtown for dinner.

April 5, 1932 – Tuesday

The weather was cloudy, warm, and the wind was calm.

I flew for one hour today in the NC122K "Prep" making four flights, 180° turn and land (fine), 360° turn and land (better), figure eights (great) and the stalls were a breeze. When practicing the stalls I was tempted to let the ship slide into a tailspin. But the test is to prove you can prevent the ship from spinning during a stall. Your life may depend on your ability to prevent the ship from stalling in the first place. A stall is simply the ship losing flying speed. If the plane doesn't regain flying speed quickly by pushing the throttle wide open and diving, the controls become useless and the ship will fall like a rock and crash.

When the controls feel "mushy" and the ship does not respond it is a sure sign the ship is stalling. If the plane stalls close to the ground, like making a climbing turn on take-off before the ship has gained plenty of flying speed, it is gonna crash for sure.

The takeoffs and landings were fine. Flying is becoming more automatic, especially take-off and landing. If the left or right wing is low, or the nose is high or low, I just naturally make the necessary correction with the controls. It is a nice comfortable feeling flying the "crate" up there doing the different maneuvers; and this is different feeling that just happened. Up till now I might have been putting myself under a little stress trying to learn the different maneuvers perfectly.

My crew did such a great job yesterday cleaning the hanger and workshop so all we had to do was straighten up the joint. We chased everyone out of the office and cleaned it from top to bottom. By this time it was getting late so we parked all the ships in the hanger and headed downtown for supper.

After supper I went up to my "cubby hole" and spent the evening studying for the exam. All this studying is starting to hurt my brain. I got a letter from Pearl today.

April 6, 1932 – Wednesday

The weather was cloudy, rainy, warm, and the wind was slightly rough.

I flew one hour and ten minutes today in the NC110K, Velie M5 Prep making four flights. I breezed through the 180° turn and land, 360° turn and land, and figure eights, but the verticals were terrible so I spent most of the time on them. I lose altitude whether it's a left or right vertical no matter what I do.

Clyde H. Beyer

I have to solve it before the exam or my goose is cooked. I am using the controls properly but there is just something wrong. Joe can straighten me out at the next check time I hope.

The rain didn't cause me any trouble it was more of a drizzle and the visibility was good. The take offs and landings are becoming so routine I decided to come in high over the wires and sideslip in for a landing. For some reason sideslipping seems to come naturally to me. At the last second I leveled the ship, headed west down the runway and didn't even feel the wheels or tailskid touch the ground. You can't land any smoother than that.

Of course Joe would be standing on the hanger apron, and when I walked up he asked me if I wanted to be grounded again cause if I sideslipped in again before the exam that's what would happen. As I was going in the hangar I heard him say that was a perfect sideslip landing.

Even Harvey, the school's licensed mechanic, doesn't have enough work to keep him busy so he has started to prepare for a L.C. license. That's another feller trying to squeeze in line for flying time. He has a Private Pilot's license.

We all went out on the runway and filled in the chuckholes, ruts, and ditches with gravel in between the students landing and takeoff.

After dinner they to have decided to have school with Harvey teaching us Lesson No. 9 "The Wings".

The modern airplane wing has a fairly uniform arrangement of members regardless of materials used. The most common wing today is the wood fabric but it may be metal fabric in a year or so

There are two long beams running the length of the wings. These are known as beams or spars and are the backbone of the wings. They are usually made of plywood box construction, called box spars. Running at right angles across these spars are long pieces of the thin plywood (built up in triangular sections) called intermediate ribs, which are cut to the thickness and chord of the airfoil or wing sections. These ribs serve a double purpose as they not only brace and strengthen the spars but they also hold and secure the wing covering.

Since the wing must hold its proper top form it has to remain strong and rigid in order that it may withstand the great forces put upon it: this is done by bracing. Internal bracing consists of wires running from one compression rib to the next compression rib diagonally across the bay. The large steel tubes are called compression tubes, if reinforced ribs are used. They are called compression ribs, and the space between them and the front and rear spar, is called a bay. Compression ribs themselves

258

are really intermediate ribs with compression rods or bracing to make the wing frame rigid.

Figure No. 3.   Curtis-Robin Left Wing Panel with Aileron removed from false spar.

Spars are usually made from a single piece of spruce and routed (carved out) into sections between points where compression ribs are connected to them. Wings have two spars, a front one and a rear one, and a false spar placed behind a rear spar.

All material used in plywood should be quarter sawed. All plies used in one piece of plywood should have the same specific gravity. High-grade animal casin glues shall be used on clean bare wood.

There are three classes of ribs, compression, intermediate and false,

The compression ribs are steel tubes used to space the spars and strengthen the wing frame.

Intermediate ribs run parallel to the compression ribs and are spaced to give rigidity to the wing structure.

False spars are capstrips running from the top of the front spar to the leading edge and at times back to the bottom of the front spar. Capstrips are strips stripes of wood placed on the edge of the ribs to offer more surface against the fabric so that it will not be cut or torn.

All ribs except the false ribs are made in three sections known as the leading web, center web, and trailing web.

Questions to Lesson No. 9

1.   Of what does internal bracing consist?
2.   What does routing out mean and accomplish?
3.   Tell in your own words how plywood is made.
4.   What is a capstrip? Why is it used? Where?

5. Describe the construction of false ribs and tell where they are used.
6. Define the following; bay, compression tube, web, spar, and gusset.

I bummed around town after supper to clear all the cobwebs out of my mind. That sure was a tough lesson we had this afternoon. I needed that lesson before we rebuilt the wings on those Waco's, Eagle Rock, American Eagle, Fleet, Monoprep, Monocoupe, Curtiss Wright, Stinson, Curtiss Robin, Ryan, Travelair, Cloudcoupe. Harvey told me what to do and how to do it but I never knew the purpose of what I was doing until now. I wrote Mom a letter.

April 7, 1932 – Thursday

The weather was warm, cloudy, drizzly, and the wind was breezy.

I flew 40 minutes today in the NC110K, Velie M5 Prep making 4 flights. The 180° turn and land, 360° turn and land, figure 8's, and spirals were pretty darn fair. I only overshot once and the landings were smooth. I have to practice these maneuvers every time I go up because they, and a few others, like spins and verticals, are part of the flying test.

As I taxied up to the hanger Joe came up to the ship and told me we were going back up for "check time". I took off and went through the maneuvers. During the spins I had trouble coming out of them in the same direction that I went in. Joe told me I should start coming out if the spin one quarter of the turn sooner.

I made a left vertical and lost more altitude than I usually do on account of Joe's additional weight. I felt a poke in my ribs and looked over at Joe. He was MAD, and he wanted to know what the heck I was trying to do, kill him? Two or three more inches and we would have been upside down, and that would have been all she wrote. You can't do a 100% vertical power turn in a ship like this; it doesn't have enough horsepower or speed to keep it from losing altitude. And if you stayed in the turn long enough the ship would lose flying speed and fall, he said. He was surprised that I hadn't fallen out of control before now. He took the controls and made a left vertical and a right vertical without losing any altitude. The ship was about 80% vertical which he said was the best this plane could do safely. And don't ever let me catch you setting this crate on its ear again.

After 45 minutes of check time Joe told me to take the ship down and call it a day. We taxied up to the hanger, shut the engine off, and sat in the cockpit while he told me the other maneuvers and landings were very good because I didn't think they were that good.

As the hanger was in good shape, my group cleaned the workshop. Then we started cleaning and waxing the ships, inside and out, one at a time. It is quite a job, but it helps to protect the ships from the weather. By the time we get the last ship waxed it will be time to start all over again.

I watched Al Thockle zigzag down the runway splashing water and mud all over the crate. He finally got airborne in time to clear the treetops, and when he came in to land I sure thought his wheels slid across the high electric lines. A couple inches lower and he would have been a dead duck. He leveled off about 2 ft. above the ground, pulled the stick back, and the ship hit the ground – "ka boom"! It had to jar his eyeteeth.

We put the ship in the hanger, locked up the joint, and left Marlin Christenson to watch the place while we went down to Barber's for supper. John is staying over tonight.

I wrote Bert a letter and spent the evening studying. It's been so long since I've gone out with a girl I'll have to ask her what to do next.

April 8, 1932 – Friday

The weather was warm, clear, and the wind was strong. Of course, it being a nice day I didn't get to fly today.

Some of these "fair weather" student pilots are going to have a hard time trying to fly in rough weather. I guess I'll just have to pray we have bad weather more often. Frank Peternell and John Yellitz haven't been here as long as I have, and yet they have as much flying time. I gave both of them their first airplane ride. I've given every new student, since I got my Private Pilot's license, his first airplane ride. I hope they will get some girl students.

My group worked on the runway, filling in the ditches and potholes, at the same time keeping our eyes on those goofy fellers taking off and landing,

To give us something to do they decided to have school after dinner, with Harvey teaching us *Lesson No. 2. Aerodynamics.*

The study of aerodynamics, as applied to aviation, is simply the study of how the air affects the plane and how the plane affects the air.

The first principle of aerodynamics – Motion is relative.

The second principle of aerodynamics – a curved surface going through the air, with pressure underneath, has a strong lifting effect, caused by a partial vacuum on an area of reduced pressure over the curve of the surface. This curved surface lifting force may be as high as 75% of the entire force that lifts an airplane, only 25% of it coming from the "kite"

261

effect caused by the pressure of the air against the underside of the wing, as it is forced through the air by the engine.

The four principal factors determining lift:

1. The angle of incidence of the wing
2. The velocity of the wing
3. The curvature of the wing
4. The area of the wing

The angle of incidence is the number of degrees that the front or leading edge of the wing is elevated above the rear or trailing edge.

The velocity of the wing is the thrust, or speed with which the wing is forced through the air, and is the principal factor affecting lift. A plane rises from the ground when its speed had created enough lifting force to overcome its weight (or the pull of gravity, which is the same as its weight).

Camber is the curvature of the wing, and it differs in nearly every kind and type of ship. The greater the camber or upward curve, the higher the air is thrown up over the wing, and this means a greater partial vacuum area, which has much to do with the lifting and flying of the plane. If you were more concerned in speed than lift, you would cut the camber down, so that it would offer less resistance to the wind and still maintain the desired lift.

A biplane is an airplane with two sets of wings, one above the other. The distance between the two sets of wings is known as "gap".

The gap must be great enough to keep the action of the air over the surface of the lower wing from interfering with the flow of air under the surface of the upper wing, and vice versa. Sometimes this controlled by placing the upper wing forward of the lower wing, which is known as positive "stagger" of the wings. And in the other cases the lower wing is placed ahead of the upper, then it is called "negative stagger".

The area of the wing (the span times the chord, as well as its camber has much to do with its lifting force. The lift generated by the wing is in exact proportion to its area. If we wish to double our lift, then we must double our area.

Questions to Lesson No. 2

1.   *What is the meaning of the word aerodynamics?*
2.   *Name the factors determining lift, and state which of these is the most important.*
3.   *What is meant by camber of the wing?*
4.   *How does camber affect lift?*
5.   *Define angle of incidence.*
6.   *What is meant by "positive stagger", "negative stagger", and gap?*
7.   *What effect does speed have upon the rising of the plane from the ground?*
8.   *What effect has wing area upon the lift and in what proportion is area to lift?*
9.   *Why does an airplane fly?*

We can't get all the ships in the hanger at the same time so we tie the rest down at quitting time.

There were six new men at supper tonight. We found out there are men, who are working on the new Hanover Hill road, staying at Richart's and Spillman's Rooming House. I guess these guys are part of that bunch.

I got a letter from Fred Moss today. Spent the whole evening studying aerodynamics and then hit the sack.

April 9, 1932 – Saturday

The weather was clear, warm, and the wind was gusty. I didn't get to fly today.

Joe was busy giving dual flying instruction to the students who have not soloed. As each student finished his lesson another took his place and Joe would take off again in the NC110K. After several lessons Joe and the student would get out of the NC110K. Then Joe and another student would jump in the NC122K, which we had serviced, and take off.

While they were flying, we checked the NC110K and filled it with gas so Joe and student could transfer from one ship to the other without losing any time. He switched planes a number of times when the gas gauge registered less than a quarter tank of fuel. This kept my group busy and gave us something to do. They are having a hard time trying to keep us busy. My Cheve is one thing that is always busy, especially at night and not by me. I'll have to give it a good check-up one of these days. If the Cheve could only talk.

While Joe was giving a flying lesson to the last student we parked the ships in the hanger and tied the rest down on the apron. He taxied the ship up to the hanger and after letting the engine cool off, we covered it with a tarp, tied it down, locked up the joint, went downtown, and ate supper. The same men were eating supper so I guess they will be eating here till the new hill road is finished.

I spent the evening loafing up town. The sidewalk and stores were crowded with farmers and their families. Don't try to get a haircut on Saturday. I never came in contact with farmers, nor knew much about them until I came to Madison. Since then I have met a lot of real nice farmers, and they have taught me a lot.

April 10, 1932 – Sunday

The weather was cloudy, cool, and the wind was wild. I didn't get to fly today.

I had an early breakfast and went out to the airport with my group to open up the hanger. This is the last day we have to take care of the hanger until May 2nd. We took Joe's Waco "Peg" out of the hanger and filled it with gas. He usually takes sightseers up for a short flight on Sundays.

I went back to the boarding house, washed, shaved, took my one and only suit out from under the mattress and went to Sunday School and church with the Sachleben family.

Mildred Ricketts invited me, during Sunday School, to meet her family on Tuesday evening, and I thought that was a good idea. As I was leaving the church, Jesse Ricketts, Mildred's younger sister, invited me to meet her parents on Thursday evening, and I thought that was a great idea. They live up on the East End so I I should be pretty safe. You don't go way down the West End alone to meet a girl without taking a chance of getting beat up.

I had a great Sunday dinner and then went out to the airport to see what was going on. There were quite a few sightseers and Joe was busy taking them up to see the town, Ohio River, Cragmont, and Clifty Falls Park from the air. I helped the passengers get in and out of the cockpit while Joe and Carl S. lined them up and collected the money. We only had to stop to refuel the ship one time. It was a busy afternoon and I had a great time. The passengers sure got a kick out of their first airplane ride. As the last passenger left we parked the Waco in the hanger. Art Prosek is staying at the hanger tonight.

I walked across the field with Joe & Carl S. to their farm house and ate supper with them. Joe, Carl, Fred, Pa, and I shot the breeze while Ma washed and put the dishes in the kitchen cupboard. We spent the rest of the evening taking turns playing euchre. We closed out the evening with a tin cup of Pa's "homebrew". Great stuff. Jumped in my Cheve and high-tailed it down to the boarding house and hit the sack.

April 11, 1932 – Monday

The weather was cloudy, cool, and the wind was rough. I didn't get to fly today.

I took Sam, Linn, and John out to the airport in the Cheve. The rest of the fellers went out in the old International flat bed truck. We pushed and pulled the planes out of the hanger and lined them up beside the apron. The fellers checked the gas, oil, and water, and then cleaned the ships while waiting for Joe to show up. I went in and organized the tool room. With nothing else to do I cleaned my Cheve inside and out so those "odd-balls" would have a spiffy car to caterwaul with the girls.

Joe showed up with a new student and told us he was raised in a convent up north by a group of nuns who paid for his tuition. Matthew is his handle and I guess he is an orphan. Joe told me to me to show him around and kinda look after him while he was giving the fellers their flying lessons.

After showing the feller around and explaining how the school operated, we spent the rest of the afternoon watching those "clowns" trying to kill themselves. Sam earned the Chinese nickname "One Wing Lo" nickname. He couldn't keep the wings level no matter what.

We locked up the hanger, Sam, Linn, John, and I went down to supper in my car. Joe took Matthew in his car and the rest went in the truck. William is staying the hanger tonight.

I spent the rest of the evening studying Aerodynamics after writing a letter to Pearl. I think I am going to get stuck with a "doozy".

April 12, 1932 – Tuesday

The weather was cloudy, cold, and the wind was very wild. I flew for thirty minutes today.

Marlin, Pete, and David S. went out to the airport with me after breakfast. The windsock was whipping around like crazy and, for a time, I thought it would tear off the hanger and all. This was the kind of day I should get to fly. Sure enough, Joe, asked me if I wanted to take Matt up and do a few maneuvers. I never said no when it came to flying and I wasn't about to start now.

I checked the NC110K, filled it with gas and cranked up the prop. When the engine warmed up, cackling like a contented rooster, Joe strapped the new student in the seat beside me. I disconnected the dual control and gave Joe the signal to pull away the wheel chocks. Facing the wind, I pushed the throttle open and pushed the stick slightly forward to raise the tail. The ship rolled slowly down the runway just a short distance due to the strong headwind, when it practically jumped off the ground and we were flying; that is, sort of.

The wind bounced like a cork in rough water and I was afraid it would slam us back on the runway as fast as it lifted us up. But we cleared the trees and finally reached a safe altitude although the wind kept bouncing us around pretty darn hard. I looked over at Matt and he looked like he was going to be sick. I tried to do a few maneuvers, like flying figure eights, etc., but the wind was too wild, and it kept changing directions. So I decided to give up and go down.

I came in high over the high lines so the wind wouldn't blow me into them, and that made me come in high over the runway. I sideslipped in and just as I pulled the stick back to set her down the wind blew us up about 15ft. and off the runway. I pushed the throttle wide open, pushed the stick forward and to the right and left, and used the rudder pedals to gain control of the ship and regain flying speed to get up over the trees.

We flew around again, came in high, and as I sideslipped in, the wind slammed down on the runway with a bang jarring my eyeteeth. I knew this was it, and quickly idled the engine, pulled the stick back to my belly and held on. My passenger was white as a sheet but at least he didn't get sick and never said a word while we were in the air. Several of the fellers came out and steadied the wings so I could taxi up to the hanger. We inspected the "Prep" from top to bottom and that ended the flying for the day. This was probably the day to say no – I think.

They are going to have a problem with that new student. He told me he had never roller-skated, rode a bike, or drove a car. We took him out in the field and played a game of softball. He was terrible; he couldn't hit or catch the ball worth a darn, and he sure didn't have any coordination at all. When the game didn't help I cranked up the old truck, took in out in the field where he couldn't hurt anyone, and tried to teach him how to drive. He scared me to death so I gave up before one of us got killed, or the truck got ruined. Besides, I had to be in good shape to visit Mildred.

I had a nice time at Mildred's home. I think she lives on Ferry Street not far from the river. We sat in the living room shooting the breeze with her family. She seems to have several brothers and sisters, but I didn't see hide or hair of Jessie. She has a real small blond brother who kept causing a ruckus so I was glad when it was their bedtime.

Then her Pap brought out the "home-brew", gave me a full tin cup and told me to wait a minute before drinking it. He took an iron poker and put the tip in the fireplace. When the tip got cherry red he put it in my brew for just a split second. It made a sizzling sound and steamed. I never saw that done before. My brew wasn't very cold and tasted like it had been in the cup all day. I didn't like it but I never let on as he said that was the only way to drink it.

I now have six different ways to make "home-brew" and am getting to be quite an expert. When I get a place of my own it is the first thing I am going to make. It is against the law but almost everyone makes it, and they would have to put the whole town in jail.

April 13, 1932 – Wednesday

The weather was cloudy, cold, and the wind was not so rough. I flew for one hour and fifteen minutes today.

I must have got out of bed on the right side. The takeoffs were ok as usual. The figures 8s were smooth as were the power turns. The wingovers and the stalls were pretty darn fair. The left vertical was good but the right vertical wasn't, I am still losing altitude. I even gained a little altitude doing the left vertical. My left spin was fair enough but the right spin was poor, the ship came out of the spin 90 degrees past the starting point.

I always have more trouble with the right maneuvers than the left. The 180-degree and 360-degree turn and land were pretty normal.

I overshot and undershot once, but the three-point landings were just about two inches off the ground, although one was a wheel landing. If I could only set her down under one inch, now that would be perfect.

I helped Harvey Kattelmann work on the "Fleet" rigging. I hope to get to fly her one of these days. They say she is pretty "hot", meaning her glide ratio, without power, is not as good as a lot of planes, but she flies faster. I think the factory sacrificed lift in favor of speed.

Matt has a real bad habit and if he doesn't get rid of it he will run into a lot of problems. He constantly licks his nose with the tip of his tongue. I don't know how he does it, and I think it is automatic, like blinking his eyes. We tried to talk him into breaking the habit; that didn't work so we made fun of him. We really want to help him so we tried to shame him, and even yelled at him. It must be beyond his control cause he keeps right on doing it, and it really gets to me. There is no way, no how, he will be accepted, or fit in, at work, or anything else unless he changes.

It is my turn to stay at the hanger tonight. I got in the cot, put the 45 cal. pistol on the floor beside me and studied aerodynamics. I finally turned out the light but had trouble going to sleep thinking about Matt. I feel sorry for him as he has a lot to overcome.

Just as I dozed off the wind picked up and the hanger began to creak and groan. The hanger's sliding doors rattled as though someone was trying to break-in. I turned on the light, "Mr. 45" and I" checked the doors and windows, then went back to bed and finally fell asleep.

April 14, 1932 – Thursday

The weather was cloudy, cold, and the wind was kinda rough. I didn't get to fly today.

When the fellers showed up at the hanger I went downtown to the boarding house and had a late breakfast. I washed and shaved then decided to write Mom, Pearl and Helen (Pearl's daughter) a letter before going back out to the airport.

I took the Dayton School of Aviation book to study while everyone was eating dinner down at Barber's. When they came back Joe Vail didn't show up. So several of the fellers pitched horseshoes, several washed and waxed the "Prep", and the rest of us filled the ruts and chuckholes in the runway. Harvey Kattelmann tried to keep busy double-checking the "Fleet". It is almost ready to fly, but it first must be okayed by Joe Shumate, the inspector before it can be flown. Matt has been hanging around Harvey. I spent the rest of the day studying airplane construction, as there was nothing else to do.

I wrote Marlin Christianson a letter after supper. I then shined my shoes, shaved, threw on some sweet smelling stuff and drove down to Jessie's house. I had a real nice time. Just about like the other night when I went to see "Millie". Her Pap brought out the "home-brew" and put that darn red-hot poker in my cup again. "Millie" was nowhere in sight. Finally, Jessie suggested that we drive down along the river, and I seconded the motion. There wasn't any traffic as we drove along Front Street but we did see a headlight going up the Milton, Ky. hill. We even saw a car crossing the Ohio River toll bridge.

We wound up in front of the Ricketts' house turned off the engine and lights and started to pitch a little high class "woo" when the porch light came on. Jessie said the girls were to go in the house when the porch light on so she gave me a quick smooch and took off running. !Yipee! I think "Millie" is a little old for me; she is about two years older than I am. I would rather go out with Jessie; I think she is about my age. I cranked up the old Cheve and drove home, trying to think of a way to do it without upsetting the apple cart. With that worry on my mind I hit the sack.

April 15, 1932 – Friday

The weather was warm, clear, and the wind was quiet. I flew for an hour today.

Harvey told us that Joe would not be coming out to the airport until late afternoon. He told me to inspect the ship, go up and practice the maneuvers for an hour and no monkey business. He said that for the benefit of the students, as he and I started with the school about the same time and are real good friends.

I had been waiting for this for a long time. I inspected the ship extra carefully, filled the tank with gas, cranked her up and took off. Instead of turning north I fly straight ahead to Clifty Park. I wanted to find out why Joe spent so much time chasing buzzards, and now I know. It was a poor day for buzzard chasing, they are not very active when the air is real quiet. They prefer strong and rough wind, which makes it easier for them to soar without having to flap their wings all the time. I imagine it takes a lot of energy to flap those enormous wings.

Just as I was about to hurry back to the field a buzzard soared up out of the valley in front of me. I pushed the throttle open and went after him. The buzzard took off like a scared rabbit, climbing almost straight up. I was not even close when he circled back and I had a chance to gain on him. I finally got close; at least I could see him, when he suddenly banked sharply to the left. They can't fly very fast but they sure can maneuver. With the engine running wide open I pushed the stick to the left and kicked the left rudder pedal hard (too hard). Instead of following him the ship skidded to the left and headed in a different direction. It was like a car sliding on ice. !PHEW!

I quickly pushed the stick forward and to the right slightly passed neutral, and pushed the right rudder pedal slightly past neutral cause I was about to lose control of the ship. Regaining flying speed I took one last look at the buzzard, he was circling high above me laughing up a storm. They are not pretty, but to me they are the ballet dancers of the sky; gracefully they twist, turn, spiral, circle, twirl, soar and glide, slowly moving their wings to the tune of an unseen orchestra.

I high-tailed it back to the airport. This ship just doesn't have the horsepower or speed to try maneuvers like this but I'll bet the "Fleet" does.

My hour was almost up so I quickly made a left and right vertical. The verticals were not so good as I still lose too much altitude. I spun the ship about fifteen times letting it wind up three times in each direction. I love it. The right spins were not so good. I came out of the spins too soon, and not soon enough. The left spins were nigh perfect. I overshot the line twice and undershot it once; three were normal. The three point landings were pretty darn fair. Without looking at me Harvey said, "that was a long hour", as I walked past him.

Bob Parker is as reckless as ever. I watched him go down the runway, and as usual, he tried to lift the ship off the ground before it had flying speed. The ship left the ground a foot or so then settled back down, and by the time it reached flying speed he was at the end of the runway. He was lucky to get up over the trees, and if the engine had coughed or misfired he would have been a goner. He came in too fast and had to use the entire runway to get the ship stopped. What a terrible landing.

I got even with Pete Nemetsky this afternoon. He borrowed my car, went out to Cleone's house and tried to get her to go out with him. I heard about it and blew my stack. That was an insult to my intelligence – borrow my car to go out with my girl. I had been trying to get her to go steady with me for several weeks and all the fellers know this.

The school's policy when two students get mad at each other is to settle it with boxing gloves. So they brought out the gloves, marked the shape of the ring in the dirt and Harvey was the referee. I really gave him the works and he quit after the first round. I had expected to cut loose in the third and last round but he didn't give me a chance. We had to shake hands but he can't borrow my car anymore, he does have a doozy cut lip.

Art Prosek walked Cleone Grimes, that's her handle, home tonight.

I spent the evening reading the Dayton School of Aviation book, but I better start doing some serious studying.

April 16, 1932 – Saturday

The weather was cool, rainy, and the wind was rough. No flying today.

I hauled a ton of coal on the old International truck from the coal company on Front Street to the airport. When it stopped raining I had to shovel the coal off the truck close to the workshop.

Joe finally showed up and had Matthew with him. As it was raining we were in the hanger goofing off. Joe told us that Matt was going back home and he wanted to say "so long". After all the fellers shook his hand and said "so long", I went over and after shaking his hand I said I hoped he would get home okay. He seemed kinda dejected so I asked him what he thought of the exciting flight we had. He sorta grinned and said it was really great. I told him to be sure and tell his friends all about it. Joe took him into the office and we all wondered what the heck was going on.

It being Saturday, we all loafed around town the rest of the day.

I really feel sorry for Matt, but he has to quit that awful habit or it will ruin him and people will avoid him.

I went to the show down on Main Street tonight hoping it would get Matt out of my mind, and it was the worse show I've ever seen. It was a Wild West picture of course; its Saturday night.

I might not be an expert at flying airplanes, but I am getting to be an expert at flying a coal shovel. Geez! What a day.

April 17, 1932 – Sunday

The weather was cool, clear, and the wind was pretty stiff. I flew for 50 minutes today and I did poorly.

I went to Sunday school with Joe and Carl Sachleben this morning. I said hi to Millie Ricketts and waved to Jessie Ricketts. Surprise, Cleona Grimes was there and that was the first time I ever saw her in Sunday school. Millie and Jessie left before I had a chance to ask either one for a date. I invited Cleona to come out to the airport after dinner and I would show her around.

It was unusual to be flying on Sunday as Joe usually takes sightseers up for a plane ride. Harvey ran the joint, as Joe never showed up. He never bothered to give me any instructions except to put in 50 minutes flying time.

After the usual pre-flight check, I took off west and as I circled the field I looked down and saw Cleona loafing around with the fellers. I spent the entire 50 minutes practicing 180 and 360 degree turn and land maneuvers to show off a little bit, but that was big mistake. Terrible. I overshot the line twice, undershot it once, and one was normal. I set them all down pretty rough. I must have been thinking about Cleona with all those fellers rather than flying. I resolved not to spin, my favorite sport, until I set the ship down right. It begins to look like I'll never spin again.

I showed Cleona the planes, hanger, workshop, and had her sit in a cockpit explaining it in great detail. In fact, I explained everything in great detail to make it last as long as possible. She is the best-looking gal I have ever seen, so far, and is she ever stacked up! All good things have to come to an end. She didn't seem to be as interested as I thought she would be. We joined the fellers and watched as they took their turn taking off and landing. I took a picture of **Art Prosek taking off**.

I helped unload two truckloads of the Dayton School of Aviation stuff tonight at the 216 E. Second St. office.

April 18, 1932 – Monday

The weather was cool, clear, and the wind was pretty stiff. I flew for one hour today. Still terrible. Getting rotten generally.

I overshot the line three times; two landing were normal. I set the ship down pretty rough. The 180-degree and 360-degree turns were normal, but it was the landings that were rotten. Verticals and landing near the line are the only maneuvers I have to keep practicing. All the other maneuvers have become automatic. But those verticals are giving me fits, the harder I try the worse they get and Joe says that's my problem, I try too hard. I know what to do and how to do it, but I think I am in a rut and can't seem to break the bad habit. I will try to get Joe to go up and straighten me out on this maneuver. I hate to do that because he got mad the last time and really chewed me out, but I have to do something.

The fellers spent the morning horsing around but I helped Harvey piddle around in the workshop until Joe showed up. I am not very good at horseplay. I have a short fuse.

Joe got us together in the workshop and told us about Matt. The school decided he would never have the ability to fly and didn't think he would master any of the other professions in the aviation industry the school had to offer. Mrs. Barber told Joe the regular boarders wouldn't eat at the table with Matt because his bad habit ruined their appetite. She didn't have time to fix meals just for him, and his habit bothered her and the help. Amen! Joe took him to every "greasy spoon" in town with the same result.

The school got in touch with the convent and explained the reason they were sending Matt home. They also sent the convert his tuition fee. This was the first feller the school turned down. We didn't know his last name or exactly where he was from, only that the convent was up north. It could even be Canada.

I spent the afternoon helping Joe, Harvey, and Art blow tree stumps out of the ground with dynamite so we can make the runway wider. Gee, it's a lot of fun. It was hard work getting rid of the stumps. We had to load them on the truck and haul them to the dump. We had to blast the stumps out after Howard Brennan, and then Art Thockle took off, and before they came in for a landing.

I spent the evening studying *Lesson No. 48 – Meteorology.*

April 19, 1932 – Tuesday

The weather was clear, warm, and the wind was fairly rough. I flew for 50 minutes today.

I barely overshot the line four times, slightly undershot it once, and hit it right on the nose seven times. I set the ship down pretty darn nice each time. I breezed through the figure 8's, 180 and 360-degree turn and land, stalls, and made decent verticals. It must have been the warm weather. If I do as well tomorrow I'll begin to practice spins again.

They ought to take movies of the fellers taking off and landing; it sure is sumpin. It seems as though each one tried to invent a weird way to takeoff. And the landings, you would bust a gut laughing if it wasn't so serious. Jack Yellitz almost landed on his left wing. "Bus" Kirsch made a giant "frog" landing, Pete Nemetsky tried to land in the hanger, and made us all scatter like a covey of quail. Dave Swett touched down, zigzagged down the runway, went in the left field, and finally stopped in the right field.

Those ships sure take a lot of punishment. No wonder the landing gear and tailskid need daily maintenance.

We cleared brush from all the fence rows, made big piles of it and had a huge bonfire.

We went to school tonight and Harvey reviewed Lesson No. 20, Rigging. Now I think I can rig a ship if I have the proper blueprints. Herb Schwartz wrote me a goofy letter. I felt bad all day.

April 19, 1932 – Tuesday

The Dixie Flying Service and the Dayton School of Aviation Correspondence Course of Dayton, Ohio officially consolidated today. The school is now "the Dayton School of Aviation", and the office is at 216 East Second Street, Madison, Indiana.

Joe D. Vail, President and Flying Instructor
Robert R. Foster, Vice-President
Harry L. Vail, Secretary
John A. Naill, Treasurer

Employees
Edward D. (Bud) Guild, Transport Pilot, Cincinnati, Ohio, Flying Instructor
Harvey Kattelmann, Mechanic and Airplane and Engine Instructor
Zelda Kellems Gercken, Office Secretary, 1932- 1935
Florence Johns – Office Secretary
2 Office Secretaries from the Dayton School of Aviation, Dayton, Ohio.

Robert R. Foster was Vice-President of the Dayton School of Aviation Correspondence Course at Dayton, Ohio for the past three years, and will continue in that capacity. He has 14 years experience in schoolwork. He and four other office workers are to move to Madison from Dayton, Ohio and operate the Dayton School of Aviation Correspondence Course.

Robert R. Foster and Major Robert L. Rockwell, were two of the original founders of the Dayton School of Aviation, which is a national correspondence course and very successful. Major Rockwell was an early pioneer in military aviation during World War I. Foster was experienced in advertising at that time.

Consolidate Dixie And Dayton Flying Schools

4—19—1932

Madison today became the headquarters of one of the nation's foremost schools of aviation.

The consolidation of the Dixie Flying Service, headed by local men, and the Dayton School of Aviation of Dayton, Ohio, became effective this morning. A merger of both services, the school is henceforth to be known as the Dayton School of Aviation, it was announced by Joe D. Vail, formerly vice president and general manager of the Dixie Flying Service and now president of the new school.

R. F. Foster, for the last three years executive vice president of the Dayton School of Aviation, continues in that capacity, Mr. Vail announced. Mr. Foster, he said, has had 14 years of school work and he will directly supervise the large volume of aviation correspondence courses which the former Dayton institution is bringing to Madison. Mr. Foster will make his home here.

Other officers of the new school are prominent local business men. The secretary is Harry L. Vail, a member of the firm of the Vail furniture store. The treasurer is John A. Naill, president and general manager of the Madison Lumber Co., and associated with other local business enterprises.

Students in every state of the union and several foreign countries are being trained as aviation mechanics by means of the Dayton School of Aviation's correspondence courses. Consolidation of the Dixie Flying Service, which has been giving actual flying lessons, with the D. S. A., as it is nationally known, will provide an unusually well rounded instruction service for young men, in the opinion of Mr. Vail. Numerous D. S. A. students have already signified their intention of coming to Madison for ground work and actual flying

April 20, 1932 – Wednesday

The weather was cool, cloudy, and the wind was pretty rough. I didn't fly today.

We washed and waxed Joe's Waco "Peg" and one of the "Preps" after we gave them a thorough inspection. The rest of the fellers leveled the runways with truck by dragging a skid we made loaded down with creek rock. They then filled in the low spots with gravel.

Joe brought his girl out to the airport this afternoon. Her handle is Mariana Kruger. He had John Hoover, Linn Wright, Sam Poletunow, and Art Prosek lined up for dual instruction. His girl got tired waiting before he finished with the fellers so he asked me to take her home. She's pretty decent.

When I got back to the airport I helped Harvey fix a window that was broken by someone who wanted to rob the joint. I guess they didn't know that a student stayed at the hanger every night. Harvey and I straightened up the workshop, swept the floors, and called it a day.

I went out to the Sachleben's this evening. We shot the breeze, played euchre, and had a tin cup of that good old "homebrew". Had a great time.

I am around them so much that Fred, Carl, and Joe call me their brother. Ma and Pop treat me as a member of the family. I like that. They are the salt of the earth.

So far I haven't had any luck with Jessie or Millie Ricketts, and I haven't been able to get Cleone Grimes to go steady with me.

April 21, 1932 – Thursday

The weather was clear with a few showers, warm, and the wind was pretty tricky. I didn't fly today.

I helped Harvey take the Velie M5 engine off the NC122K "Prep". The thrust bearing went phooey and tore up the engine. We took the engine all apart and while Harvey made a list of the parts that had to be replaced, I cleaned the parts with gas. I scraped the carbon from the cylinders and pistons, and then ground the valves and seats. The intake and exhaust valves and seats turned out to be in perfect condition. It took just one lousy bearing to practically ruin the engine. The bearing checked out okay at all the inspections and maintenance, but those things will always happen. It will be several days before this ship is ready to fly; it all depends on how long it will take the new parts to get here. I like this, it beats loafing, washing and waxing ships, filling chuckholes, cleaning out fence rows, cleaning the hanger, etc.

Art Prosek and I spent the rest of the day studying Lesson No. 44, Aerial Navigation. Art is one of my favorite students, although I like them all, including me. I loaf around a little more with him than the others, and he is my bodyguard although he doesn't know it; he sure is big. He is quiet, always willing to volunteer, and is not interested in horseplay.

Joe, although he is the instructor and part owner of the school, is another one I like a lot. I can't figure out what he thinks of me cause he is usually bawling me out when we are flying together. I hope it is not personal.

I loafed down at the Maddox Garage for a while after supper and saw, for the first time, a one cylinder, 2-cycle gasoline engine run.

I got a letter from Max C.F. Sittner, Hudson, Kansas, also one from Mom, and just in time. I owe Mrs. Barber 2 week's room and board. The only money I have is what Ma sends me. Paul, Charles, Earl, Ralph, Pearl, and Bert send me a little every now and then. I try to do my part and spend only what I absolutely have to. If the fellers don't keep the Cheve full of gas I walk. As part of my tuition I have to work for the school when I am not flying. Besides, there are no jobs of any kind, part time or otherwise available.

I really hate to take the money from home; they are having a hard time due to the depression. They own 2 bakeries, my brother Charles runs one and my Dad runs the other, but they are losing money and it won't be long until they go broke. The chain grocery stores sell their bakery products for less than my Dad pays for his supplies. I'll bet as soon as they run my folks out of business they will raise the price. There is no comparison as to the quality and taste of the baked goods, but the neighbors have to save every penny to get by.

April 22, 1932 – Friday

The weather was clear, warm, and the wind was mild. I flew for forty-five minutes today.

I overshot the line about a foot twice, and hit it right on the nose once. I made mighty fair landings. My spins were ragged and the right verticals were rotten. The left verticals and the figure eights were okay.

The Velie M5 engine cut out on me twice. The first time the engine cut out (misfired) for just a revolution or two, then started running again. I thought it might have an air bubble or a drop of water in the carburetor preventing the gas from reaching the cylinders. But a short time later the engine cut out a second time and I thought sure it wouldn't start running again, so I looked for a place to land. The airport was too far away to make a "dead stick" landing there. Just as the prop. Was about to quit turning the engine coughed and began to run normally again. Once the prop stops turning there is no way to start the engine except by hand cranking it, and I wasn't about to get out and crank it 400 ft. in the air.

I hightailed it back, while it was still running, to the airport and made a nice 3-point landing. I told Harvey about it and we checked the fuel system, took the carburetor apart, cleaned and replaced it. We checked the ignition and everything checked okay. Finally, not finding anything wrong Harvey decided to drain the gas out of the system including the gas tank. We filled it up with fresh gas using the chamois covered funnel as usual. We chocked the wheels, tied down the tail, tied the "joy" stick back against the seat, and let the engine run at fast idle for an hour. It never missed a beat.

Bob Parker had a close one. He was flying the NC122K at about 400 ft. right over the airport when the engine suddenly stopped. He made a decent "dead stick" landing. Some guys have all the luck. I helped Harvey take the engine off the ship and then we took it into the workshop and took it apart. The crankshaft broke and tore up the engine.

We all went out to the airport after supper and reviewed *Lesson No. 28 – Carburetors*.

I waited until all the fellers left then asked Joe about that job Mr. Oglesby had promised. Joe acted surprised and said he didn't know anything about it. I told him that was why I came here after reading a small advertisement in the Popular Mechanics magazine. The ad said that the first student enrolled in the Limited Commercial Flying Course at Dixie Flying Service, Madison, Indiana, and graduate with an L.C. flying license would be guaranteed a job. I also told him before reading that ad I had planned on going to a flying school in Cleveland.

Joe wanted to see the ad because he had never heard of it. But I had given it to Mr. Oglesby when I signed the tuition form and he kept it. The biggest mistake I made was not getting it in writing, but I was brought up to trust a man at his word. Another lesson learned the hard way.

Joe said he would look into it, but it seems as though I don't get the job. It's a bad way to do business. I am upset, disappointed as I had my heart set on the job, and mad. After all the extra work, using my car and gas to run errands, taking students to Seymour for their physical exams, etc.

I didn't even raise a lot of fuse when they are giving the other students a lot more flying time because they have to pay their room and board; but not mine. I'll not blow my stack until I get my L.C. license.

April 23, 1932 – Saturday

The weather was warm, clear, cloudy, rainy, and the wind was mild. I flew for 50 minutes today.

I overshot the line slightly one time, and hit it right on the nose twice. The landings were barely fair. I leveled off too high and it seems as though I'm getting a little ground shy. I've never had this trouble before. I am still hot under the collar and that could be the problem, I've got to get it out of my mind until I get my license but that won't be easy. My left verticals were okay, but the right ones were rotten. The spins were sloppy, but the figure 8's were okay.

Joe didn't mention the job and I am not going to bring it up again; yet.

I spent the rest of the day helping Harvey finish taking the NC122K, Velie M5 engine apart. We washed and inspected everything, including nuts, bolts, etc. When we get the parts in we will have to rebuild the engine "from the ground up". We have some of the parts in stock, and when we get done it should be the same as a new one.

Those guys are still rebuilding the Hanover Hill road. They said it had been a gravel road until 1927 when it was blacktopped, and now they are tearing it up and replacing it with concrete. Their language is very colorful, to put it mildly. Mrs. Barber has rules of behavior and they stretch it to the limit. I take great pride in my eating ability, but they make me look like a piker, they sure shovel it in. Mrs. Barber isn't going to make any money off those odd balls.

Al Thockle and I loafed around town this evening. The town was crowded, as usual, with farmers and the girls were of particular interest. Some were tall, short, big, little, fat, thin, and some were downright cute as a button.

If I don't start doing some serious studying I am going to be up the creek without a paddle. I got a letter from Red Sandstrom and one from Paul.

April 24, 1932 – Sunday

The weather was cool, cloudy, drizzly, and the wind was fairly rough. I did not fly today.

I went to Sunday school with the Sachleben's this morning. Cleone G. wasn't there; actually I've only seen her in Sunday School twice. Millie and Jessie were there but I didn't get a chance to talk to Jessie alone. I got the impression they were avoiding me so I guess dating both sisters won't work. Don't give up so soon.

I went riding out in the country with Al Thockle this afternoon. The airport was closed, bad weather. About half of the fellers went out with their girl friends, and the rest spent the afternoon in the living room shooting the breeze.

I tried to do some studying this evening but couldn't concentrate so I loafed around town awhile. I gave that up and went back to my humble abode, read parts of the Dayton School of Aviation Correspondence book, then hit the sack.

April 25, 1932 – Monday

The weather was cool, cloudy, rainy, and the wind was rough. No flying today. It was kinda strange not hearing the engines' cackling sweet music.

Clyde H. Beyer

Harvey and I worked all day trying to repair the oil pumps on the Velie M5 engine that was damaged when the crankshaft broke. The pump still didn't work when we quit for the day.

We really need a spare Velie M5 engine. Then when something like this happens we can keep the ship flying while the damaged engine is being repaired. Right now we have only one trainer ready to fly.

The other crew spent most of the day rearranging the ships in the hanger, and then washed all the windows. They made sure the engines on the ships parked outside were covered securely with the waterproof tarps.

In addition to a little horsing around my crew helped, and checked all the tires, landing gears, and tail skids.

Art Prosek and Howard Brennen walked downtown with me and I picked up my pictures at Inglis' Drug Store. While there we flirted with the gals who worked behind the soda fountain.

I wrote Mom, Herb Schwartz, and Pyatt a letter.

April 26, 1932 – Tuesday

The weather was cold, cloudy, and the wind was rough. No flying today.

A new feller came in from Canada yesterday. Alan Lindstrom is his handle. I have to take him to Seymour tomorrow for his physical exam.

We fooled with the oil pump most of the morning. We tried everything; changed gears, valves, gaskets, and shims, but it still won't pump oil. It can't be fixed. I wish the parts would come in so we can rebuild the Velie M5 engine. This loafing around is ruining me; pretty soon I won't be worth killing.

During dinner we argued back and forth about which cars were better, Cheves or Fords. As usual I bragged about my Cheve and feeling pretty safe as I was the only student that had a car. "I would race and beat any Ford that came down the pike".

It was almost 2 o'clock when Bob Parker showed up at the airport. He was driving a 1931 Ford roadster and was grinning like a silly cat that had just swallowed the canary. It was Gladys Gourley's car; she is the school's secretary. With all the fellers standing around admiring the Ford he slyly yelled over at me "Well Friday, do you still want to race a Ford?"

I knew right then and there I should have been listening instead of bragging. My Cheve is about seven years old and I had converted it from a coupe to a sedan. But with all the fellers watching me I wasn't about to eat "crow", so I said, "you better believe it".

We agreed to start the race at the top of Michigan Hill, about three miles south of the airport on S.R. 29, and the first car that turned into the airport was the winner. Art Prosek rode with me and Al Thockle rode with Bob Parker. There were just a couple of cars that got in our way but that really didn't make a difference in the race. Right off the bat that Ford took off burning rubber and left my Cheve in a cloud of smoke. Bob whupped me for sure, and my poor Cheve was really embarrassed, but I couldn't think of an excuse. The fellers didn't razz me very much cause they know which side of their bread is buttered, my Cheve is still the only car they can borrow.

Several of my crew, including me, pitched horseshoes the rest of the afternoon. The rest, and the other crew played softball. I am not a very good horseshoe pitcher. I don't think I ever saw a horseshoe before coming to Madison.

Harvey bought a 1927 Cheve coupe yesterday and he tinkered with it the rest of the day.

April 27, 1932 – Wednesday

The weather was cool, clear, and the wind was rough. I didn't fly today.

We watched as Joe and Frank Peternell went up for a check flight. Frank still has trouble staying on the runway, but he did make a nice landing. Next up was George Hayden - he went straight down the middle of the runway, but he has a tendency to takeoff with the right wing a little lower than the left. He came in a little low and had to gun the engine wide open to get up over the high wires. That made him land pretty hot, and by the time the ship stopped they were at the end of the runway near the trees. As usual he came in with the right wing a little low.

Walter Etchison was next and he made a nice takeoff and three-point landing. They didn't use the line in the landings, and I don't know how they did in the maneuvers. Howard Price was the last to go up before dinner and the takeoff and landing looked okay to me. He must have done something wrong during the maneuvers cause they sat in the cockpit on the runway for quite a spell.

After dinner George Dystra went up with Joe. George and Linn Wright are the only fat students, and when the runway is wet and soft, or muddy, they have a hard time getting the ship off the ground.

I didn't watch George and Joe fly. Instead I helped Harvey take the engine out of his Cheve. I kinda supervised it as I had the experience of taking my engine out when I changed the bodies on my Cheve. I think I am somewhat of an expert on 4 cylinder Cheves, having worked on every part of mine from the back bumper to the front bumper, and everything in between. I have tightened, loosened, removed, or replaced practically every nut and bolt so often, they seem like old friends.

It was a dirty job; the engine was covered with cruddy oil. We spent the afternoon taking the engine apart, cleaning each part with gas (while we were smoking) as we took it off the engine block, just in time to close the airport and go down to the boarding house for supper.

Al Thockle and Alan Lindstrom left the school, so I didn't have to go to Seymour.

Robert R. Foster, V.P. of the Dayton School of Aviation, took pictures of us eating supper. Then, after we finished eating, he took several pictures of us in the parlor of our boarding house. These pictures are to be used in a pamphlet promoting the school.[32]

THERE'S NO TIME LIKE THE DINNER HOUR

What a happy time it is when the boys gather around the abundantly set dinner table. It is a revelation how Mrs. Barber finds time to cook such excellent food for the boys. Every day she has standing orders for special delicacies from those who are completing their training and are about to go out on the job. You will find Mrs. Barber a marvelous hostess and will enjoy every minute of your stay with her.

In this photograph you see some of the boys in the living room of their home. We say home because Mrs. Barber, who keeps the boys has turned her entire 13 room house over to them. In this photograph you see a group of them around the piano and believe me they spend mighty happy evenings under the supervision of Mrs. Barber right here at home.

Art Prosek and I went to roller skate at the Riviera Gardens. It is a brick building on the river side of Highway 56, west of the city limits and east of Clifty Falls State Park's south gate. It is about 2 ½ miles west of our boarding house. I was having a swell time until a big guy began bumping into me, trying to make me fall. So I blew my stack, and just as I was about to get my ears pinned back, Art skated up to us and saved my

---

[32] Pamphlet is at the end of the book

hide. That bozo took one look at Art and scrammed. I skated close to Art the rest of the evening.

I wonder what Joe said to Howard Price while they were sitting in the cockpit out on the runway this morning?

I got a letter from Rodney Pert today. I wrote Paul a letter and then hit the sack.

April 28, 1932 – Thursday

The weather was warm, clear, and the wind was pretty rough. I didn't fly today.

The parts for the NC122 engine came in yesterday. Harvey, Art, and I began rebuilding the engine as soon as we got to the airport this morning. We put the rebuilt engine in the ship after dinner and tied it down. We started the engine and let it run all afternoon to break it in. It sounded great.

Art, Harvey, and I double-checked the Fleet rigging and it was okay. It sure looks pretty with a new coat of paint.

We finished the day pitching horseshoes while keeping on eye on the NC122. The rest of fellers got some flying time, while I filled in the ditches with cinders.

I wrote Art Sandstrom and Fred Moss a letter. I studied for the Limited Commercial test till the wee hours of the morning.

April 29, 1932 – Friday

The weather was cool. Cloudy, and the wind was pretty quiet. I flew the NC122 for 15 minutes today, and the engine never missed a beat. The 180-degree maneuver was pretty darn nice, and the landing was right on the money (on the line). The ship spins terrible. It didn't fall into a spin during the stall, even though I idled the engine and pulled the stick back. The ship lost flying speed, the nose dropped and started into a steep dive. I pushed the stick and rudder all the way to the left, but it didn't wind up (spin). So I pushed the throttle half open keeping the stick & rudder to the left. As the ship started to spin I pulled the throttle back to idle. When I tried to pull the ship out of the spin it kept spinning a couple of turns before it quit. Then I pushed the throttle open, put the rudder pedals in neutral, gradually pulled the stick back and leveled off.

I tried the same maneuver to the right with the same result. So I stalled the ship, pulled the stick back, pushed the right rudder and stick all the way to the right with the throttle half open. As it started to spin I idled the engine. This time it would not stop spinning when I tried to pull it out, and I began to think it never would. After trying every maneuver I was taught, and the ground rapidly approaching, I reversed the controls, pushed the stick all the way forward, pushed the throttle wide open and held on. We were headed to earth in a hurry when the ship finally stopped spinning. I gradually pulled the stick back, keeping the wing and rudder neutral, leveled off and pulled the throttle back to cruising speed. It is going to be awhile before I make this crate spin again.

Actually, a ship that doesn't fall into a spin during a simple stall is much safer than one that does at the drop of a hat.

We have just about finished the Fleet. What a ship! We watched Sam Polentunow and Linn Wright practice takeoff and landings. They are getting pretty good. The runway is awful soft and Linn used up more runway taking off than Sam.

We all went to school at the airport after supper. Joe wasn't there. We horsed around all evening but he never showed up.

April 30, 1932 – Saturday

The weather was clear, warm, and the wind was slightly rough. I didn't fly today.

I went up on the hanger roof and took the windsock down and replaced it with a new one. It was about ready to fall apart. While on the roof I checked and cleaned the weathervane.

Usually on Saturday the only fellers at the airport are those who are going to fly, unless one of the training planes must be repaired. As I didn't get to fly I loafed around and watched the others. If you watched each student fly several times you can tell who is in the cockpit without seeing him, as they all fly in a different manner.

I worked on my Cheve this afternoon. I cleaned and adjusted the sparkplugs, cleaned and timed the distributor, then cleaned and adjusted the points. I also adjusted the valves, and tightened the steering. The engine isn't running very good.

Linn Wright, Sam Polentunow, and I walked around town this evening with the rest of the farmers.

May 1, 1932 – Sunday

The weather was cool, cloudy, and the wind was awful rough. No flying today.

I went to Sunday School this morning and I learned a lot and it was not just about the bible. The teacher is trying to keep a lid on the conversations and there was not a dull moment. Mildred and Jessie were there and so was Cleone. I've just about given up going out with Jessie or Millie so I will try to talk Cleone into going steady with me. It seems as though every boy in town wants to go out with her. She is very pretty, stacked up, and very popular (I think she knows it).

I loafed at the airport all afternoon. There were very few sightseers taking airplane rides, as the wind was awful rough. So I spent most of the afternoon "shooting the breeze".

When I finally had a chance to put in my "two cents worth", I told Joe and the fellers a small plane the size of our trainer set an altitude record of 10,000 ft. I asked Joe if he thought one of our "Preps" could go that high. Everyone had a different opinion, and after batting it around, I suggested there was one way to find out. As I was the most experienced student weighing less than most of the fellers, I should be the one to try it.

Joe didn't seem to be very excited about the idea but said that Harvey and I could prepare the 110K the first day the weather was ideal, and give it a shot. The ship will have to be in "tip-top" shape, and the wind blowing strong but steady, not calm or gusty.

Geez, I sure would like to have a job here teaching students to fly.

I waxed the top of my Cheve after supper. I was going to drive downtown, but the car won't start so I wrote Mom a letter.

May 2, 1932 – Monday

The weather was clear, warm, and the wind was a little rough. I flew for one hour today and it sure felt great sitting in the cockpit again. It is the greatest sensation in the world once you are in the air.

The left and right spins were okay. I went into the spins with the nose of the ship pointed towards the river and came out in the same direction. That ain't easy, depending on the way the wind is blowing. The figure eights were fair but the left and right verticals were not so good. Still doing them too steep. I set the ship down once right on the line, but overshot it five times. Made one decent three point landing but the others were pretty bad.

I am beginning to wonder how in the world I ever managed to pass the Private Pilots flying test. I am still shook up because I won't get that job, and Joe's attitude seems different since I asked him about it. It might be my imagination. I think all of this might be affecting my flying and if it is I have to snap out of it. They told us that to be an excellent pilot you must control your emotions, disappointments, aggravations, temper, and have patience while flying. In other words, have a one-track mind totally devoted to the business of flying. I am weak on emotions and temper but I'll solve them.

My group cleaned the hanger, cleaned the cockpits, and rearranged the ships at quitting time.

The fellers pitched horseshoes until it was their turn to take a flying lesson.

I cleaned my Cheve engine with gas, took the cylinder head off, installed new gaskets, replaced the cylinder head, and timed the distributor. Then I took the starter off the engine, replaced the brushes, cleaned the armature, and when I put the starter back on it wouldn't turn the engine over. I think it is bolted too tight against the flywheel gear. So I'll have to hank crank it and take a chance on breaking my arm until I have time to change it. After all that work I discovered the cylinder head is cracked.

May 3, 1932 – Tuesday

The weather was cool, cloudy, and the wind was rough. I didn't get to fly today.

The rest of the fellers kept both "Preps" busy. As soon as one "Prep" landed the other one took off. The student was sitting in the cockpit waiting for Joe to climb in. The students are sure getting better. Most of them land less than 3 ft. off the ground; most of the time.

Each ship had a quick checkup after every flight. I checked the engine, Frank Peternell checked the controls, Jack Yellitz checked the landing gear and tailskid. I made sure there was plenty of gas in the tank and then parked the ship on the hanger apron. When we saw the "prep" approaching the airport runway, we cranked up the ship, the next student sat in the cockpit and waited for Joe to climb in. We repeated these inspections until all the students had received dual instructions. On one inspection Jack Yellitz found several landing gear bolts bent and loose. He replaced them in no time. All the other inspections were trouble free.

My group took care of the hanger today. They cleaned all the windows and swept the office and workshop floors. They also checked the tires on every ship and made sure each tire had the proper air pressure. At quitting time they parked all the planes and locked up the hanger.

A new student came in today. Bud Lawrence is his handle and he is from St. Mary's, Ohio. He seems okay. As usual I gave him his first airplane ride and he got a big kick out of it. I get a big kick out of it because no matter how many or what kind of mistakes I make they think I'm perfect.

I don't understand why I can't enter this time in my logbook. It would build up my flying time, but I am just happy to get the chance to fly.

I stayed in again this evening and studied Lesson No. 37, Aircraft Navigational Instruments. I hope tomorrow is the day.

May 4, 1932 – Wednesday

The weather was cool, clear, and the wind was very strong from the west. I flew the 110K for forty minutes today.

It was a perfect day to see how high the "Prep" would go. Harvey and I checked the ship from "stem to stern", tuned up the engine, adjusted the carburetor, and checked the controls. Between the two of us, we decided to put just enough gas in the tank to get me up there and back. In that way se would eliminate some weight.

I sat in the cockpit and Harvey cranked up the prop. After the engine warmed up Harvey pulled the wheel chocks from in front of the wheels and waved me down the runway. All the fellers watched as I pushed the throttle wide open; we were on way headed west. I banked the crate to the right and began a series of upward spirals. I tried to make the circles about a mile in diameter, as I had to keep the airport in sight at all times, Joe said. The right spirals soon got monotonous so I reversed directions spiraling to the left.

With the throttle set at two-thirds open we gained altitude with each spiral. At about 8,000 ft. the farms looked a lot smaller, about the size of checkerboard squares, more or less alike, and we weren't climbing as fast as we had been. I could tell the difference in the air, as it was getting thinner. As we continued to climb I kept inching the throttle forward to keep the rpm's steady. At 10,000 ft. the throttle was wide open but the rpm's stayed the same. I couldn't see the airport; everything was so tiny and blended together. I was glad the Ohio River was there although it was real small too. The engine was laboring due to the lack of air and I was breathing pretty hard myself. The engine carburetor was now out of balance as it was getting more gas and less air. As the density of the air became thinner, the exhaust gases, at the same time, began to change from clear or white to light bluish or light blackish.

I began to think we were not going to reach at least 11,000 ft. The gas gauge was dropping faster than normal, and the controls were now beginning to get sloppy. There was a very slight delay between the time I moved the controls and when the ship responded. I kept the nose slightly above the horizon, but not so steep. I didn't want to take a chance that the ship might fall off into a spin. I made the circles real shallow for the same reason.

The ship finally reached 11,900 ft. that is, if the altimeter was accurate. With the throttle wide open, the engine was really laboring, and he rpm's were beginning to drop. The engine was simply not powerful enough, in this thin air, to create the lift necessary to raise the ship any higher. I kept the nose up and the ship slowly struggled up maybe a foot or so, then the nose slowly dropped down a couple of feet. The nose continued to slowly rise and fall, but I didn't gain any altitude. It felt like being in a canoe and riding in the swells created by a big boat. I looked down, but couldn't see the houses or roads; everything was so small. Geez! It was a long way down and I didn't have a chute on.

I had reached the absolute ceiling of this ship, and the gas was running low, so I decided to get down as fast and safe as possible. I thought about spinning down, which would be the fastest, but I didn't know how the ship would react in the thin air. So I eased the nose down slightly below the horizon, kept the throttle wide open and flew in a straight line until I felt the controls respond in a positive manner to my touch. As the ship was now flying, more or less normally, I pulled the throttle back to cruising speed. Then, as we were losing altitude fast, I banked the ship toward the river and began looking for the airport.

We continued to spiral down, and when the farms, roads, and the river became recognizable, I found myself over strange country. I stayed over the river, as continued to lose altitude, but I couldn't find the bridge. I banked the ship and flew down the river and finally saw Madison and the bridge. We had plenty of altitude as I flew over the airport and cut the throttle to idling as the gas gauge showed "empty". As we spiraled down I started into the 360 degree turn and land maneuver. We lined up with the runway heading west, the engine sputtered so I quickly pushed the throttle wide open but it belched and stopped.

I finally got my wish, a "dead stick" landing. It was stranger than I expected and more complicated. First there was the deathly silence, except for the wind whistling past the ship. Then the prop stopped straight up and looked just like a cemetery monument.

We came in high over the high lines "hot" (too fast) and at a steeper angle than usual to make sure we maintained flying speed. If I undershot the field, without an engine to help," it would be all she wrote". I flew the ship right down on the runway, as I didn't have the time to let it touch down normally. The ship still had too much lift and would leave the runway, if I pulled the stick back so the tailskid would drag and slow the ship down. I kept the stick in neutral; when we left the end of the runway and headed towards the trees I pulled the stick all the way back and held my breath.

The landing gear got tangled up in the underbrush and stopped the ship just as a big cedar tree stared me in the face. Geez, if we had put just a couple of more gallons of gas in the tank, I wouldn't have been in this mess. When the fellers showed up in the truck to tow me back to the hanger, and heard me "cussin" up a storm cause it ended up this way, they knew I was okay. The only damage was some scratches on the landing gear and front of the ship.

We went into the workshop and I told them we got as high as 11,900 ft. We talked about how the ship flew at different altitudes, and how it reacted to differences in the atmosphere, especially near the ceiling. I never looked over at Joe, but as we were leaving I heard him say, "Friday, I don't want to ever hear anymore of your bright ideas, or any other "cockamamie" suggestions you might think up, NO WAY, NO HOW, PERIOD, AMEN." His exact words, and it was a one-sided conversation. I guess he forgot he gave his okay.

While I was flying my group cleaned up around the outside of the hanger. They also leveled the hanger apron with creek gravel. We parked the ships in the hanger and tied the rest of the ships down on the apron at quitting time.

A new student came in today and I took him up; it was his first airplane ride. Daniel Booth is his handle and he is from Tower, Indiana.

I took Bud Lawrence to Seymour for his physical exam. I stayed in tonight and studied Aerial Navigation.

May 5, 1932 – Thursday

The weather was cool, clear, and the wind was kinda rough. I flew the 122K for 15 minutes today.

I flew over the new Hanover Hill road and saw 3 fellers, who eat at our boarding house, working on top of a bluff. The bluff seemed to be about 25 ft. above the roadway. I spiraled down to treetop level to wave at them. As I headed toward them I pulled the throttle back to idle to slow the ship down, and for some reason the engine backfired. Maybe I pulled the throttle back too fast.

I guess they thought I was going to crash into them cause all three jumped off the bluff. I pushed the throttle wide open and flew out of there hoping they didn't get hurt or get the ship's license number. After a few maneuvers I made a mighty nice 3-point landing.

I took Dan Booth to Seymour for his physical exam. My group policed the hanger, then washed and waxed Joe Vail's "Peg". The other group was busy taking dual flying lessons.

I took the Cheve starter loose and adjusted it properly against the flywheel gear. Now I don't have to hand crank it. I am having trouble finding a used Cheve cylinder head that isn't cracked.

The pictures V.P. Robert Foster took April 27 were very poor.

We have finally learned to stick together. Most of the local jokers still give us a hard time every chance they get, and would like to see us disappear into the wide blue yonder.

When we went in to eat supper the three fellers who jumped off the bluff on the Hanover Hill road were there. I was pretty sure the students wouldn't tell them anything. They didn't seem to have any broken bones but they were sure mad, and said a pilot chased them off the bluff on purpose. They wanted to know who flew the ship, and when they found out, they were going to wipe up the floor with him. I thought they could and would. We all denied knowing anything about it and used every alibi in the book. They did seem kinda stiff and sore when they left still bellyaching. Phew!

I studied *Lesson No. 36, Aircraft and Engine Instruments* all evening.

May 6, 1932 – Friday

The weather was clear, warm, and the wind was rough. I didn't get to fly today.

Joe had a full schedule of students posted on the bulletin board to take flying instruction. He wants to have the fellers ready to take their Private Pilot's test on May 17. Evidently he thinks Frank Peternell, John Yellitz, and I are ready to take the Limited Commercial Pilot's test on May 18, cause our names were not posted today. But there's plenty of time. So far we are the only ones qualified to take the L.C. test.

I watched as Bud Lawrence and Dan Booth took their first flying lesson. Joe took with Bud, circled the field and made a nice 3-point landing. They taxied back to the apron, turned around, and then Joe let Bud handle to controls with his assistance. They rolled down the runway at half-throttle; they just tried to keep the wings level and the ship in the middle of the runway. They practiced this maneuver three more times. The third time Joe took off circled the field and made a nice 3-point landing.

They taxied back to the apron, Bud climbed out of the ship and Dan climbed in. As they were taking off a gust of wind almost blew the ship off the runway. The left wing tip hit the runway but Joe managed to level the ship and takeoff. He landed the ship, got out and examined the wing tip. It was not damaged so they repeated the same maneuvers as with Bud. As I watched them I remembered what Joe said to me as I took my first lesson, and no doubt told Bud and Dan. "Put your feet and hands lightly on the controls so you can follow my movements. But don't ever 'freeze' or prevent me from using the controls or I'll bounce the fire extinguisher of your noggin".

My group and I hauled several truckloads of cinders and spread them on the hanger apron and parking lot. We parked all the planes and locked up airport at quitting time.

Fred, John, and I spent the rest of the day studying Elementary Meteorology and Navigation. Although it is not part of the L.C. written test, it is part of the Transport Pilot test that we hope to take someday. The cross-country flight test is the only additional test we would have to take to pass the Transport test, plus 200 solo hours.

I went out to the Sachleben's after supper and had a nice time playing euchre, also enjoyed the tin cup of home brew. While playing cards I told them my Cheve cylinder head was cracked and I couldn't find a good used one. A new one was way too rich for my blood. "Pop" said he had an old 1918 Cheve sitting in the barn and if the cylinder head fit my Cheve engine I was welcome to it. Joe S. said he would help me.

May 7, 1932 – Saturday

The weather was clear, hot, and the wind was very rough. I didn't get to fly today.

Art Prosek washed out the landing gear on the 122K today. I didn't get to see it happen but the fellers said he came in kinda crab-wise. Maybe he tried to sideslip in and didn't get the ship straightened out in time. When the wheels hit the runway the landing gear folded over like an accordion and broke the propeller at the same time. It shook Art up quite a bit and sure made Joe mad. Well, the saying goes, "the landing was good one if you can walk away."

We put the front end of the ship on the back of the truck, put a dolly under the tail and towed it back to the hanger. Harvey, Joe, Art, Bud, Dan and I wrestled the ship into the hanger. We unbolted the landing gear from the fuselage, and on inspection found the fuselage connectors were not damaged. What luck! It could have been a major repair job. I felt sorry for Art; he is a swell feller and my favorite student, along with Joe and Harvey. We examined the ship from stem to stern and it seemed to be okay.

Howard Brennan flew the 110K and made a fair landing. As he taxied back to the hanger we heard the engine make a loud knocking sound, then silence. Howard turned the engine off as soon as he heard the knock. We pushed the ship in the hanger and that ended flying for the day.

Dave Swett, Harvey, and I worked on the 110K. We took the engine cowling off and turned the engine to find which cylinder was making the noise. It was the number one cylinder so we took the can (cylinder) off and discovered the piston pin had come loose and scored the cylinder wall pretty bad.

We replaced the cylinder, piston, pin, rings, valves, and sparkplugs; the connecting rod was okay. We put the new can on the engine, adjusted the valves, and cranked up the engine. When the engine warmed up we readjusted the valves and installed the cowling. She is ready to fly.

The other group took the 122K landing gear apart and will rebuild it Monday. Harvey was pretty smart; he had stockpiled a lot of landing gear parts and propellers. I guess he had a good idea what we would break.

While my group was policing the hanger and grounds I went over to Sachleben's barn, which is on the north edge of the airport. Joe S. had removed the cylinder head by the time I got there. We took it to the hanger workshop and cleaned it. It was sure a job as it had been sitting in the barn for years. We finally got the sparkplugs and valves off the head. The valves and guides, to our surprise, just needed a good polishing. We will grind the valves and valve seats, polishing them with grinding compound. I hope it fits my Cheve as it is in good shape with no cracks.

Those odd balls are still singing the blues because they can't find out who chased them off the bluff. They were lucky and landed on a pile of loose dirt where the roadway had been bulldozed. Its strange, Joe has never mentioned it.

I got a dandy letter from Paul and Mom. Mom sent me a package of medicine, but the bottle of peroxide was broken. I got a swell notebook from the Standard Oil Company.

The fellers have been spending a lot of time lately swimming at the Nor-Rose rock quarry.

May 8, 1932 – Sunday

The weather was clear, hot, and the wind was pretty rough.

I went to Sunday school with the Sachlebens this morning. Cleone never showed up. Mildred and Jessie look right at me and acted like I wasn't even there. So that's that. It is just as well as I have to catch up on my studying. I had a nice time and I am absorbing a little religion at the same time. It could come in handy if I ever get in trouble while flying.

I flew for forty-five minutes today. Frank Peternell went up with me. The figure eights, power turns, climbing turns, wingovers, and the 360-degree turn and land were pretty darn good. I undershot the line twice and the rest were normal. All the landings were very good.

The verticals were very poor. Joe sure was mad and wanted to know many times he had to tell me the verticals were too d--- steep. I am having a terrible time breaking the habit. As soon I get the wing toward the ground I can't resist until the wings are perfectly perpendicular and I am on my side looking straight down at the ground. When I do this the ship loses altitude fast and can't complete the maneuver. I have to do it right or I'll flunk the test.

Joe S. and I ground the valves and seats on the 1918 cylinder head, and replaced the rusted valve springs with springs from my cylinder head. We put the head on my engine block and it fit perfectly. We put the engine back together, manifold, hoses, etc., filled it with water and cranked her up. The Cheve runs good.

I went over to Nor-Rose and watched the fellers swim. I took a picture of Linn Wright diving.

We policed the hanger and grounds. This is the last day my group will have to do it for a while. I studied *Lesson No. 54, Air Commerce Rules and Regulations* this evening

May 9, 1932 – Monday

The weather was clear, warm, and the wind was very rough. No flying today.

We pushed the ships in front of the hanger and washed each one, inside and out. We looked for scratches, cracks, holes, and any other defect in the fabric. We also checked the tautness of the covering. While the fellers were horsing around I took the cracked Cheve head to the junkyard and sold it for 50c. I split it with Joe S.

I sure wish there was a way to get $400.00 or $500.00 without robbing a bank. I can buy one of our rebuilt planes for that and would be able to fly anytime the weather was fit. It would also give me a chance to accumulate flying time a lot faster. At the rate I'm going it will be forever before I get the 200 hours I'll need for a Transport license. This is a pipedream cause even with the ship; there is gas and oil to buy, license fee, hanger rent, insurance, and maintenance to pay for. I loafed around the airport the rest of the day. I am going to Seymour tomorrow for the Limited Commercial Pilot's physical exam.

Art paid my way into the picture show to see the "Big Parade". It was a fair picture. I promised to take him to the show if I passed my L.C. test.

Mrs. Barber and her daughters are very nice people and I am lucky to be staying here. Mrs. Barber is a nice little "mother hen", but Inez is a young "daughter hen" type that always seems to be running around the chicken yard taking charge, flapping her wings and cackling as she goes. She treats me like a younger brother although she is only a couple years older. But, geez, she gives me all kind of advice and her opinions on every girl I go out with. Thankfully, she hasn't been doing any "cackling" for a while as I have been concentrating on the big test. I am sure she will the first chance she gets. I am not serious about any of the girls; I just want to have a nice time. Flying is the only thing I am serious about. (I think I could get serious about Cleone)

May 10, 1932 – Tuesday

The weather was cloudy, cool, rainy, and the wind was very rough. I didn't get to fly today.

Frank Peternell, John Yellitz, and I went to Seymour this morning to take the Limited Commercial Pilot's physical examination. It took most of the day, but we all passed. The doctor seemed to spend a lot of time on my eyes but he didn't say anything. He is a man of few words; its either you passed or failed period. It took me exactly one hour to travel the 42 miles.

We had school at the hanger tonight. It was a lot of questions they think will be asked when we take the written test. I could answer a fair number of them.

May 11, 1932 – Wednesday

The weather was cool, cloudy, rainy, and the wind was rough. I didn't get to fly today.

I went up with Frank Peternell and his maneuvers were fair. His verticals were not near as steep as mine. Personally I wouldn't even call them verticals but I must be wrong. Before we knew it we flew into a rainstorm, no lightning or thunder, just sheets of rain. The windshield wipers couldn't wipe the rain off fast enough so it was hard to see where we were going. We found the airport and didn't waste anytime setting the ship down.

I waxed the Eaglerock Biplane. It sure is a lot of work. Art and I went roller-skating at the Riviera Gardens after supper. We had a good time, but we had to run most of the way home in the rain. It is 2 ½ miles from the Riviera to the Boarding House.

May 12, 1932 – Thursday

The weather was cool, cloudy, and the wind was a little rough. I flew for forty minutes today.

All the figure eights were fair. The right verticals were pretty fair, but I still lose out on the left verticals. I am beginning to get the idea, Not So Steep. The spins were normal so I am going to spend more time on the other maneuvers. The wingovers were okay as were the stalls. The 180 and 360-degree turn and land were normal but I did overshoot the line once. All the three-point landings were pretty darn fair.

I waxed the Travelair while the fellers were taking flying lessons.

The fellers rebuilt the landing gear on Monday, May 9th. After installing it on the 122K Harvey taxied up and down the runway a dozen times, then said she was ready to fly.

I tightened the connecting rods and cleaned the oil system on my Cheve. I got a letter from Mom, and then I studied a little this evening.

Inez B. is sure sweet on Harvey. When he walks into the dining room its as though the rest of us don't exist. I can see this young "daughter hen" fluttering around making sure he gets his share of the food. He doesn't need any help; he is 6'2" with long arms, and big hands.

Harvey is a swell feller and I think he kinda likes Inez. As a matter of fact he has taken her out a time or two. But he has a strange way of going with Inez, he acts as though she is a sister rather than a girl friend. The only way I can describe it is he acts kind of bored about the whole thing and is letting her do the chasing. I never saw a feller as nonchalant about girls as he is. He is the most private person I ever knew. He will talk and listen on any subject, but never one word about himself. I know just as much about him now as I did the first day I met him.

May 13, 1932 – Friday

The weather was clear, warm, and the wind was pretty fair. I flew 35 minutes check time with Joe today.

I got an awful bawling out and it all started after I took and began the maneuvers. During the figure eights he yelled that they were sloppy, that the stalls were too steep, and the more he yelled the more mistakes I made. And the more mistakes I made the louder he yelled until I thought he might "kack out" or sumpin'. I completed the spins and wingovers and he slowly began to calm down. I was afraid to do the vertical maneuvers and he didn't mention them.

As we circled the airport he suddenly yelled "Beyer" (this was the first time he called me by my last name so I knew I was in trouble), "there is only one thing the matter with and your flying. You keep complaining that the landings were only fair, when the wheels and tailskid are 3 or 4 inches off the ground when landing, till I want to blow my stack. And the maneuvers, they are only fair, why ain't they good or great once in a while? Why does everything have to be absolutely perfect all the time? THAT is your problem. You are trying so hard to be perfect that the stressful pressure make your efforts worse, rather than perfect. You have to get use to the idea that no one is perfect, and just try to be the best. You work on your car all the time for the same reason, but it will never be a Stutz Bearcat".

He pointed down at the airport as I circled the field, so I cut the engine, and glided toward the runway. I'll never know how I managed to land the ship. Joe shocked me when he, in a calm and quiet tone, asked me, "Friday, be honest with me, what do you think of my ability as a Pilot?" I told him I thought he was an excellent Pilot. As we taxied up to the hanger he said he was going to take the ship after I got out. And I was to stand right there and watch him very carefully during the takeoff, maneuvers, and landing.

He gunned the engine; the ship started down the runway, and as the ship picked up speed the tail left the runway in flying position. Then I noticed the wheels barely left the runway for about 10 ft. or so, and settled back down. The ship continued to gain speed and finally took off. It seemed to me that Joe tried to take off before the ship had enough flying speed. As for the maneuvers, he was too far away from me to judge how well he was doing. Joe approached the runway with the engine idling, in a long gentle glide. I thought he was coming in a little low, and sure enough, just before he came to the high wires, I heard the engine beller wide open as he climbed up over the wires. He set the ship down on the runway from about 8 to 10 inches off the ground.

Joe taxied back to the runway like he was going to a fire. It would be "Katie bar the door" if I ever taxied like that. He shut off the engine, climbed out of the ship and wanted to know if I thought the takeoff, maneuvers, and landing were perfect. I tried to be very diplomatic when I told him what I thought. I was in enough hot water. He said, "Friday, you told me I was an excellent Pilot, and now you tell me the takeoff and landing was not perfect. This is the way I fly and it sure is not perfect." I am going to watch him fly again cause I think he flew that was on purpose to prove something.

The company took a mess of pictures today. I fixed the timing gears on my Cheve so they wouldn't rattle.

We went to school tonight out at the hanger. The subject was different types of magnetos.

Splitdorf Dixie Magneto was extensively used, being an extremely simple unit. It developed four sparks per revolution of the magneto shaft. The revolving poles rotated between two horseshoe magnets creating waves of electric current through a primary coil.

Berling Magneto operates similar to the Dixie Magneto.

Splitdorf S. & S.S. types are featured by their rugged construction, simplicity, and protection against dust, oil, and water.

Splitdorf VA-type Magneto is virtually two independent magnetos built into a single housing, and requiring only one means of drive and control. Aircraft engines employ dual ignition for safety and improved power output.

Scintilla Magnetos were originally built in Switzerland but they are now manufacturing in this country and are perhaps more extensively used at the present time than any other type. They are featured by their dependability, simplicity, accessibility, and light weight.

Bosch Magneto is manufactured in Germany and many are imported into this country. The company is the oldest and most experienced builder of high-grade magnetos in the world. These magnetos are very efficient, and require little attention.

Delco Ignition System is a generator-battery ignition system. In this system the current is furnished by a generator with a regulator to keep the electrical output practically constant over a wide range of engine speeds. A special battery supplies ignition current for starting, and for operation in case the generator fails.

May 14, 1932 – Saturday

When I arrived at the airport this morning Robert R. Foster, Executive Vice-President of the Dayton School of Aviation, asked me to take him up and demonstrate some of the maneuvers we were taught. All the maneuvers were normal, but then I really scared him with several steep verticals. He decided it was time to go down, and the landing was pretty nice. He seems to be a real reserved cookie, but then he is an expert, and highly respected in his profession. We were up twenty minutes.

I cranked up another plane and flew for 60 minutes. The three spins in each direction were just right. The figure eights were what the doctor ordered, and the verticals were normal (not so steep) for a change. The wingover, stall, and climbing turn were okay. The sideslip was perfect, and I hafta admit that I am the best at performing this maneuver. The first two landings should have been better, but the last two were pretty darn nice.

As I was taxiing back to the hanger I was suddenly surprised when the landing gear collapsed and the ship stood on its nose. I automatically shut off the ignition switch before I even knew what happened, and there I was hanging by my seatbelt against the dash. I was still cussin when the fellers came out and turned the ship upright. The only other damage was broken prop. They put the tail up on the bed of the truck, put a dolly under the broken landing gear and towed the ship back to the hanger. Just as I expected Joe was awful sore. Two days in a row. If I had taxied like Joe the ship would have landed upside down and would have had a lot more damage. Joe gave me the broken prop, and someday I'm going to make it into a clock.

As we were taking the landing gear apart I noticed Joe cranking up his Waco "Peg". I watched him from inside the hanger so he wouldn't notice. He wasn't "putting on" the other day when he asked me to watch him fly. That's the way he really flies, and I had never noticed. I am going to take Joe's bawling out seriously, no more trying to be absolutely perfect, and no more beefing about how poor I did. But, I am always going to try to be the best.

The landing gear seemed to be okay during the preflight inspection, so we are going to try to figure out what made it collapse.

I knew I should have stayed in bed all day yesterday, Friday the 13th. Things generally happen to me on Friday the 13th, like yesterday. When I was in grade school my older brother Earl and I joined the YMCA. One Friday the 13th, after school, we went down to the Y swimming pool. When the lifeguard blew the whistle that it was time to get out of the water, everyone made a dash for the locker room cause the last one out got riprapped on the rear with a wet towel. As I was running past a couple of kids rassling, they knocked me into the deep water. That is the last thing I knew.

Everyone got in the locker room and my brother wanted to know where I was. He ran back to the pool and saw me lying motionless on the bottom. He let out a yell and dove in. I came to throwing up, water running out of my mouth and nose. I was awful sick and tried to make them leave me alone, not knowing what they were doing. Earl saved my life. I still can't swim.

A new feller came in today, he is from Bangor, Michigan, and his handle is George Hayden. Howard Brennan and I loafed around town tonight.

I took George Hayden up for his very first airplane ride and he sure was impressed. I'll probably take him to Seymour for his physical exam on Monday.

May 15, 1932 – Sunday

The weather was clear, warm, and the wind was fair. I didn't get to fly today.

I took my Sunday-go-to-meeting suit from under the mattress and the cardboard fell apart. Sometime tomorrow I'll go back down to Vail's Furniture Store and bum Mr. Harry out of 2 empty large cardboard boxes (furniture is shipped in them). Two sheets of cardboard under the mattress sure keeps my suit pressed and it's cheap. I brushed the small pieces of cardboard off the suit, shined my shoes, got all slicked up and went to Sunday School with the Sachlebens. You never know whom you might meet; I hope. Cleone and the Ricketts didn't show up. I met a keen girl and exaggerated a tiny bit to impress her, about my experiences as a pilot. She told me her Uncle was real interested in flying. I kidded her about piloting a plane for her for $35.00 a month and room and board. She took it to heart and is trying to get him to buy her one.

I went out to the airport and messed around with the fellers. I took pictures of the Fleet and some of the fellers. George Hayden took a picture of Cleone and me. We went over to the Nor-Rose quarry and I took some pictures of the fellers and "Boots" diving.

I went out to the airport after supper, as it was my turn to stay at the hanger all night. I spent the whole evening studying "Airplane Construction and the Theory of Flight." I won't have any trouble with the engine section. I think Frank, Jack, and I will be taking the L.C. written test in a couple of days. I still need one hour and twenty minutes flying time to qualify.

May 16, 1932 – Monday

The weather was cool, cloudy, and the wind was fair. I didn't get to fly today.

I've got to get that one hour and twenty minutes pretty darn soon. Joe keeps scheduling me off the flight list so he can get some of the fellers ready for their Private Pilot's test.

I took George Hayden to Seymour for his physical exam this morning. The Cheve made the trip in sixty minutes. The school gives me enough aviation gas to make the trip.

The fellers repaired the landing gear and installed it on the ship while we were gone. We put the new prop on last Saturday.

After the exam I dropped George off at the airport and noticed the fellers were playing softball till it was their turn to fly. I drove on down to Vail's Furniture Store and Mr. Harry gave me two dandy cardboard boxes. I cut the biggest pieces that would fit my cot and Mr. Harry said he would get rid of the rest. He sure is a mighty fine man. I put my one and only suit between the sheets of cardboard under the mattress. I'm back in business. I wonder if could get a patent?

I loafed and studied the rest of the day. I picked up Cleone, then stopped at the boarding house for "Boot" this evening, and took them for a ride out in the country. For some unknown reason I can't seem to get a chance to be alone with Cleone. On the way back to Madison we ran out of gas and had to walk home. But we had a real swell time. I walked Cleone home, and as I was leaving she gave me her cameo ring so I gave her my high school pin.

I wouldn't run out of gas on purpose, cause if I had I sure wouldn't take "boots" along. I walked down to Maddox Garage (a good half mile) and talked Clyde Weber into taking me and a couple gallons of gas to get my Cheve. He laughed all the way and said I was kinda dumb for taking "Boots" along.

Those highway construction workers haven't mentioned jumping off the bluff lately. The day they move out of town I just might confess.

May 17, 1932 – Tuesday

The weather was clear, warm, and the wind was fair. I flew the one hour and twenty minutes this morning.

The figure eights couldn't be better, the verticals were fine, and the spins were as easy putting a record on the Victrola. The stalls were normal, and the 180 and 360-degree landings were really good. Every landing was a three pointer. I don't even think about the takeoffs anymore. I have taken Joe's advice and it sure makes flying easier.

Frank, John, and I spent the rest of the morning studying for the L.C. written test.

Joe Shumate, the Inspector for the Department of Commerce flew in this morning. He gave the Private Pilot written test to Pete Nemetsky, Bob Parker, Art Prosek, Linn Wright, Sam Poletunow, John Hoover, and Clair L. (Bus) Kirsch. "Bus" failed the written test so he doesn't get to take the flying test. He is the last one I would have expected to fail.

The inspector then gave the flight test to Bob Parker, Art Prosek, Linn Wright, Sam Poletunow, John Hoover, and Pete Nemetsky. Pete failed the flight test. Pete and "Bus" will have to wait 90 days before they can retake the test. I thought Bob Parker might get in a little trouble cause he is as reckless as a bedbug, and that is reckless.

The Inspector gave the Limited Commercial written test to Frank Peternell, John Yellitz, and me this afternoon. It sure was tough and it took two hours and thirty minutes. It was just about as I expected. I breezed through the engine section, and no doubt all the work I put in on the Cheve helped some.

The section on airplanes was a totally different story. I did okay on the questions regarding the rebuilding and repairing of the fuselage, wings, ailerons, stabilizers, elevators, and rudder. Also, I knew most of the questions on Air Commerce Rules and Regulations. But the questions on history, theory, and meteorology sure made me sweat and I didn't do as well.

So, I got a good grade on the engine section, and my grade on airplanes was good enough to pass but nothing to brag about. Frank and John had trouble passing but it was the engine section that bothered them. They were graduates from the Dayton School of Aviation Correspondence Course. I am sure I am going to make it now.

Joe asked the three of us to come into the office. He said we were to go to Bowman airport in Louisville, Kentucky tomorrow and take the L.C. flight test. We will toss a coin to see who will fly the ship Joe to Bowman field and who will fly it back to Madison. This sure was good news.

He then told us a stranger had stopped at the office on East Second Street looking for licensed pilots. He wanted to hire pilots who have at least a L.C. license, or better, and he is willing to pay $600.00 a month. Wow!

The contract is:

You will be flying for the Chinese government in China. They will pay the transportation from your home to China. You must stay two full years with a chance to stay a third year. They will keep your monthly salary on deposit for the two years, but if you leave before the two years are up you will forfeit it all, and you will have to pay you own way home. They will furnish the room and board and give you all the living expense you need. If you go home after two years you will get the two years salary and they will pay your transportation home. These same conditions apply to the third year.

Joe said this man is traveling all over the country trying to hire licensed pilots. He said that if we were interested, to let him know and he would get in touch with the man and arrange an interview. We all said it sounded great, but that we would have to think about it.

On the way down to the boarding house Frank and John said flat out they were not interested in leaving the country. I said maybe, as a last resort, if there were no jobs in this country. But I would have to think long and hard before I made the jump. I wondered if a feller could go two or three years without sparkin' with a nice American girl. Frank said he heard that after you were there a while all the girls begin to look pretty darn nice.

But first, I am going to talk to the office and the officers of the school about that job I was promised.Pete Nemetsky and I went out tonight and had a good time. Now that the written test is over I am going to renew my social life. Fit as a fiddle and ready for love.

May 18, 1932 – Wednesday

The weather was cool, fair, and the wind was exceedingly rough. I flew for an hour and forty minutes today. I didn't sleep very well last night. I counted the hours until it was time to get up. I had been looking forward to this day since the day I soloed. I ate a quick breakfast, jumped into the Cheve and high-tailed it out to the airport.

When Joe showed up, I told him I was ready; so let's go. He told me to keep my drawers on as the ship had to be inspected and filled with gas. Frank P, John Y, and I went in Joe's office and he told us that we were well prepared for' the test and what to expect.

We inspected the ship thoroughly cause we didn't want anything to happen that might mess up our chance to take the test. After we filled the tank with gas, Joe told us to change our clothes, get spruced up a bit, and come back to his office. It didn't take us long to get back. He took a dime out of his pocket and told me to call it as he tossed it in the air and I called "tails". When it hit the floor and stopped bouncing, tails it was.

Joe said he would fly with me to Bowman field; Frank and John would drive my car and meet us there. They will fly the ship back to Madison after the tests, and Joe and I would drive my Cheve back home. I wondered why they didn't drive Joe's car. I kinda wanted to fly the ship back but it was really six of one and a half dozen of the other. I got to fly all the way there and they each got to fly halfway back. Frank and John filled my Cheve with aviation gas and drove down to Bowman Field in Louisville. KY.

Joe cranked up the ship, climbed in and charted the course we flew. He said we could just follow the Ohio as Louisville is right on the river, but it has so many twists and turns it would be a waste of time.

I pushed the throttle wide open and at 1,500 feet we made a beeline for Bowman Field. Joe reminded me to keep a sharp eye out for potential emergency landing fields. He never touched the controls but made sure I followed the course he had mapped out. We crossed the Ohio at Madison, at Hanover and again at Charlestown. I saw, for the first time, all the numerous twists and turns of the Ohio. It flows straight west a very short distance from Madison then flows straight south. When it reaches the vicinity of New Washington, it flows southwest to Louisville.

The engine never let out a peep, just the reassuring cackle of the 5 cylinders keeping perfect rhythm. It took us 40 minutes and we hit the town right on the nose. As we approached Bowman Field, we scanned the sky to see if there were any planes in the area. As we had the sky to ourselves, I circled the field to let them know we were preparing the land. I made a nice landing; even Joe seemed satisfied. He told me to go up and practice landings a few times while he took care of all the details. I didn't see my Cheve as I prepared to take off. I had to keep my eyes peeled for any planes in the area, as there are bound to be a lot of them in a town this size.

I made several swell landings and I noticed my Cheve parked beside the hanger. I parked the ship, went over and joined Frank and John. They said the Cheve ran great.

Finally, Joe and Joe Shumate, the Inspector, jointed us and said John would be the first to take the test. John made 3 fair landings, the inspector climbed in the ship; they took off and made the required maneuvers. They landed, and while we were filling the tank with gas, the Inspector told John that he had passed the test.

I went up and made 2 mighty fine landings. As I took off for the third time, I thought once more, this is going to be easy. I made the 180 degree turn and looked down it suddenly occurred to me that wasn't a dirt runway, it's concrete. As I approached the runway, I kept thinking "holy mackerel" I never took off or landed on concrete (I know now, I should have concentrated on landing on the white line instead of being distracted by the concrete runway; consequently I messed up. I came in about 5 ft. above the line so I gunned the ship and went around and approached the runway. As I got close to the line I was still too high and had to go around again. I looked down and saw the Inspector waving me in. I knew what that meant so I decided I was going to do something right. As I approached the runway, I deliberately came in too high. I kicked the rudder to the left, moved the joystick to the left, and slightly toward the seat. We sideslipped in and made a nice 3-point landing right on the line.

The Inspector didn't have to tell me I flunked, but he did. He told me that I could come back in 90 days and take the flight test without having to retake the written test. I told him I'd be back come hell or high water, as I looked over at Joe. As I turned away he said, "of course you know that side-slip is not permitted during the test". I was almost out of earshot and barely heard him continue, "but it was a dandy". Coming from a man of very few words, it could have been considered praise, but I felt awful and had to stand there and watch Frank pass the test.

Frank and John filled the gas tank, took off and headed for Madison. I took pictures of Bowman Field, and then Joe and I climbed in my Cheve and headed home. The only comment he made during the trip was, "what in the world happened up there?" The only thing I could think to say was "I just didn't know," and I don't. We both were silent the rest of the trip. At least he didn't blow his stack.

Frank Peternell & John Yellitz going to the "Prep" to fly to Madison from Louisville – May 18, 1932

I dropped Joe off at the airport. Frank and John were there so I drove downtown. I didn't feel like eating supper, or facing the fellers, but decided to take the bull by the horns. So I walked in the dining room with a chip on my shoulder, but was surprised when all

the fellers actually tried to cheer me up. They talked me into going with them to the Carnival down on the river front. But it didn't do any good; my mind kept trying to figure out how and why I messed up that third landing right after making five swell ones.

I finally hit the sack but had a terrible time going to sleep. I am ashamed, disappointed, unhappy, disgusted, and I feel awful. If it were possible I would kick myself in the behind. Tomorrow morning I am going to the office on East Second St. and talk with them about that job no one seems to know anything about.

I sure ain't fit as a fiddle or ready for love tonight.

May 19, 1932, Thursday

The weather was clear, warm, and the wind was fair.

My world came to a screeching halt this morning. I ate breakfast with the fellers and instead of rushing out to the airport, I went down to the office at 216 East Second St. hoping to catch Joe before he went out to the hanger. They were all there, the office girls, Robert R. Foster, V.P., and Joe. I asked him if they had ever found the Popular Mechanic advertisement regarding the job guarantee, or any evidence of such a guarantee. I reminded him that I had talked to him about it, and he told me he never heard of such an advertisement. I told them I had never heard of the Dixie Flying Service until I accidentally read the advertisement in the Popular Mechanic promising a job to the first student who signed up the Limited Commercial Pilot's Course. I answered the advertisement right away and Richard Oglesby sent me a pamphlet describing the school

I said my mother and I met Mr. Oglesby in his office, I signed the LCP Course contract, paid the $398.00 tuition fee, and gave him the advertisement I cut out of the magazine. He assured me I would get the job. Incidentally, when we saw the airport and planes behind the Sachleben farmhouse, I was so disappointed, especially since I saw the airport and school at Cleveland, Ohio, that we went back down to this office. I told Mr. Oglesby I wanted my tuition, but he talked me into staying, and said I would have a job and prosper as the school prospered. The office girls said they searched through all the office records and could find nothing. One of the girls wondered whether the information could have been among Richard Oglesby's files they sent to his estate. Right off the bat, Mr. Foster said he was not involved; that he has been primarily wrapped up in the Dayton School of Aviation Correspondence Course.

I asked Joe whether it would make any difference if I could find the magazine and advertisement. He said he didn't think so because there was not enough work to keep even him busy, and Edward D. (Bud) Guild, a Transport Pilot, as you know, had been laid off. Harvey doesn't have enough work to keep busy, but the law says the school has to have a licensed A. & B. mechanic. I said I had been treated very poorly. I spent more time than all the other students combined, working on things that had nothing to do with aviation. But I was glad to do anything because I thought I had a job to look forward to. I never got my share of regular flying time, and it took me a lot longer to reach 50 solo hours than Frank and John. I think that, and the controversy over the promised job, might have played a part in my flunking the flying test.

Finally I told Joe I wasn't blaming him, that I liked him a lot, and always thought he reminded me of my oldest brother. I wondered if there was some way I could get the flying time to take the test in 90 days. I hemmed and hawed, like maybe, perhaps, accidentally, possibly, probably, hopefully, perchance the school might want to give me one of the cheap planes we rebuilt. He said the school couldn't afford to do anything like that, but he would talk to the other officers and try to work something out. He would talk to me in the morning. I asked if it would be ok if I talked to Mr. Harry (his Dad) and Mr. Naill, and he said it would be ok with him.

I went down to Naill's Lumber Yard and told Mr. Naill that I was a student at the Dixie Flying Service. I said that I had just talked with Joe V. at the office, and then I told him the whole story. He seemed concerned but said he don't know anything about the operation of the school. He seemed real nice.

Then I went up to Vail's Furniture Store and I didn't have to introduce myself as I had met Mr. Harry several times. I told him that I had just talked with Mr. Naill about the conversation I had with Joe at the office. He also seemed concerned but said he didn't take an active part in the operation of the school. I thanked him for listening.

Frank and John went home today. Before they left we talked about all the swell times we had, and wished each other lots of luck. And, let's never forget, when in trouble, get the nose down and fly right on. I promised Frank I would look him up sometime, as he doesn't live that far from me. He lives at Meadowlands, PA, which is due south of Pittsburgh and near Washington, PA.

I went out to the Sachleben's for supper and it was a fine typical farm meal. After supper while Ma was putting the leftovers away and washing the dishes, Pop, Carl, Joe and I sat around the table shooting the breeze.

Fred and his gal Lizzie Dierkes were downtown "sparkin". They couldn't believe I flunked the test. Joe and Carl said they would bet that of all the students, I would be the one sure to pass, and had planned to celebrate. Pop was surprised and wanted to know what I was going to do. I told them the whole story and that I had spent the entire day talking to Joe and the office staff. And that, I talked to Mr. Naill and Mr. Harry to try to find a way to retake the flying test. I said Joe was going to talk with me in the morning.

Ma said she was sorry to hear the bad news but was sure everything would turn out all right.

Pop filled his corncob pipe with "long green". I rolled a cigarette with the same tobacco and we spent the rest of the evening playing their brand of Euchre. When we quit playing cards, we wet our whistle with a nice cool tin cup of good old "home-brew". They keep it, milk, butter and other stuff cool in the springhouse. We ended the evening with a tiny glass of Pop's special beer wine.

I think that had been the hardest day's work I'd ever done, and my brain and jaws were worn out.

May 20, 1932 – Friday

The weather was cool, fair, and wind was rough.

I crawled out of the sack, went downstairs and had the first quiet breakfast I can remember, as the fellers had left for the airport. Inez's little sister, Opal, was there, and I think she goes to grade school. That is a strange name, but there are a lot of strange names around here. Mrs. Barber wanted to know what my plans were in regard to my room. I told her that I was meeting with Joe shortly and would let her know.

It had only been two days but I had a strange dismal feeling of being left out; that they would go on as usual without me and as the old saying goes "out of sight, out of mind". I sat opposite Joe as he leaned back in his office chair and put his feet up on the desk. He is usually all business, no monkeyshines, although on rare occasions during evening school he will cut up. But this morning he seemed friendly and relaxed as he said "Friday, I talked with all the officers and after a long discussion, we all agreed we will do everything we possibly can to help you."

I thought this sounds encouraging. "But as I said yesterday, we simply can't give you a plane or a job." My ticker dropped down to my boots. "If it was possible, you would be the first person I would hire, but surely you can see, we are not doing very good right now." I never said a word, but I had noticed the school spinning down since Richard Oglesby died.

"We are going to guarantee you will get your Limited Commercial License. If you fail in August, we are betting you won't, we will keep trying until you do. Well Friday, what do you think?" My ticker came up to my hips and I said "I realize you can't give me a job if there is no job to give so what are the requirements?" "They will be the same as they are now, and you will pay your room and board as usual. You will get all the flying time necessary to pass the test and we will pay for your physical exam. When not flying, you shall help Harvey K and also work around the airport."

"We think you should get away for awhile and visit your folks. You are to report back at the airport on July 30, 1932 ready to fly. I gotta get back in the saddle, Dan Booth is out there dying to get in the crate and screw up. The secretary will type up this agreement and give you a copy; then both of us will be sure to have it this time. I know you want to talk to the fellers and your friends before you leave, so Friday, believe it or not I'll miss you. And you better not be late reporting in July 30th." He shook my hand as though he meant it and my ticker returned to normal. He left and his cute secretary Gladys Gourley gave me a typed copy of the agreement.

I went out and said "so long" to each student, and said that I would be back by July 30th. It was kinda hard cause I know several of the students will be gone by the time I come back and I probably will never see them again. Harvey said I could count on him being here when I return. Of course there is a good possibility we will meet again in that magnificent airport in the heavens. George Hayden, Waiter Etchison, and Clair Kirsch signed my autograph book. All the other students including Harvey and Cleone have signed it.

I walked across the airport to the Sachleben farmhouse. I didn't go into the details but told them the school wants me to go home for a visit, and then come back on July 30th so I can get prepared to take the flying test again. They seemed really glad to hear the good news; I asked them to tell Fred S, he works downtown at Bob Greene's Trucking Co. (That's where his gal Lizzie works).

I drove downtown to Barbers Room & Boarding House and told Mrs. Barber the same story I had just told the Sachlebens. I asked her to save my room for me beginning July 29th. She said she would, and in the meantime will change it back into a broom closet.

I spent the rest of the day saying "so long" to Penn Maddox and Jim, Bob Jones, Clyde Weber, Ed Becht, Bob Shucks, the Ricketts, Mr. Naill, Mr. Harry V, Robert Foster and the office stag and many others. I sure hated to tell Cleone I was leaving tonight but I would be back in July; she said she would write to me. I can just picture in my mind, all the students lined up and knocking on her door. One thing for sure, they are going to miss my Cheve.

I checked the Cheve from stem to stem, and filled her with gas, oil, and water. I ate an extra helping at suppertime so I wouldn't have to stop on the way home. You can't find a place to eat in the middle of the night anyway. I paid my room and board for the full week - $5.50, loaded the Cheve with all my belongings and about 9:00pm with barely enough room to sit, I climbed in my car, waved good-bye, and slowly headed north on Michigan Road. At Versailles I headed east on S.R. 50 and arrived in Cincinnati, OH about midnight. I filled the tank with gas in Norwood and continued east on Hwy. 22 through Wilmington, Washington Court House, Circleville and Lancaster.

I had all the roads to myself, and all the sidewalks were rolled up in these towns. I have to figure out the best way to tell Mom and Dad my situation without lying. I bought gas at the only station I found open, when I drove in Zanesville, then continued east on Hwy 22 through Cambridge. At Steubenville I crossed the Ohio River on the toll bridge, 25c, into West Virginia at Weirton, and 13 miles later I was in Pennsylvania. A short time later I crossed the Ohio River at Pittsburgh, and since it was Saturday morning, there was little traffic, so it didn't take me long to park the Cheve behind the bakery at 9:00 a.m. (15 gal. gas and 3 qts oil $3.45).

May 21, 1932 – Saturday

The weather was warm, cloudy, and the wind was calm.

I opened the back door; Dad seemed kinda surprised to see me as I said "Hi" and shook his hand. I went right back out and unloaded the Cheve cause they will steal everything that ain't nailed down, before the car even cools off. By the time I finished, Dad was ready to go upstairs.

We went up and when Mom saw me she sure was surprised. I didn't write to them cause I didn't know how it would turn out. I gave her a hug and as they were about to sit down to eat I grabbed a plate and sat down with them. It had been quite a while since I ate at home and it sure tasted great. Mom was curious as to why I didn't let them know I was coming home. Between bites I started by telling them the first time I asked Joe about the job. Mom reminded me not to talk with my mouth full. I was expecting that, so I decided to concentrate on eating, and continue after Mom was finished with the dishes.

When I told them Joe said he had never heard of the advertisement that guaranteed a job, Mom wanted to know why Mr. Oglesby didn't straighten it out and give Joe the clipping we gave him. I said Oglesby had died and the school couldn't find any evidence of such a deal, and even if we could find the magazine with the advertisement, Joe said it wouldn't make any difference. The school had been going down the tube ever since Oglesby died, and no one seemed able to fill his shoes. If they didn't make some radical changes soon, the school was going to be long gone. They laid "Bud" Guild, the other instructor and a transport pilot, off. Harvey, the *A&B* mechanic didn't have enough work to keep busy and neither did Joe. I know they can't hire anyone now, but Joe said I would be the one they would hire if it were possible. Mom said, "Mr. Oglesby convinced us to stay, so he had to be a fine promoter, and I was sure the school would be a big success. Why don't they hire someone like him, and what are you going to do now?"

Dad got up and said he was tired after working all night. I seconded the motion as I had been up since 7:00 a.m. yesterday. I went upstairs and went to sleep in a full size bed for the first time in quite a spell and it sure felt great. I was asleep before my head hit the pillow. I finally woke up and heard Mom and Dad talking so I went downstairs.

They were getting supper ready, as Dad didn't do any baking on Saturday night. Ralph came in just in time to sit down at the table and we passed the time of day. He told me he was about to graduate from the Industrial High school next month as a mechanic and Dad said he was a good one too.

Ralph took off right after supper and I dried the dishes for Mom while Dad read the paper. After Mom finished tidying up the kitchen and I finished reading the funnies, we went in the dining room. I started things off by saying that I had really messed up and that was the reason I came home. Dad asked, "what do you mean, you messed up?" I said, "don't ask me why or how cause I don't know, but I flunked the L.C. flying test which I was expected to pass easily. The fellers all thought the written test would be my downfall cause I didn't have any previous schooling on aviation." They both looked at me as though I had done something terrible, and I thought "Katie Bar The Door".

I told them that I had talked to all the officers who told me they didn't take an active part in the operation of the school. But Joe told me he had a meeting with the officers and they decided to, at least, guarantee that I would get a Limited Commercial license. He gave me a copy of the agreement and said I was to report back at the airport on July 30th ready to fly. If you flunk a test, you have to wait 90 days before you take another one.

I gave the copy to Dad, and after reading it he handed it to Mom. Dad said he always believed that everything turns out for the best, and at least, I would be getting some extra free flying time and experience. Mom agreed, but the room and board, etc. had to be paid.

Dad said he understood the problem the school was having as he was facing the same situation. After all those successful years, they were now in serious financial trouble. The bank where they had a savings and checking account closed, and they were using up the money they had in the B&L. I promised to try to get part time work when I went back, but there were no jobs there, and if one did turn up, they favored the "locals". I told them I could get by on $10.00 a week.

It was time for Lowell Thomas and the news on the radio. Following the news, we listened to "Amos and Andy" and they sure are funny. Mom said Earl was out riding with his girlfriend. I finally said "good night", picked up a book to read and went upstairs to bed. I'll talk to Dad tomorrow about the China deal.

May 22, 1932 – Sunday

The weather was warm, partly cloudy, and the wind was brisk.

I rolled out of the sack this morning and was glad to be awake. I had a real bad dream last night; that I was flying and couldn't get the ship down close enough to land. I went downstairs to the kitchen, Mom, Dad, and Earl were eating breakfast so I sat down and ate a bowl of "rolled oats", a donut, and drank a cup of coffee. We were not much for "chit-chatting" when we first wake up so I just said "Hi" to Earl between bites. Mom went to the foot of the stairs and called Ralph. He never answered so she went up, and then he finally came down. He sure hated to get up and was in a bad mood.

We finally were all ready to go to Sunday school and church. I had the honor of getting the Marmon out of the garage and driving them there. That car was the "cats-meow" and made mine look and act like a "tin-lizzie". Of course there was about $3400.00 difference in price.

Ralph and I went in our classroom, everyone was already there including Herbie Schwarz, Don and Jack Bicht, Earl Seitz, Helen Milliron (the preacher's daughter), and Eleanor Yeakel. Mr. W. Haller taught the short Sunday school lesson, after which in talking about the flying school, I forgot to tell them that I flunked. I just said that I would be going back to continue my training in a few weeks. As we were going in the sanctuary, I saw Paul, his wife Caroline, Charles, his wife Gladys and we all sat in our pew. After the service a lot the members came by and said "hello".

We went home and had a mighty fine pot roast dinner and a day old peach pie for desert. We never ate fresh baked goods; it's for the customers. Mom, Dad, Earl and I jumped in the Marmon after Dad glanced through the Sunday paper, and Earl and I read the funnies. We headed north for Bakerstown and visited Mom's relatives. On the way home, I eased the Marmon up to 70 MPH and Dad quietly reminded me that we didn't have wings on this car. We had learned to take his quiet suggestions, cause the next suggestion is given in German English and then you have a problem.

We had a late, left over supper, then went in the dining room and listened to the radio. Dad went to bed as he starts baking sometime during the night so I'll talk to them tomorrow about China. Earl works at a bank in E. Liberty and it wasn't long until he went up to bed. Ralph was still messing around somewhere. I told Mom "good night and don't let the bedbugs bite", went upstairs and hit the sack. She almost threw a book at me for making that crack. I read for a little while then went to sleep thinking about the fellers and the airport. I hoped I wasn't missing anything exciting.

May 23, 1932 – Monday through May 28, 1932 – Saturday

I overslept Monday morning and had to fix my breakfast; Mom was taking care of the store until Anna showed up. I ate two donuts and drank a cup of coffee. Then went out on the back porch and quickly smoked a cigarette. I was half awake by this time so I went down in the shop to see

Charles before he left. They were just about ready to quit for the day. I spent the morning getting the shop ready to begin baking tonight.

After dinner I got the old bike out, pumped air in the tires, checked the horn and bell to be sure they worked, that's the law, and went for a ride. I pedaled southeast out Liberty Ave. to Baum Blvd., went northeast to Penn Ave., rode around Peabody High School, my alma mater, then went west to Main St., and home. It was quite a trip; it had been some time since I rode a bike.

Don't ever, walk or ride a bike downtown on Penn or Liberty Ave. I had that painful experience when I was about 10 years old. Although I went to school from Kindergarten through High School with colored kids, and a few belonged to our church, it wasn't a good idea to walk or ride a bike in colored town.

I washed, inside and out, and waxed, the Marmon. It was an all day job, but it sure looked great. Dad had a Masonic emblem about the size of a saucer and I talked him into letting me put it on the front of the Marmon. He wasn't sure that it was a good idea as our neighborhood was about 80% Catholic and it might hurt our business.

I washed and waxed Charles' Whippit on Friday. As a senior in high school, our class collected cans of food for the poor people. I volunteered to deliver some of the food and asked Charles if I could use his new Whippit. *And* of course, after making my first delivery, I backed into a telephone pole, and made a big dent on the back fender. That ruined the rest of the day as I wondered how I was going to tell him and what he would say. I was shocked when he told me that it was ok, he would get it repaired soon.

I went across the alley and visited Tony Fellow and his family at their Laundry and Dry Cleaning Store. We chitchatted for quite a spell then Tony took me down the cellar and we sampled several different kinds of wine. Saturday evening Dad & I took off up Main St. and as soon as we turned right on Penn Ave., Dad lit his "stogie" and I lit my pipe. I'm sure Mom knew he smoked cigars; it leaves a strong odor clear across the room. But I guess she figured as long

as they played this game, at least there would be no smoking in the house or in her presence. We walked along "Gasoline Alley" as usual, and as we walked past we saw the Marmon "16" sitting in the showroom.

Dad helped Mom close the store; I picked up a book, went upstairs and hit the sack. I was going to start studying for the Transport Pilot's written test just in case I got a chance to accumulate enough flying time to qualify. The flight test is not a whole lot different than the L.C. test I was going to take. I was going to pass the L.C. flight test, come h--l or high water cause I was not going to make that same mistake ever again.

Sunday May 29, 1932 through Wednesday July 27, 1932

We all flew to church this morning, I piloted the Marmon, Dad was the co-pilot because he owns the car, buys the gas, and pays the bills. Mom and Dad went into the Adult Class, Earl went into his class, and Ralph and I went into ours.

No one prepared me for the new member in our class. She and the Preacher's daughter are cousins; her name is Ruth Milliron. I had been smitten with the Preacher's daughter once, with no luck, but this girl is a dream come to life. She is a "humdinger", no that doesn't do her justice, she surpasses any movie star I've ever seen. According to our high school art class, clinical, analytical diagrams of the human body, her features are well nigh perfect. We used these diagrams to draw the human form in the exact proportions. Naturally I can't vouch as far as the proportional area is beneath the clothes, but I am willing to make an educated guess. Had a real nice time.

After dinner and before anyone could leave, I said that I needed some advice. Mom, Dad, Earl., Ralph, and I went into the dining room. I brought up the China flying proposition and explained the requirements, stipulations and rewards. Ralph said, "What's the problem? That's a lot of money". Earl said, "I don't think it is worth it, it's only temporary, and there is a lot of turmoil going on over there". Mom said, "I'll agree to whatever you want to do, but I think Earl is right, their culture is different, as is their language." Ralph popped up with, "and there won't be any Ruth Millirons either". That crack almost settled the whole question.

Dad said, "I faced a somewhat similar decision when I was sixteen, but with totally different aspirations. My sister, living in Pittsburgh, offered to pay the ship's passage across the ocean for my brother William. He changed his mind and didn't want to go. I immediately volunteered to go in his stead because I wanted to take advantage of all the opportunities of a lifetime that America offered. It seems to me there has to be something brewing over there to make such an extravagant offer". I had to agree, especially when you can rent a plane here for $7.00 an hour. I added, "What if they decide to refuse to pay you at the end of the two or three year term?" Earl wanted to know about the flying part of the deal; he thought it would be for the government. You might find yourself in the middle of a war. Well, I'm not about to go around killing people that's for sure. The flying duties weren't spelled out in great detail. I finally decided to take my chances on finding a flying job here even if I have to join the Army Air Force.

Instead of the customary rite of taking a Sunday afternoon drive out in the country, Dad settled in his chair and proceeded to read the paper. Earl went out somewhere and Ralph went out someplace, to each his own. Mom turned on the radio. I asked Dad if he was finished with the funnies, and said that I wanted to take them for an airplane ride one day. Morn seemed surprised and said she hoped so. Dad said that would be fine, you have taken us flying several times in the car.

I decided to clean the bakery shop every day and have all the ingredients prepared for Dad and Charles to begin baking. I put the rod back up in the doorway I used it to do chin-ups every day from about the fifth grade through high school. Every time I walked through the shop I would pull myself up a few times. I was not the best at any sport in high school, but I won the Physical Efficiency medal of the Pittsburgh high schools, my junior year. You had to participate in every individual sport, you earned points in each event, and as I accumulated the most points I received the medal. Football, baseball and basketball were not included, they were team sports, but there was a football, baseball throw and basketball foul shooting. All running, field and gym events were part of the program that took about a month. I'll never forget, I was last to do the chin ups, I easily passed the highest number and wanted to find out how many I could do, but the referee finally said I could quit any time, and stopped counting. I'm sure the number of chin-ups helped me win the medal I still have.

Mr. W. W. Haller had started teaching the Sunday school lesson as Ralph and I arrived a tad late. All our friends were present including Ruth M. I didn't remember the lesson but I sure remember her. She is beautiful but I just as well write her off my list, she lives in a different circle than I. Now on the other hand, those gorgeous gals in Madison definitely live in my circle.

We drove out Highland Ave. to the Highland Park Zoo on Sunday afternoon. It is about a mile north of my high school. We practiced running from the school to the park, around the reservoir, and back. It took me several times to get over being self conscious about running through the upper crust neighborhood, in my gym shorts. The weather was warm, sunny, and the wind was breezy. We parked the car in the parking lot and started walking up the winding path toward the zoo. It sits on top of a large grassy knoll. Ralph looked up toward the zoo and yelled, "hey you", and began running up the grassy slope. Earl and I looked up and ran after Ralph. There was a man standing up there with his pants down around his ankles and he didn't have any underwear on. By the time Ralph got up there, it was quite a distance; the guy had disappeared in the crowd.

Ralph was still mad when Mom and Dad joined us at the entrance into the zoo because he didn't catch the guy. He told Dad he wanted to "cold cock" that "freak", but Dad told him to forget it. He said, "don't you remember what happened to Charles that Halloween night when he knocked that man down for stealing his wallet"? He was arrested, fined and sued because "he" happened to be a woman wearing a man's costume. The judge said Charles didn't have to knock the "man/woman" down to get his wallet back.

We walked through the zoo, all the animals, birds and fishes were beautiful, but the building that housed the big cats stunk. I don't consider reptiles beautiful, but to be honest, if you examine them artistically, which I don't, you have to admit they are. They were small cages with a concrete floor. The big cats' endless pacing back and forth with head moving from side to side was hypnotic. It was fascinating to watch their powerful muscles ripple as they paced, but the stench chased me out. We left at closing time, and Ralph grumbled all the way home. After supper Mom, Dad and I listened to the radio and read the paper. Dad and Charles have to bake tonight. Earl and Ralph disappeared.

Ralph and I fooled everyone, including ourselves, as we were the first ones in Sunday school class. All the faithful members came in, including you know who. I felt a mite sorry for the Preacher's daughter, all her admirers have abandoned her since her cousin joined the class. It's downright embarrassing the way those young turks fawned upon Ruth M., they could be a little more subtle. I'm just jealous. All the fawning doesn't seem to affect her one way or another and that's in her favor. Being a licensed pilot doesn't seem to make any difference either. Well, it sure does in Madison, In. I'm afraid the teacher is wasting his breath as long as she is in class.

Our Sunday drive took us to the Pittsburgh airport. We toured the place and it is much larger than Bowman Field, naturally. Pittsburgh is much larger than Louisville. I pointed out the planes that were the same as ours and explained what the pilots were doing or should be doing. And I observed them more carefully since that bawling out lecture Joe V. gave me and it continues to prove him right. However, I am a little disappointed to find out that some of them are quite sloppy. We left and returned home for supper.

Dad and his cigar, me and my pipe, continue to take our Saturday evening walk out to gasoline alley.

Ralph and I sat back in our Sunday school chairs and watched those young dudes currying her (you know who) favor without success. Ralph is not a bit interested in the outcome. Mom would be surprised at how much Bible teaching we ain't hearing. I'll bet Dad wouldn't be surprised, he ain't blind.

We returned to the customary Sunday drive out in the country. We drove up to Connellsville to visit some of Mom's relatives, and it is a real small quiet town. I do mean small, quiet, tranquil and very pretty. It is so small it has a weekly newspaper and the most exciting news is when we visit. One of our relatives works for the paper, she always reports in the paper that the Beyer family visited Mom's folks this past Sunday and sends us a clipping.

Every time we drive up to Connellsville I am reminded of the time we made the trip in our Oldsmobile touring car. It had to be before 1920 because I was so young I didn't realize the serious situation we were in. We got lost and were driving on an unpaved road when a car ran into us. There were six of us in our car, Dad and the other driver talked for a long time, apparently the cars were ok because the other car left and we proceeded down the lane. Smoke began to come out of the engine compartment then the engine quit. By the time we all got out of the car it burst into flame. We all stood there, Mom carrying Ralph and watched the car burn to the ground. That's all I remember but that image is just as vivid today as it was then. Probably someone saw the smoke and helped us. Dad never drove another cat. He claimed he put our lives in peril.

This is turning out to be a long summer and I am getting homesick. I got a letter from Rodney Pert, Max Sittner and Chet Mckinley. The letters were short; none of them were working in or out of the industry.

We all performed the Sunday ritual and went to church, either that or else. I personally don't mind, for one thing it helps to keep track of the rest of the week, and you might miss something interesting or learn something. Those young chaps can't take a hint, they are still acting ridiculous, and I am learning what not to do. It is kinda entertaining, but if I wanted to act like a hypocrite I might say it was interfering with the lesson.

Our Sunday drive took us across the Allegheny River via the Washington Crossing bridge and out to the Bettis airport. There is a PT-I Army Plane, French Spad, and Pirate Bomber in the hanger. Jack Morris is still there. I have visited this airport quite a few times. In a way, it is similar to the Madison airport. We looked the planes over and watched them fly for a spell, then returned home. I am grateful Joe opened my eyes. I have been watching planes fly but never really seeing the skill or expertise. I watched them land but never noticed if they were three pointers or whatever. If they took off smartly with wings level or struggled, etc. Now I not only watch or look, but I see and observe.

There is an ironclad rule in this house, on Sundays everyone goes to church. There are two ways to escape this rule, up and die, or move out, so we go to church. Curiously, Pearl, Bert, Paul and Charles are married and have their own homes, yet they still go to church.

Nothing new in Sunday school, I practiced the observation mode this morning and noticed Ruth M. doesn't use makeup, it would be dumb to conceal beauty.

Our Sunday drive took us across the Monongahela River to the south side, where we visited Dad's sister. She owns a grocery store; naturally it is closed on Sunday. Like ours, the store is part of the house. It is interesting, listening to the conversation, the English with the strong German brogue; it sounded nice. (I took German in high school).

What a surprise. Ralph and I were lounging in our chairs, the male students were ogling you know who, and out of the clear blue sky Ruth asked me if it was difficult to fly an airplane. I should have prepared for such an eventuality; instead I made a complete jackass of myself. If words were feet I would have tripped all over the place. I don't know what I said, but Ralph said it sounded like gibberish to him, and he didn't let me forget it. That Sunday school lesson is a complete blank.

Our Sunday drive took us across the Allegheny River to the north side and we finally arrived at the Rodger's airport. Bob Dake is the pilot there. I loaf there a lot and it is like the Madison airport, has grass-landing areas. He was taking quite a few sightseers up for a short spin. After watching him and Jack Morris, two transport pilots, I am trying to convince myself that I might not be so bad after all. The sun was going down so we flew home.

There are only three girls and six of us boys in our class. Since the advent of Ruth, her cousin tries to act nonchalant about the obvious distraction. Eleanor Yeakel, she is the true intellectual in the class, seems truly indifferent to this juvenile activity. I was prepared to take center stage but the opportunity never presented itself.

We fired up the Marmon and drove downtown, we drove around and visited the Fort Pitt Stockade and the point where the Alleheny and Monongahela rivers meet and form the Ohio River. It has been a long time since I have been at the point, the railroad tracks are there. It didn't take long to drive home.

I told the Sunday school class I was leaving to further my education and bid them a fond farewell. I looked Ruth M. straight in the eyes as I said that I hoped to see them all again.

I've been studying the Meteorology lesson but so far it seems difficult to retain, so I might have to memorize the entire lesson. I know the rest of the required written test to obtain a Transport License, and will have plenty of time to brush up on it. I have no doubt whatsoever about passing the L.C. test.

I answered Rodney Pert, Max Sittner and Chet McKinley's letters. I wrote a letter to Mrs. Barber and told her I would be there sometime July 29, 1932.

I tried the chin up bar quite a few times and I need to spend more time at it, those muscles when downhill.

I sure enjoyed sleeping in my nice big bed but I'll be glad to get back to my "hole in the wall".

July 28, 1932 – Thursday

The weather was partly cloudy, hot, and the wind was calm.

I worked on the Cheve all morning. I checked her from stem to stem and everything seemed to be ship-shape. Next I washed and waxed her and cleaned the inside including the windows. Then I filled the tank with gas, thanks to Dad, and Earl gave me 3 quarts of engine oil. Now all I have to do is load her up and head southwest toward Madison.

I "woofed" awhile with the old "gang" and told them I would be leaving for Madison early in the morning. They haven't been able to find jobs since high school. Tony gave me 2 bottles of grape wine and 1 bottle of anise wine.

I called Fred Moss but he wasn't home. I guess he was still going to Pitt U. to learn to be a doctor. I hope he stays with it; maybe he will give me free medical care, ha! I told his mother I was leaving for Madison early in the morning and would write to him.

I called Earl S. and he happened to be home. He sure is a go-getter, although he hasn't been able to find a steady job. He is always out trying to earn a buck. From selling car-cleaner wax at stoplights to collecting "dead beat" rent. I went with him one time to try and collect overdue rent. Not for me, I don't know how we got out of that area alive. It didn't bother him a bit, he just told them that when he came back, they better have the money or he would throw them out the door on their rear ends. I told him I was leaving for Madison about 4:00 a.m. and that I would write to him sometime.

Pearl, Bert, Paul, Charles, Earl and Ralph each gave me $1.00 and promised to help pay my living expenses. Dad gave me a handful of 'stogies", Ralph gave me 5 packs of pipe tobacco. Mom gave me $10.00

for my room and board (she was the Treasurer in the house) and a big box of cookies and cakes. Before Mom went to bed, she put my breakfast on the table.

July 29, 1932 - Friday

I was so anxious to go back to Madison I couldn't sleep so I went down in the shop and aggravated Dad and Charles while they worked. I finally gave up and went up to the kitchen, ate my breakfast, changed clothes, put all my clothes, cookies, and tobacco in the car. I put a towel around the wine and put it under the seat with the oil. I went in the shop, said so long to Charles, and shook Dad's hand as he said, "so long skipper". I went out, jumped in my "crate", fired her up and took off for Madison at "perzactly" 4:00 a.m.

As I drove through Weirton, WV, I thought of Bert who lives a few miles down river in Wellsburg, WV. I crossed the toll bridge and filled the gas tank at the same station in Steubenville, OH.

I filled the tank with gas at the same station in Lancaster, OH and added a quart of oil. The Cheve burped along on 4 cylinders as usual, eating up 40 miles each hour. The towns and countryside looked a lot different in daylight. My previous trips were at night.

I filled the tank with gas at the same station in Cincinnati, OH and added another quart of oil.

I finally arrived at Barber's Boarding House at 4:00 p.m. The trip took 12 hours, 15 gal gas, and 2 quarts of oil.

I walked in the kitchen, Mrs. Barber and Inez were cooking supper, and they seemed glad to see me. No, they didn't jump up and down nor clap their hands in glee, but they did give me a nice warm smile and said they missed me. We talked as they continued cooking the food and Inez said quite a few students finished their training and had gone home. I sure hated to hear that. Mrs. Barber said my "room" was still upstairs so I unloaded the Cheve and was glad to see the cardboard was still under the mattress.

I called Joe V and said "Friday is back in town and I will see you at the airport at the crack of dawn". His response was "Oh yeah"!

I went in the bathroom and got all "spruced up" before the fellers came in. There is only one bathroom and my room is the farthest away. It is wild, you never have the bathroom to yourself, and there is usually 3 or 4 jammed in at the same time, especially in the morning. Luckily some of the students get up real early and some are sleeping at Richert's Rooming House, around the corner on Main St. Sometimes I wonder whether I am shaving myself or one of the other students.

I went down in the parlor just as the crew came in like a cyclone, as usual. After we said, "how-de-do" to each other, we made a mad dash to the dining room. One thing never changed, everyone snatching and grabbing the food at the same time. But I noticed one difference; I had a strange feeling, that this time I was the new student. The conversation only concerned what had happened during the day, and I felt kinda like an intruder.

Geez, I was only gone a few weeks, but I guess that old saying "out of sight, out of mind" could be true. Harvey did ask me if I enjoyed my "vacation", but that was it, he never "chit-chats" or starts a conversation. To this day I don't know him personally although we have worked and lived together since the school started. Surprise, I didn't have to hide my empty pie plate, Inez slipped me a second piece of apple pie.

I joined the fellers in the parlor and Art P told me Lynn Wright, Bob Parker, Marlin Christianson, John Hoover, Pete Nematsky, Dave Swett, Sam Poletunow, George Dystra and Fred Works completed their training and had gone home. I sure hated to hear that as we were kinda like brothers. We slept and ate in the same house, (most of us), flew and worked together every day. Of course, some were better and some were worse, but you don't expect a brother to borrow your car and try to steal your girlfriend. My brothers at home would never do that.

I went up to my "room" early and it was hot, there was no window or air circulation. I'll see if Mrs. Barber can get me a fan tomorrow. I hit the sack, as tomorrow was the big day.

July 30, 1932 – Saturday

The weather was partly cloudy, hot and the wind was calm.

I got up early and left the bathroom just in time, cause here they came. I went down in the dining room, ate my breakfast, climbed in my Cheve and flew out to the airport, at 40 mph, that is. When Jack, the pup saw me he came running, jumped up on me and darn near knocked me down. He is one big dog. Howard B, he stayed at the hanger overnight, told me everything was just the same since I left and it had been boring. By this time the rest of the fellers showed up and Howard climbed in the truck, went downtown, and ate breakfast at Barber's. I was so glad to see my old friends again, (the ships) and I patted each one several times on the back and said I was sure glad to see them.

Harvey opened the hanger and we pulled both "Preps" out on the apron. Just then Joe drove up. He got out of his car and looked like something the cat dragged in. He evidently does a lot of "cattin" around at night. They say he is quite a jokester, and is the life of the party.

He saw me and mumbled, "Friday let's go in the office". We drank a cup of coffee; he brightened up a bit, and said he was glad to see me. He asked about my "vacation", and said things had been awful quiet and monotonous around here. He hoped it would stay that way and that I would disband the "Grounded Club". I told him that I'd never forget that "lecture" he gave me, and that he was absolutely right. I visited several small airports at home and watched several pilots flying that proved his points. And now I have a different perspective toward flying, and the only thing on my mind is to pass that test. "Whew" he said, go out there and crank up the "110", you have 35 minutes." As I left he said "Friday, we really did miss you". I didn't know exactly how he meant that.

I rushed over to Harvey and said I was to take the "110" up for 35 minutes. He wanted to know if Joe was going to check time me. I said he never mentioned it and I wasn't going to take the time to find out. Harv told me to, at least, take the time to do the preflight inspection before I took off. I quickly inspected the ship from stem to stern, checked the oil, filled her up with gas and had Harv man the cockpit. I cranked the prop and the engine coughed, belched, snorted, and hiccupped till I thought the ship would shake apart. Just as Harv was about to shut off the ignition switch, the engine began to run smooth. He climbed out and thought we should check the engine before taking off. I suggested we check it when I got back. I climbed in the cockpit, fastened my seat belt and gave him the thumbs up. He pulled the chocks away and waved me off. I opened the throttle and as we rolled down the runway, I let out a "war-whoop" as I told "her" how happy I was to be sitting here. I leveled off, banked right and made the figure eights around the round barn and the hanger. Went up and completed 3 nice spins to the left and right, and stalled the ship without any problem. I made 3 decent verticals to the right and left before I had time to think about it. The 180-degree turn and land was normal, and the two 360-degree turns and lands were pretty darn close to the line. I chuckled as I taxied up to the hanger because all the maneuvers were great and I knew Joe would be watching me. I told Harv the engine sounded ok.

While the other students were taking flying lessons, I shot the breeze with Harvey and Art P for a spell. I then walked across the field and visited the Sachlebens for a short time. It being Saturday, as soon as the last student parked the ship, we closed the airport and went downtown for supper. I walked down Main St. tonight and met quite a few friends. Bud Lawrence stayed at the hanger tonight.

July 31, 1932 – Sunday

The weather was partly cloudy, hot and the wind was pretty calm.

I ate breakfast, went back upstairs, washed, shaved, and even shined my shoes, you never know who you might run into. Satisfied that I was presentable, I went to Sunday school and church with the Sachlebens. Jessie and Mildred Ricketts were real friendly and introduced their local boyfriends. I didn't see any new "frauleins". Cleone wasn't there so I guess I'll have to go visit her out on N. Walnut St., that is, if I can catch her at home. I'm glad she doesn't live at the west end of town. Everyone acted as though I had never left. They are nice people and it is a nice church.

I drove out to the airport after dinner to help Joe as he gave visitors airplane rides. There weren't very many sight-seers just then so Joe said that since I was there I might as well fire up the "110", and take her up for 45 minutes and 4 flights. It didn't take me long to inspect the ship, fill her up with gas, and as Art was just standing around I asked him to man the cockpit. I cranked the prop and after the engine warmed up, Art pulled the wheel chocks away. As Joe had taken off with a couple of passengers, I had to be sure it was clear to take off. This was a new experience for me; usually there was only one plane in the air at the same time.

He was nowhere in sight so I took off and had to constantly scan the sky because he never takes the sight-seers very far. Hopefully, he would be looking for me too. I flew a little further north than usual to do my maneuvers as he usually flies close to the river, Hanover College, Clifty Falls, and Cragmont State Hospital. I zipped through the figure eights, stalls, nice verticals, very nice spins, even did a few extra.

I went up higher than he flew the sightseers and it was the first time I watched a ship flying below me. It was kinda strange and gave me another perspective on flying. I watched Joe make a 180-degree turn and land. I noticed, that before he crossed the high lines approaching the runway, he opened the throttle all the way because I saw the exhaust belching a lot of smoke. I would bet he undershot the runway a bit, but I'll never mention it that's for sure. He took off again and I had to quickly dive down and land, as he doesn't give very long flights. As I had to make four flights, and my 45 minutes were about up, I decided to make "touch & go" landings.

That is, as soon as the wheels touch the runway, the throttle is pushed wide open and the ship takes off. I could still see Joe so I quickly made my second 180-degree turn and "touch & go". This time I went back up and make several "wing-overs" and watched Joe land. I saw Howard B help the passengers in the front cockpit so it didn't take Joe long to take off again. I went down in a hurry and this time I made a 360-degree turn and "touch & go". I took off and quickly made my second 360-degree turn and land and taxied to the parking area before Joe came in and landed.

I helped Howard take care of Joe's passengers and helped close the airport at suppertime. I spent the evening goofing off.

August 1, 1932 – Monday

The weather was clear, hot and the wind was quiet. I didn't get to fly today, flying two days in a row is about all I could expect. It was ideal flying conditions so the "fair weather pilots to be" were scheduled to fly.

It is certainly different now, there are only ten students left including me. Art Prosek, Howard Brennan, Bud Lawrence, Dan Booth, George Hayden, Clair "Bus" Kirsch, Walter Etchison, Howard Price, Dick Boepple. That Walter is something else, he always comes out to the airport wearing a necktie, rain or shine. I am glad Art is still here, we are pretty good pals even though I accidentally busted his lip while boxing (he could have wiped up the floor with me). He always helped me work on my Cheve although he very seldom borrowed it. (He hasn't found a steady girlfriend yet.) He is my idea of a typical, good, trustworthy friend, and a mighty good bodyguard. The local "tuffs" won't tackle him; I can vouch for that.

Joe showed up just as Art finished the pre-flight test on the "Prep". After talking to Harv he told Art to take off.

It is comical to see Art trying to get in the "Prep" as he is tall and lanky. By the time he twisted his long arms and legs into the cockpit around the "stick", I thought he was trying to imitate an octopus. I almost busted a gut laughing. I watched as he went down the runway, he made a nice smooth take off.

It took a little time to soak in as I took off last Saturday that the trees on the west end of the runway were gone. So now we don't have to worry about trees at either end, just the high lines on the east end.

Joe came over and told me to take the landing gear and tail skid off each ship, one at a time. Take them completely apart and show Harv any part I thought was defective and needed to be replaced. He will make the decisions and will check on you occasionally; he said he trusts your workmanship. And he will tell you which ship to work on. We want all the ships in tip-top shape so they can be licensed. We are going to make a vigorous effort to sell quite a few ships. Harv will spend as much time as possible teaching the students the principles of aviation.

I went over to Harv and asked him what was "cookin". He just repeated what Joe had told me, and said Joe will stick to flight instruction. He said

I could start working on the other "Prep", and to look him up whenever I needed help. As we got the ship in position, and raised the front end up, he slyly said, with a straight face, "this project should keep your mind and muscles busy, and out of trouble". We put wood horses under the fuselage to support it as we removed the landing gear. I acted insulted and told him he should feel sorry for me as my heart was broken cause I flunked the test. All he said was "join the club" and walked away. Just then Art landed, taxied up to the hanger apron, un-wound his arms and legs, and struggled out of the cockpit after he hit his baldhead on the doorframe. I had to laugh. I didn't get a chance to work on the landing gear, as it was time to close the airport for the day.

I spent the evening "shooting the breeze" with the fellers down at Maddox garage.   Mr. Maddox caught me again. He said, "Friday how are you"? I said I was pretty good, and as usual he said, "you might be good, but damned if you're pretty". Every time I see him he always asks me the same question, and regardless how hard I try, I automatically say the same thing.

When I went upstairs, I saw a nice big fan in the doorway of my "broom-closet". At least it will circulate the hot air. It is hot.

August 2, 1932 - Tuesday

The weather was stormy, rain and muggy, the wind was very gusty. No flying today.

We opened the airport as usual. Harvey told Howard B to help me work on the landing gear. The rest of the students went in the workshop with him. We took the landing gear assembly off the "Prep" and took it completely apart.   We washed everything with gas and examined thoroughly each strut, absorber, wheels, bronze bearings, axles, nuts, bolts, and the connectors on the fuselage. The wheel bearings and axles appeared to be in good condition; so did the absorbers. The tires looked ok, but the air pressure was low. Harv came out and after testing the absorbers said everything was ok. We painted the assembly and will put it back together when it dries.

Howard and I joined the other students in the workshop. While I was gone, Harvey made a replica of an airplane wing, and obtained a radial air-cooled engine, to use in teaching the students. Today's lesson was engines, and I had been through all that so I daydreamed a bit, then listened awhile to catch him making a "boo-boo". I even dozed off, almost fell off the chair, and interrupted the discussion. When the discussion resumed, I mentioned that the school had a lot of material, pictures, and information on the Packard diesel air-cooled radial engine. And as I don't know a darn thing about diesel engines I would like to attend a class devoted to it.

We closed the airport for the day and I jumped in the Cheve. But before I could start it, Harvey leaned in the window and said, "I wish you would do your sleeping in bed". I saw he was slightly aggravated so I said "Harv, old pal, you are a mighty fine teacher". I think a lot of Harvey, but I don't think I will ever know what he thinks of me.

I drove across the field to the Sachlebens before he could answer that. I invited myself to supper and spent the evening playing cards. So far, I have resisted the temptation to join Joe, Carl, and Fred in arguing over the games. It seems to have become a ritual to end the evening with a tin cup of "home-brew" and a tad of "beer-wine"; just enough to wet your whistle. I think that's nice. Joe S. told me "Pop" breaks out the liquid refreshments only when they have visitors.

I drove downtown to the boarding house, climbed in bed and slept like a baby.

August 3, 1932 – Wednesday

The weather was partly cloudy, quite hot, and the wind was quiet. Well, I flew for one hour today.

Joe showed up bright and early for a change. As a matter of fact, he was quite chipper. From what I've read about the army, he usually acts more like a drill sergeant than a flight instructor. He came over just as Bud Lawrence and I were starting to put the landing gear back together. The paint job looked great. He told me that I was scheduled to fly one hour this afternoon, provided we got the "Prep" in flying condition.

Bud and I didn't waste any time, but we didn't take any shortcuts either, in putting the landing gear together and installing it on the "Prep". We used new bolts and nuts and cotter keyed them so they wouldn't come loose. We put grease on the axles, put the wheels on, and put air in the tires and I told Harv the crate was ready to fly, except I thought the pad on the tailskid should be replaced. As it was dinnertime, Harv ok'd the landing gear and said we would put a new pad on after I flew this afternoon. During dinner I looked around and noticed the Hanover Hill workers were not at the table. I guess they have left town; I hope.

After dinner we removed the wood supports and lowered the "Prep" down on its landing gear. Harv, Bud and I pushed it out on the apron and I gave it a good preflight inspection. I was to make six flights so I took off and zipped through all the maneuvers. The steep and gentle figure eights were normal, the stalls ok, the right and left vertical power turns were passable, the sideslips were automatic, and the climbing turns were fine. The spins were just what the doctor ordered, went in and out in the same direction. Not too bad after several weeks layoff. I spiraled down from 2,000 feet twice with engine throttled, and landed within 50 feet of the line. I made the two 180-degree turns and land maneuvers and landed within 50 feet of the line. I made the two 360-degree turns and land maneuvers and landed within 50 feet of the line. All of the landings were tolerable.

I taxied up to the apron, turned off the ignition switch and asked Harv for help in lifting the tail so I could put a wood saw horse under the fuselage. He sent Dan B and George H out, they lifted the tail and I slid the sawhorse under it. I took the tailskid pad off; it was practically worn out, and replaced it with a new one.

I spent the rest of the afternoon watching the students practice the maneuvers, take offs and landings. They have improved a lot, but there was an occasional "ground loop" and "leap frog". Those poor "Preps".

Instead of eating supper at "Barber's", I drove out Walnut St. to see Cleone. I thought it would be the best time to catch her at home. They were just about to sit down at the supper table and they invited me to join them. Cleone introduced me to her mother, aunt, and little sister. I was a little uncomfortable at first as I had never met any of them, but they were very friendly and made me feel right at home'. I had a nice time although I didn't get a chance to pitch a little "woo".

I drove back to "Barber's" and got there before "curfew". I went to sleep thinking maybe flunking that test might not be all that bad, but it still bothers my pride.

August 4, 1932 –Thursday

The weather was cloudy, pretty hot, and the wind was calm. I flew for 45 minutes today. I wonder if they are anxious to get rid of me.

I was scheduled to fly first this morning and make four flights. After opening up the airport and lining up the ships, I warmed up the "Prep" engine. As I started to roll down the runway, I caught a glimpse of "Jack", our dog starting to chase me. He never chased the cars out on the highway and this was the first time I saw him chasing a plane. Thankfully he didn't get in front of me or we would have had a lot of dog meat flying around. As I flew back over the field, I saw several students chasing him, so he was ok, at this time.

All the maneuvers went as planned, even the "vertical" power turns. I wonder what would really happen if I made "real" verticals. After the test, and if I get a chance, I am going to make one, and see what happens.

I spent the majority of my time on spins and they were ok. I made two nice 180 degree and two nice 360 degree turn and land maneuvers, and each time set the ship down close to the line, in three point configuration. One never knows where the Inspector is going to place the line, but I'll let him worry about that. I parked the ship and left the engine idling. The engine seemed to run ok but I had a feeling it wasn't quite up to par, it seemed to have a "lisp" once in a while. I asked Harv to come out and check it, he thought it sounded pretty good but we decided to check the spark plugs and they were pretty well carboned up. I installed new spark plugs and it solved the problem. When my car's spark plugs got fouled with carbon, I cleaned them with spark plug equipment. It used a combination of sand and air to blast the carbon off the points. Usually, the only time car spark plugs are replaced is if the porcelain is cracked or the points are badly burned.

They tied "Jack" up and decided that he would have to be kept tied while the ships were flying.

It looks like I've graduated at the school of "Gravel and Cinder Hauling College". "Bus" Kirsch hauled creek gravel and cinders all day, and Walter helped spread the stuff. Yes, he wore a necktie. Otherwise things are back to normal; once again I am the "old man" of the school. As I didn't want to disturb Harvey's class again, "Jack" and I watched Howard and Dick try to wreck the "Prep". One of these days they probably will. "Jack" tried his darndest to get loose and chase them.

After dinner, Harv and Walter helped me hoist up the front end of the other "Prep" and put wood supports under the fuselage. Walter and I took the landing gear assembly off the "Prep", completely disassembled it and after examining it thoroughly, had Harv ok it. We cleaned the assembly and gave it a coat of paint. Walter has to do something with that necktie; it is beginning to bug me. I don't know how he keeps from getting it messed up and work at the same time, but somehow he manages to do both.

I went down to Ed Becht's house and spent the evening trying to learn how to play the Hawaiian guitar, but am not making any progress. I think it makes better sounding music than the other type. Bob Shucks stopped by and got in the act.

August 5, 1932 – Friday

The weather was partly cloudy, quite hot, and the wind was still. Yep, they want to get rid of me for sure, this is the first time that I flew three days consecutively. I flew six flights for one hour and 15 minutes today.

Walter and I were just starting to assemble the "Prep" landing gear when Joe showed up and said I was scheduled to fly first after dinner. He wanted me to perform all the maneuvers over the airport so he could watch me as though he was the Inspector. We assembled the landing gear, attached it to the fuselage and Harv's official opinion was "it will pass", so we lowered the ship down on it. The paint job looked pretty good but it won't last long after it rains.

I installed a new tailskid pad after I finished flying. As it wasn't quite time for dinner, I slipped quietly into the workshop so I wouldn't disturb Harvey's class, but those jokers were just goofing off, so I joined the fun.

Well, I found out at noon, it's official; those road workers are gone. I don't know why they got so mad, I was just trying to be friendly, and no one got hurt. I'll admit they could have.

I checked and warmed up the ship, then told Joe I was ready to go up. He came out on the runway and I took off. I tried to imagine Joe as the Inspector, and that was easy as Joe has a short fuse. I started from 1,500 feet with my engine throttled, made a 360 degree turn, and a 3 point landing about 50 feet from the line. I took off, and then from 1,000 feet I idled the engine, made a 180-degree turn, and a 3-point landing almost on the line. As I took off, I hoped Joe watched those landings because they were pretty smooth. I made 5 gentle figure 8's at 800 feet and 3 steep figure 8's at 1000 feet. Then from 2,000 feet I idled the engine and spiraled down right, and made a 3-point landing slightly in front of the line. This was my lucky day; all 3 landings were pretty darn nice. I don't quibble whether the landings are 1 inch or 2 inches above the ground any more.

Normally, this is when the Examiner climbs in the ship, but I was surprised when Joe motioned that he wanted to climb aboard. Before I started down the runway he told me that he was not going to say anything, except to tell me what he wanted me to do. The first maneuver I was to perform was a left vertical power turn, and as I completed it, I noticed a slight frown appear. So I made sure the right vertical power turn was not as steep, and the frown disappeared. Next I made a nice recovery from a stall, and made a 3 turn spin to the left, and one to the right. I was really surprised when, as I was making a climbing turn, he suddenly throttled the engine, and said I had to land without using the throttle; and I did. As we sat there on the runway he told me the Inspector would do something unexpected, such as he just did, but not necessarily the same maneuver. He said to take off again heading south as the wind was from the west, so that I would be taking off and landing in a crosswind. I sure didn't expect this, at least the runway was dry and hard and the wind was pretty calm. But the runway was so short, and I guess Joe weighed about 160 pounds.

I taxied as close to the north edge of the runway as I could, turned the ship south, held the stick back against the seat, and pushed the throttle wide open. When the engine reached maximum rpm's, I pushed the stick slightly forward to raise the tail. As we picked up speed, I thought, if he was willing to take a chance on getting up over that wire fence and telephone wire, I sure wasn't about to chicken out. The ship picked up speed quicker than I expected and I didn't notice the crosswind. As we approached the fence I tried to coax the ship up off the runway and I knew Joe was doing his share of coaxing. Finally the ship wanted to leave the runway, so I steadily pulled the stick back and cleared the wire easily.

At 1,000 feet he told me to make a 180-degree turn and land maneuver, side-slipping in. This created a different problem, there was very little wind, and a cross wind at that, so there was less air resistance to slow the ship down. As I lined up with the runway, I decided that I had to sideslip down as soon as I crossed over the north fence, and pull the stick back until the ship stopped rolling; which I did. Joe then told me I had just enough time to take off and make a normal landing into the wind. We took off heading west, made a 180 degree turn, and as I flew over the high lines, for just a second, I thought I would have to side-slip in, but was able to set her down within the line limits.

As we taxied up to the hanger, Joe suddenly said "Beyer", my ears perked up and I thought uh oh, he continued with a suspicion of a grin, "Friday, you passed the test today, and did very well. I don't think you will have any trouble with Joe Shumate unless you just plain flat out screw up. By the way, that side-slip, was as good as they come". I let out a real big sigh of relief. We parked the ship. That one hour and 15 minute test left me with a mental perception that I had just finished a hard half-day's work.

It was a nice change of pace to go back to work on the "Prep".

I spent the evening playing "checkers" with William B, Bud L, and Howard B and won more than I lost. Louis B and Don V borrowed the Cheve to double date.

August 6, 1932 – Saturday

The weather was cloudy, pretty hot, and the wind was breezy. I flew for forty minutes and three flights today. I was so surprised yesterday when Joe complimented me on my flying that I never said a word, didn't even thank him.

It being Saturday, there was no school, or work, just those students scheduled to fly, and I was one of them. Harvey was fooling around in the workshop and I went in to help him fool around. I happened to mention again, that I couldn't figure out how I flunked that flying test. He said it happens to the best, in fact Joe Shumate told him he fails over fifty percent of the students the first time. Now I understood what he meant the other day, when he said, "join the club". He told me he failed the flying test during the time I was in Pittsburgh. Well at least I wasn't the only one that messed up; it did make me feel a little better.

I asked him how his love life was coming along, and that mine was at a stand still. He didn't bother to answer me, and for a young twenty one year old fellow, he sure doesn't' show much enthusiasm about dating. He lets Inez do the courting.

At last, it was time to fly. I went out and helped refuel the "Prep", cranked her up, and took off heading west. I wondered if the "air pocket" was still near the railroad station and decided to fly over there. Sure enough, when I flew into it, the ship dropped like an elevator, but quickly recovered. I've experienced it several times in this area and don't know why; neither does Joe. I performed the obligatory maneuvers in decent fashion. I made one spiral landing, one 180-degree turn and land and one 360-degree turn and land. All three were close enough to the line, and were 3-point landings. Hopefully, I've put those wheel landings behind me.

My Cheve seems to be percolating pretty good so I quit trying to turn it into a "Stutz Bearcat". I guess as that old saying goes, "if it ain't broke, don't fix it" (might be right, but it is a hard lesson to learn. Walter told me he had a date lined up, but was scheduled to stay at the hanger tonight. Before I could express my sympathy, he asked me to take his place as night watchman. I didn't have a date, so I said ok. Then he added "insult to injury" when he asked to borrow the Cheve since I wouldn't be needing it. He sure pressed his luck, but I said ok.

After supper I went back to the airport. Walter was waiting; I gave him the keys to the Cheve and reminded him not to forget the gas tank. "Jack" and I went in the workshop and settled down for the evening. I listened to the radio for a while, read the Dayton School of Aviation book a spell, and finally turned out the light.

Lying in the dark, listening to the hanger creaking and groaning, I heard a strange sound; the darn dog was snoring. Now wide-awake, I thought my mind must be scrambled by all the spinning I have been doing lately. I can't seem to get rid of the thought that strangers suddenly came into my life. Then I live, eat, work, and socialize with them every day for a short period of time and consider them close personal friends. Just as suddenly they pack up, and walk out of my life forever. Never to be seen or heard from again as though_they never even existed, and this really bothered me. It is a good thing I am not a Physicist, Physic, Philosopher or any other kind of "shrink" or I'd probably go nuts. I had a hard time going to sleep; I should be concentrating on flying and girls.

August 7, 1932 – Sunday

The weather was hot, cloudy, rain, and the wind was fairly strong. No flying today.

Walter, and necktie, showed up right on time in my Cheve. I was in a hurry to go down and enjoy breakfast. But he insisted on telling me all about the great date he had, that the Cheve was in fine shape, and that he took care of the gas. He was still talking as I drove out of the airport.

It didn't take me long to eat breakfast, I was starved, and I didn't have much time to get to Sunday school. I got there just in time but had a hard time concentrating on the lesson, as I was awake nearly all night. Cleone never showed up, and all the interesting girls seemed to have steady boyfriends.

The dinner was great as usual and I think I overdid it a little. Bud stabbed my hand with his fork as we both reached for the same piece of chicken at the same time. It was just a small cut and I got that piece of chicken. I went upstairs, turned on the fan, took off my clothes, except my underdrawers, and read an aviation magazine. Not for long, I fell asleep and the next thing I knew it was suppertime. I was still half asleep during supper and was in no decent mood to take part in the idle conversation; however, by the time I finished eating that piece of homemade apple pie, I came to life, and challenged the students to a series of "checker" games. We went in the parlor, except those that had to study, and spent the evening trying to win every game. I came very close but got kind of cute and lost a couple, Art and "Bus" beat me once.

August 8,1932 – Monday

The weather was partly cloudy, pretty hot, and the wind was breezy. I flew three times today, first time this ever happened, and I smelled a rat.

We opened the airport, put gas in both "Preps" and had just finished the pre-flight inspections on both ships when Joe drove in. We fooled around until he came out of the office with the flight schedule.

I was to take one "Prep" up for twenty minutes, stay over the field, and perform right and left vertical power turns. Joe and Dick took off right after me. After twenty minutes of verticals to the right and left, I felt like a pendulum swinging back and forth. Since I've learned not to go too vertical, I do them just fine. I hope I don't forget, after I passed the test, to try a true vertical. The runway was clear, I spiraled down and made a three point landing close to the line.

After I filled the tank with gas and checked the oil, I went in the workshop and stood quietly in the background. Harvey was teaching the students the theory of the gasoline engine and that is my strong subject. I had a strange urge to put in my two cents worth, but each time I started to open my mouth he would look over at me with a look that said, "You better not". He must be a mind reader. Joe came in and told me to fire up the ship, go up for thirty minutes, and make one spiral landing, one 360 degree turn and land, one 180 degree turn and land, and one random landing. I began to get the picture, but it couldn't be.

The "Prep" was still warm; I cranked her up and took off in a cloud of dust. I spiraled down from 2,000 ft. with engine throttled, and noticed that Joe had moved the line; I changed my approach and landed within the limit. I took off and at 1,500 ft. I throttled the engine and made a 360-degree turn and landed closer to the line. Two down, and two to go. I took off and at 1,000 ft. with engine idling; I made a 180-degree turn and landed about 50 ft. in front of the line. Joe had moved the line each time I landed, and so far, I had adjusted my approach successfully. So far so good. I had a little flying time left, after taking off west I turned east and did a little sight seeing for a change. Then I turned back toward the airport, approached the runway and noticed Joe had placed the line close to the hanger apron. There was only one way I could land within the limits of that line, I had to come in real close to the high lines and make a steep sideslip landing. I did and landed smack dab on the line. I'm too embarrassed (not really) to admit that all four landings were mighty nice.

I parked the ship, turned off the ignition switch, climbed out, and thought I did ok today. Joe met me as I went in the hanger and I thought he was unhappy with the sideslip. Instead he told me to fill the gas tank, check the oil, and crank up the "Prep" as he wanted to "check time" me for thirty minutes. Now I knew there was a rat in the woodpile. Before taking off he said we would practice each required maneuver once and call it a day. He told me what maneuvers to practice, but never criticized or praised my flying. We taxied back to the hanger, parked the ship, shut off the ignition, and as we sat there he told me Joe Shumate, Examiner, would be at the airport in the morning. He would try to talk the Examiner into giving me the L.C. flight test. He said, "So be prepared, and get a good night's sleep, in case he says ok".

After supper I went up in my private "sanctum", stripped down to my BVD.'s, it is hot, and wrote Bert and Paul a letter. Mom sent me a letter with my room and board. I am well satisfied with my flying today. I read a magazine to relax and unwind, the house became quiet and I finally fell asleep.

August 9, 1932 – Tuesday

The weather was partly sunny, still hot and the wind was slightly rough from the south. I flew for one hour and four flights today.

I ate a light breakfast as there was a possibility that I might take the flight test today.

We went out to the airport, opened up the joint, lined up the ships and filled the gas tanks on both "Preps". As usual "Jack" was glad to see us, running around like there is no tomorrow, and barking up a storm. I have to brace myself when he jumps up on me or I will hit the deck. He is huge. The night watchman (student) takes care of him, feeding, etc., and keeping him tied up when the ships are flying.

When Joe arrived, Harv went in the office with him. A short time later Joe Shumate flew in, parked his ship, and went in the office, without so much as a "hi-di-do". To kill time, we walked the east-west and north-south runways to see if there were any holes, ruts, or ditches. I was especially interested for obvious reasons. When we returned to the hanger, Harvey told the students to make the hanger and workshop shipshape. I helped them, and it seemed like forever but Joe and the Examiner came out in a short time. They shook hands with each student.

I stood in the background and was the last the shake the Examiner's hand. I had decided to try to "butter up" and treat Mr. Shumate with respect, and it was easy. I do respect and admire a pilot who is good enough to advance to the position of Inspector, who has the ability, authority, responsibility and power to pass or fail a pilot. He said, "Beyer, I believe we have met", and I said, "Yes sir". "Your instructor told me that you are prepared to take the L.C. flight test again." I said, "Yes sir", and held my breath. "It is highly unusual", he said, "And I never do this without prior preparation, but Mr. Vail talked me into staying long enough to give the test." I let out a breath of relief, and thanked him effusively. The students and Harv were all eyes and ears.

Mr. Shumate took out a pad and pencil and told me to start the test. I gave the "Prep" a thorough pre-flight check from stem to stern, I knew he would watch every move I made or didn't make. Finally he said, "Beyer, you will wear that ship out before you even get it in the air. Here are the maneuvers I want you to perform unless I flag you in. At 1,500 ft. throttle the engine and make a 360-degree turn and land as close as possible to a line I will put on the runway. At 1,000 ft. throttle the engine and make a 180 degree turn and land. Each normal landing is to be made with the wheels touching the ground in front of and within 200 ft. of the line. At 800 ft. make 5 gentle figure 8's and at 1,000 ft. make three steep figure 8's. At 2,000 ft., throttle the engine, spiral in the left direction and make a normal landing." I completed all the required maneuvers without any problems; at least he didn't flag me in. He didn't have much choice as to where to place the line, due to the length of the runway.

I taxied up to the Examiner, he climbed in the cockpit beside me, and told me to complete the following maneuvers consecutively: climbing turn to the left and right, vertical power turn to the left and right, stall and recover, three turn spin to the left and right beginning and ending in the same direction. We took off in a cross wind, one of the requirements, fortunately the wind was from the south and I didn't have to use the north-south runway. I completed all the maneuvers and never forgot that I had a 180-pound passenger. I completed the right spin facing the river and leveled off. I pushed the throttle to cruising speed; the Examiner suddenly pulled the throttle back to idle and told me I was not to touch it. I spiraled down toward the airport, he then told me to make a sideslip landing. I noticed he kept his hand on the throttle. I came in a little high on purpose, and flew over the high wires. I pulled the nose up slightly, lowered the left wing slightly, and lined the ship straight down the middle of the runway so I wouldn't wipe out the landing gear when the ship touched down. I reversed the controls and as the ship settled down on the cushion of air, I pulled the stick back. I admit I made a nice 3-point landing, not even a bump. It was quite different than the one the Examiner made when he landed "ker-plunk". I even imagined I felt the jar. I taxied up to the hanger, parked the ship, and we both climbed out.

The Examiner went in the office. I joined Harv and Art, they pretended to be busy, but it was an act. They wanted to know what was what and I told them, with my fingers crossed behind my back, that I was perfect. Then I said, truthfully, the Examiner never said anything except what he wanted me to do. Joe and Mr. Shumate came out of the office and joined us. I tried to look cool, calm and collected, but my shirt was damp under my arms. The examiner didn't waste any time, or words; he shook my hand and congratulated me on passing test. He said he didn't bring any certificates today but would send it to my home in Pittsburgh. And it will serve as my license until the Department of Commerce, Aeronautics Branch, Washington, DC sends me the permanent license. It will include my picture and license number on it. I gave him the picture I had been carrying in my wallet.

It was as though I had grabbed the brass ring on the merry-go-round. I don't know how many times I thanked him, but he finally said, "Don't thank me, you earned it". I was so pleased that I finally made it; I just had to tell the Examiner he broke my heart when he flunked me the last time. As always, I opened my mouth when I should have been listening. Without cracking a smile he said, "Beyer, I would rather your heart be broken than your neck". "Maybe I'll see you in six months to renew your license." I said, "Maybe I'll see you and take the transport pilot's test". He said, "Maybe, and by the way, who taught you how to sideslip like that"? I asked him to send me the names and addresses of all the airports and airlines. He told me to write to the Department of Commerce, Aeronautics Branch, Washington DC and they will send me a list. He walked back in the office.

I joined the students and they asked me a blue million questions. I tried to keep my feet on the ground, but was not very successful. It was hard to believe that in one hour I went from a licensed Private Pilot to a licensed Limited Commercial Pilot, and successfully completed my training. At that moment, I thought I could take off without the ship. Just then Mr. Shumate, I kind of like the guy, came out of the office, cranked up his ship, waved and took off.

Joe came out of the office and suggested he and I inspect the runways. As we walked, he said he knew I would pass the test and that he was honestly sorry about the job deal. And he wished the school would be successful and profitable, in that event, he would be more than glad to hire me. But he is worried that the school is going "kaput". He said, "You sure have changed since that day you showed up and enrolled in the Dixie Flying Service. I don't know how to say this, but you seem relaxed, confident, sort of mature and less feisty. You're real problem is, and it makes me blow my stack, that you are such a blasted perfectionist. As I said before, you have to land right on the ground, not two or three inches above. Stitching the fabric on the wing must be plum perfect. Even spreading gravel and cinders must be perfectly smooth. Painting the numbers on the ships must be just so. In other words, everything you do must be absolutely, positively perfect or you are very unhappy. But I think you are beginning to realize that it is impossible to achieve perfection all the time."

Joe was smiling as he talked, so I know he was not bawling me out like he did the other time. To rationalize my "problem", I told him that I graduated from high school and got my first job. At that time, my dad told me that as long as I performed my job better than anyone else, I never needed to worry that I would lose it. Joe said he was going to give me very good recommendations and I thanked him. Then I told him that I was not interested in that China deal, although it was a lot of money. At best, it would be temporary, and I could lose out on a permanent job in this country. He agreed and said he would never, ever consider it. When we reached the north-south runway he stopped and said, "Friday, if you can't find a flying job before next February, I want you to come back and take enough flying time to renew your license. You will only have to pay your room and board. Before you leave, I will give you the recommendations and the agreement I just made," his exact words.

He shook my hand, patted my shoulder and turned back toward the hanger. I was so surprised and shook up I just told him I wanted to walk around the airport and unwind a spell. As I reached the west end of the runway I finally realized, it was over; and, temporarily, felt like a balloon that had lost its air. Now I will be the one to walk out of their lives, probably forever, and it was not a pleasant thought. I walked every inch of the runway and visualized a mental picture, so I can recall it anytime in the future. By the time I got back to the hanger I was my old self again; and it was time for supper.

The topic of discussion at the supper table was that I had successfully passed the LC flight test would receive my license from Washington DC. Even Mrs. Barber and Inez seemed pleased to hear the news. But, as usual the main topic was eating. We all went in the parlor and discussed the flight test from A to Z for a spell. They wanted to know what my plans were and I said I was going to concentrate on finding a flying job, get a Transport Pilot's license, then buy an airport and hire all you guys. That caused a few laughs.

I was exhausted, not that I had worked that hard, just mentally and emotionally, so I went up in my humble abode. I turned on the fan, wrote Mom a letter and said I passed the flight test and would be home in a few days. I stripped down to my under-drawers, fell on the cot, turned out the light, and relived the whole blessed day.

August 10, 1932 – Wednesday

The weather was cloudy, real hot and the wind was pretty calm. Naturally I didn't get to fly today. I graduated and am a "former student". What a strange and different attitude.

I woke up and nothing had changed. The same old mad dash for the bathroom, jostled down the stairs, and bruised shins at the table. And I am going to miss it. I thought the world would stop, or at least hesitate; here I was, a licensed L.C. pilot, the students have short memories.

I went out to the airport, again I was the outsider, and the students said a brief "Hi" and promptly went about their chores.

I went in the office and Joe offered me a cup of coffee, I swear I think he is trying to poison me. He told me to go "shoot the breeze", and as soon as the secretary finished typing the recommendations and the offer he made me, she would let me know. I went in the workshop and talked with Harv and Art about some of the good times we had. The secretary came out shortly and said Joe had the letters for me. We went in and he gave me both letters and told me to read them. I read the recommendations and was so surprised that I asked him if he was referring to me. "Every word", he said. I sincerely thanked him, and then it occurred to me, he is a different person when not giving flight instructions; very personable and sincerely friendly. I asked him if Joe Shumate said anything about my performance. He said you know how he is, all he said was that you didn't make any "boo boos". He did say you evidently spent a lot of time practicing "spins" and "side-slips". I told Joe they were my favorite "sport". I read the letter regarding his offer; he spelled it out in great detail, and just as he told me.

I went out in the hanger, "shot the breeze" with the students, and they told me Dick was practicing landings. This was my last opportunity to watch the students fly. I watched him approach the airport and as he flew over the high wires, I thought, "Uh oh". His left wing was too low and his left wheel touched the runway first, the ship "ground-looped". The left wingtip scraped the ground, spun the plan around, and came to a halt. The students went out, leveled the ship, and Dick taxied back to the hanger. Before he could climb out of the cockpit, Joe opened the door for him, and proceeded to "recite Dick's history" that included all his ancestors. Joe is back, in full form. How many times was I on the receiving end? The wingtip was slightly scraped, that's all; a little paint will take care of that. I watched Art, my bodyguard and pal, fly and he is pretty darn good.

I watched Dan, George and Howard P fly and I would give them passing grades.

I had to leave sooner or later so I went in the office and said "good bye" to the cute and pretty secretary, Gladys Gourley, and to Joe; it was hard to do. I told all the fellars that I would see them at suppertime, Harv said, "Not if I see you first".

I found Ed Becht and Bob Shucks in Inglis Drug Store. I told them I graduated yesterday and was leaving town. We ordered a cherry coke at the soda fountain, and the pretty waitress was Nellie Harrod. She can put her feet under my breakfast table any time, I wish. I told her I needed a box of Bugle pipe tobacco, and to my surprise as she turned to get the tobacco, I blurted out, "My, you sure are pretty". Ed and Bob looked astonished as I stammered and stuttered. She brought the tobacco in a small paper sack and smiled as I paid her. Dave Gentrup came in as we were leaving and I told him I was leaving town. He said he had heard I passed the test and was sorry that I was leaving. It was suppertime so I said, "So long" and went back to the boarding house.

I enjoyed supper as usual and told everyone that I would be leaving in the morning. I asked Mrs. Barber if she planned a special breakfast for me. She and Inez laughed but didn't say anything.

I drove down to the Maddox garage, had the tank filled with gas, tires filled with air, radiator filled with water, and bought three quarts of oil. Bob Jones checked the Cheve while I told my friends I was leaving town in the morning because I graduated yesterday. Mr. Maddox said, "Friday, how are you?" I was ready for that and said, "I am fine, how are you?"

I said, "So long" and drove out to the Sachleben's. I was glad they were all there, usually Fred is downtown "sparkin" with his girlfriend, Lizzie Dierkes. I told them I was going home in the morning, and that I had passed the test. We sat around the kitchen table and talked for several hours. Mrs. Sachleben said I was a member of the family, and Joe said we were brothers. I said they were my "Mom and Pop" away from home, and thought Fred, Carl and Joe were my brothers. It sounded like a mutual admiration society, but we all were sincere. With that, "Pop" brought out the "home-brew" and topped it off with a spot of "beer wine". I think the world of them, they are the salt of the earth, but I finally had to leave. "Pop" told Joe to go out to the barn and get me two "hands" of their best "long green" tobacco. It was pitch dark; Joe lit a coal oil lantern, went out to the barn and brought back two nice "hands". I said, "goodbye", and Joe, with the lantern led the way to my Cheve. He gave me the tobacco, and we promised to keep in touch.

I drove down to the boarding house, parked the Cheve out front and went upstairs. It was still hot, the fan didn't cool my "cubby-hole" but it did circulate the air. I put the bugle tobacco with the "long-green" and noticed a telephone number on the bottom of the sack. Nellie H. never gave me a hint when she handed me the sack of tobacco. Darn, my last night in town and this had to happen. I am tempted to stay for several days, but that's out. What a break, she is really my idea of a steady girlfriend, pretty, a great personality, a nice sense of humor, and stacked up. I am going to save the sack, just in case. I ignored the heat, laid on top of the sheet, turned out the light, and relived an entirely different day.

Clyde H. Beyer

August 11, 1932 –Thursday

The weather was partly sunny, very hot, and the wind was calm.

I got up early so I could tell Harv and the students "good-bye" before they went out to the airport (without me). We had a fine breakfast, and as they were getting ready to leave for the airport I told Dick, Howard P, Walter E, 'Bus" Kirsch, George, Dan, Bud, and Howard B, that I hated to leave but was glad I got to know each one of them. I wished them lots of luck and success, and to never forget when in trouble, get that nose down and sail right on. Then I thanked Harv for all the help he gave me, and hoped to see him again someday. To aggravate him I said he should up and marry Inez. Before he could say or do anything, he is over six feet tall; I turned to Art and told him I considered him a true friend that I could trust. Besides being a great bodyguard. I really hated to walk out of their lives.

I haven't had to work on the Cheve this go round, and as Bob Jones said it was ok last night, I was sure we would get home without any trouble.

I got ready to leave, and decided to drive out Walnut St. and say "good-bye" to Cleone. She opened the door and I was surprised to see she was still wearing her nightgown. I told her I was going to drive around town one more time before going home. She told me to wait while she put on a light robe, and she would ride with me. Well, that sure woke me up; it generally takes me all morning. But how often do you get to see the girl you would like to woo, without makeup, or all dressed up, and still look as cute as a "mouse's ear". This is the second time in two days I was tempted to postpone my departure.

We drove down Main St., turned around at the Clifty Park entrance, went back up Main St. to Ferry St., turned south to Front St., went west to Walnut St., and north to her house. It didn't last long enough, but it had to end some time. I said, "good-bye", she jumped out of the car, waved, turned and went in the house. I felt pretty blue as I drove back to Barber's.

I put all my belongings in the Cheve, went back in the house and thanked Mrs. Barber and Inez for watching over me. They surprised me by giving me a big lunch to take along. To break the "blues" I told Inez not to let Harvey get away, that I think they are a great couple. She blushed.

I jumped in my faithful chariot, drove up Michigan Hill, and about one-half mile before I got to the airport, I blew the horn constantly until I passed the Sachleben's house. There was one "Prep" in the air, and he dipped his wings, or maybe was doing some sloppy flying.

It was quite different driving home in the daytime; the traffic was not a bit bad, even going through downtown Cincinnati, Ohio. I felt kind of depressed, but the further I drove and began to plan my next move my spirits began to brighten. I stopped and filled the tank with gas and added a quart of my oil in Norwood, Ohio. I filled the tank with gas and added a quart of my oil at the same gas station in Zanesville, Ohio. I opened my lunch as I left the gas station and ate it while I drove east. To my surprise there was a big piece of homemade apple pie, my favorite, just like my dad bakes in the shop. I made my last pit stop at the same gas station in Steubenville, Ohio and added my last quart of oil.

I arrived home at 10:00 pm, 12 hours, used 15 gallons of gas and 3 quarts of oil. As usual I parked the Cheve in back of the bakery shop. I opened the door, told Mom and Dad, "I'm home", went back and unloaded the car. Then I told them that I had passed the flight test with flying colors. The Inspector is going to send me the certificate that says I successfully passed the L.C. test. The Department of Commerce is going to send me my permanent license. I told them the Inspector told me to write to the Department of Commerce and request a list of airports and airlines including addresses. I also said that I had successfully completed the course and Joe seemed pleased with the results. I showed Mom and Dad the recommendation and the offer he made me; he wants me to come back and renew my license if I don't find a job before it expires. After they read the letters Dad said you couldn't expect anything better than that, but hopefully I will have a job by that time. I told them the license has to be renewed every six months with ten hours flying time and a physical examination.

It was late, Dad had to go down in the shop, Charles will be here shortly, and they will start baking. Ralph and Earl are still out horsing around somewhere. I was tired, I grabbed the paper and went upstairs and hit the sack. What a birthday!

Friday August 12, 1932 through Tuesday February 7, 1933

The weather was hazy, hot. And the wind was still.

I woke up sweating and went downstairs and ate breakfast. Earl went to work at the East Liberty Bank and Ralph was gone. Mom was down in the store; Dad and Charles were about to quit for the day. Ralph has been trying to get a job at the Police or Fire Department garage. He spends all his time volunteering at both places hoping to get a job. These jobs are strictly political and he has never shown any interest in politics.

Form FL-12

DEPARTMENT OF COMMERCE
AERONAUTICS BRANCH

Date, 9-18-32

(This Letter of Authority Must be Kept in Your Personal Possession)

Clyde H. Beyer
4121 Main Street -
Pittsburg, Pa -

DEAR SIR:

You have completed the tests and examinations required by the Air Commerce Regulations for a **Limited Commercial** License and are authorized to operate as such pending receipt of the actual license. Unless this authority is sooner suspended or revoked, it expires upon receipt of license, or not later than two months from the above date.

The holder of this Letter of Authority, if for Limited Commercial or for Transport Grade, is authorized to transport persons for hire in only the types and classes of heavier-than-air craft listed below for which the holder has been found qualified, as shown by the proper indorsement of an authorized Inspector for the Department of Commerce.

CLASS I	TYPES	LAND OR SEA	INSPECTOR	DATE
1,000 pounds or less gross				
1,000 to 3,500 pounds gross	A. Single engine *Airplane*	*Land*	*Jos. J. Shumate*	*8-18-32*
	B. Multiengine			
CLASS II 3,500 to 7,000 pounds gross	A. Single engine			
	B. Multiengine			
CLASS III 7,000 pounds or over gross	A. Single engine			
	B. Multiengine			

CLARENCE M. YOUNG,
Assistant Secretary of Commerce.

Base *Madison Airport*
(For Limited Commercial Pilots only)

Madison, Indiana

By *Jos. J. Shumate*
Department of Commerce Inspector.

I started the day by writing to the Department of Commerce, Aeronautics Branch, Washington, DC requesting a list and addresses of airports, airlines, factories and schools. When I receive the list, my main occupation will be to send them an application for a job. Until then I am going to clean the bakery shop every day and take over the daily decoration of the two store windows. Maybe it will entice more customers to come in and shop.

While cleaning the shop I did a chin up workout, and I soon found out I have a long way to go. I thought I was "fit as a fiddle", maybe it is the second part, that I am in shape, "and ready for love".

I continued our customary rite of attending Sunday school and Church this morning. Actually, I observed it in Madison by attending a church similar in philosophy and ritual with ours. This rite has been observed down through the ages without interruption and certainly bears fruit. Ralph told me before leaving home that I was going to be surprised. In a way I was, the class was subdued, you know who was not present and the young gentlemen were once again paying court to the reigning primadonna the Preacher's daughter. I looked at Ralph and his lips formed the words "she is gone". The lesson was moving along so smoothly and quietly I casually, accidentally dropped my license on the floor. Don Bicht picked it up, looked at it, and then passed it around to the rest of the class, which brought the lesson to a halt. I took center stage.

The lists from Washington finally arrived and I was really surprised how big they were. I'll never be able to write or visit them all, but am going to write and visit as many as possible. It was slow, Paul bought me a cheap typewriter to speed things up and it did help. I continued writing the letters all fall and winter, and sporadically would receive a reply, some said they had a long employment list, and others said they would put me on their employment list. They all thanked me for having a desire to join their company. So far, the vast majority did not bother to answer my letter. All the letters I wrote were basically the same, "that I would like to put in an application to join your company". I enclosed a copy of Joe's recommendations and my L.C. license. Joe's letter not only praised my flying ability, but also my ability to repair airplanes and engines.

I cleaned the bakery shop and decorated the store windows daily to keep from getting blurry-eyed and the writer's cramp. I took time out from writing, went over and visited Jack Morris at Bettis field. He was very pleasant and friendly, and recognized me as a frequent visitor. I told him I had renewed my L.C. license and needed a job. He said he had a second job because he couldn't make a living flying. He is a nice guy. Jack (John P) Morris

I drove out to Rodger Airport and talked with Bob Dake, I had met him several times. He is a tad more reserved than Jack Morris but still friendly. He also doesn't depend on flying for a living, and wished me luck. He said if I ever bought a plane he would be glad to rent me part of his hanger.

Morris and Dake fly Waco OX 5 planes.

I spent all morning at the' Pittsburgh airport and they have more applications than they know what to do with.

Purely by chance, I saw a small article in the paper that said a young pilot was killed in a plane crash near Washington, PA. I read the entire news item because I knew Frank Peternell lived nearby at Meadowlands, PA. I had promised to visit him the day he left Madison. I held my breath as I read the paper, but it didn't do any good, the name of the pilot was indeed Frank. I drove down and visited his family and expressed my sympathy. I told them I went to school with Frank at Madison, IN and when I told them my name, they said Frank had talked about me. His mom told me Frank bought a used Aeronca and rebuilt it; she thought it had been in poor shape. He took the plane up to test fly it, and was flying over the golf course when for some unknown reason; the plane plunged into the ground. It seemed to make them feel better talking to me about the crash and about Frank. I guess this will have to fulfill my promise I made to Frank. It also made my frustration over not getting any job offers inconsequential.

I have been getting some letters. I got a brief letter from Cleone, she enclosed a picture of herself and said not to forget her. I am not about to forget that gal; I'll answer her letter.

I got a letter from Andy Nufer and Lawrence Burklow. They are still looking for a job, and wanted to know what I was up to. I took time out and answered their letters.

My chin-ups are improving every day and it won't be long until I am back in shape. Just in time too. I met a young chap who talked me into learning the proper way to hop a freight train. The first lesson went ok until it was time to get off. When "Mike" decided it was time to get off, the train was traveling too fast and we had to stay on until it slowed down about 3 hours later. We hopped a train going back, it was dark when we arrived near the freight yard we started from. Mike jumped first and landed on his feet, I followed and ended up tumbling head over heels. We couldn't wait for the train to stop to get off cause the railroad cops are very aggressive toward hitchhikers. The horror stories they tell about those cops make your hair stand on end.

The meteorology lesson is beginning to make sense so maybe I won't have to memorize it. I've also been brushing up on the other written tests. Since I have been watching the pilots closely I am not too concerned about the flight test. The only additional flight test not given in the L.C. test is a triangular cross country flight, and land at the airport at each point of the triangle.

I had a different type of job for a spell. The woman living alone next door asked me if I would be interested in taking her to a private dance club. She appeared to be in her 30's and pleasant, although I had never spoken to her. She explained that she dearly loved to dance but had to have an escort to get into the club. She emphasized that she was not interested in a boyfriend or a date; she simply wanted to dance at the club. I was to drive her to the club, she would introduce me as her escort, and as I can't dance, I would have a table where I could listen to the music and watch the action. Yet close enough so that if her dance partner tried anything other than dancing, she could point to me as her escort. The manager was to serve me what food and drinks I wanted. It sounded down right interesting, especially since she said it was a very strict club that didn't tolerate any trouble. I was to get $3.00 and everything I would eat and drink. Naturally I couldn't drink and drive.

I gave it a try. I picked her up in the evening, drove her to the club at the edge of town, went in and she arranged a table for us in the dim light. She brought a waiter over and told him I was to be served with anything I ordered. She told me she would never leave the building without me, and if she did I was to check on her immediately. Then she walked across the floor and I had my first chance to get a good look at her. She was slightly taller than me, neither fat nor thin nor homely or pretty, just a pleasant and comfortable personality. She didn't seem to be the type that attracted men at first glance.

The waiter brought me a glass of beer, a bowl of pretzels, and a couple of big fried oysters. From now on, I am going to eat a very light supper. It was a pleasant evening, listening to the music and watching dancers of every size, shape, looks and ability. I sipped the beer slowly to make it last, but consumed plenty of oysters and pretzels. My neighbor came over just as I was getting drowsy and said she was ready to leave. I drove home and walked her to the door, she gave me the $3.00 and said to call her Laura, and would I want to pick her up Friday evening. So it got to be a habit every Tuesday and Friday evening, and the routine never changed. There were never any fights, commotions or loud arguments, like they had at the Riviera Gardens in Madison. Then suddenly she upped and moved away; I never saw her with a boyfriend other than on the dance floor. We never had a decent conversation, and it was like an employer-employee relationship. It was a unique experience.

Dad and I continue our Saturday evening walks out to gasoline alley. Those new cars are out of this world; each one has an individuality of its own. Dad seems to take a great of pleasure, and never gets tired of examining the different makes and models. We can identify the cars as they drive down the street. I would give anything if he would take up driving again.

I've rubbed elbows with Fred Moss, Earl Seitz, Tony Fellow, Red Sandstrom and the alley gang. But we were all devoting our time trying to find a job. It's hard to believe but Bill Sandstrom got a job working for the Gulf Oil Corp.

Dad decided, I guess, that I should earn my keep. He taught me how to decorate fancy cakes, wedding, etc. He's a good teacher, a mite too pertinacious; I don't know how many times I had to learn to put a dot of icing in the proper place. I enjoyed it and got good enough that he quit inspecting my work. Until, that is, he caught me improvising from the norm. I listened to his theory that each cake must be just so every time; this time however his correction didn't seem as absolute as it used to be. Since I'm a little older and have been away for a spell, I am beginning to understand him a little I think. I said, "Pop", his eyebrows shot up, "just look at how symmetrical the design is. But I can wipe it out and start over". He took a good look and told me to finish it as is, and if I got another urge to do likewise it would be ok. I'm afraid I went hog wild, but the fancy cake business increased a lot. I liked the work but there is very little money in it. It's kinda like painting with icing.

I've reached the end of the rope. I drove back over to the Bettis airport. The Army Air Force has a small corner of the hanger as an office. The dude in charge was dressed for the part. He wore hard, black, high boots, his trouser legs were down in the boots, I don't know what you call it, pantaloon? Leather flight jacket and a captain's cap rounded out his attire. He even carried a 'crop". I didn't think I would like him. I introduced myself and said I would like to join the Army Air Force. I showed him Joe's recommendations and my L.C. license. He didn't pretend to read either one, and bluntly asked me if I was a college graduate or attending college. When I said no, he practically dismissed me then and there. Being desperate, I told him I would lose my license if I didn't get some flying time. I said the government would save money because I had experience in repairing and rebuilding airplanes and engines. I didn't give him a chance to interrupt me, I said they won't have to teach me the fundamentals, (begging) and I really need flying time desperately. He interrupted me and with a bored look said not a chance. He told me they are only accepting college graduates, period, and would teach them to fly as they see fit. So far I was doing pretty good, my pressure just slightly elevated, then I asked him what made college kids so special and so much better than I. He said, with a look of scorn, you guys are a dime a dozen. That was the turning point; I blew my stack cause I didn't have anything to lose. I asked him if he knew who was paying his salary, I didn't pay taxes but my family did. Then I waved my finger at him and said I needed the government now. One of these days you will need us dime a dozen guys and I am going to tell you where to go. (Of course I didn't mean it if there was a war.) He said, "oh yeah", we will come and pick you up any time. At that I read him his history all the way back to the Stone Age. What a difference between two government employees, Joe Shumate and this pompous dude. As I walked past Jack Morris, he just shook his head and pointed a finger at his temple.

I was hot under the collar all the way home. I told Mom and Dad what had occurred and he reiterated that it would probably turn out for the best. He also said I should start looking for a job and save the money to buy flying time or a used airplane. But at the same time continue to send in those applications. A good idea.

I read the ads in the paper every day. Almost all of them wanted salesmen of all types. The few remaining ones were for professionals, the kind I could not handle. Once or twice there were requests for mechanics, I answered those, but they wanted men with years of experience and their own tools. The same reason Ralph has been unable to get a job. No one needs a "grease monkey" or office worker.

A member of our church and his family own a pipe and steel factory. The Sunday school superintendent and his family own a huge wholesale bakery with hundreds of trucks. Another member is an executive in a downtown bank. I asked Dad to use his influence on them; he did but came up empty. Each one told him they were terribly sorry, and would put me on top of the employment list.

I got a nice letter from Art P. and he remembered the times we had working on the Cheve, but wasn't particular about being chosen my bodyguard, and added just kidding. I answered his letter.

To ease the stress of constantly being rejected when applying for a job, Ralph and I took up sled riding after supper. I went down the hill and crashed into a girl on a sled. I helped her up out of the snow and brushed the snow off her coat, she just laughed. We would meet quite often when it was suitable for sledding and enjoyed talking. She was petite but pleasingly plump, good looking and a senior in high school. Ralph was kind enough to find a girl to sled with. Eventually I asked her for a date and the bubble burst. She told me she wasn't allowed to go with me because our nationality and religion was different. She is Polish and a Catholic. They must have seen the Masonic emblem on the Marmon. I told her I wasn't talking marriage I just enjoyed her company. She quit sled riding, at least here. I saw her mother up the street and if the saying is true, "if you want to know what your wife will look like years from now, just look at her mother". I don't think I'd be interested.

February 8, 1933 – Wednesday

The weather was cloudy, cold and the wind was extremely strong.

I ate a big breakfast, a bowl of rolled oats, two doughnuts, and a cup of coffee; it is a long trip to Madison. Mom gave me a lunch, a big box of cookies, a bag of candy from our store and $10 for my room and board. Dad had filled the Cheve with gas and slipped me a handful of "crooks" when Mom went to get my lunch. Ralph gave me two tins of Prince Albert pipe tobacco. Charles and Earl each gave me a dollar. I put everything in the car, went back in the house and said, "So-long". I had to go back in again, I forgot my L.C. license and log book.

I climbed in my chariot, cranked her up, waved, and took off southwest at 8:00 am. This trip is different than all the others I've taken to Madison. The excitement and anticipation is missing this time, but am thankful that I have the opportunity to renew my license. Provided I pass the physical exam, and get the promised flying time.

I stopped at the Steubenville, OH airport and talked to a Mr. Bebout about a job, and showed him my license. He told me they were not hiring anyone, not even transport pilots, anytime in the near future. I said I would take any job, even be a "grease monkey". No luck. I stopped at the gas station, and the guy recognized me, he filled the gas tank, washed the windshield, checked my tires and radiator. I didn't have the nerve to ask him to put my quart of oil in the engine so I stopped at the outskirts of town and put it in.

I took off and flew down Hwy. 22 to Cambridge and merged with Hwy. 40 then on to Zanesville. At this point I changed my normal route. I continued west on Hwy. 40 and stopped at a gas station east of Columbus and asked the guy how to get to the Columbus airport. I talked to a Mr. Boal, with the same results. When I mentioned a "grease monkey" job, he laughed and said not even a janitor job, which I would have taken.

I flew west on Hwy. 40 to Hwy. 235, turned south and stopped at the Dayton airport. I told the folks at the airport that I graduated at Dixie Flying Service, that it merged with the Dayton School of Aviation Correspondence Course on April 19, 1932, and became the Dayton School of Aviation at Madison, IN. I mentioned that Robert R. Foster was Vice-President. A Mr. Robertson said he knew Mr. Foster. They barely looked at the license and recommendation. I try to change my approach each time to make it more personal, but so far it hasn't made any difference. This was my last stop, and I still had a long way to go.

I continued south on Hwy. 48 and when it merged with Hwy. 22 I switched over onto it. I stopped at the Norwood gas station and the gas pumper recognized me. I left the gas station, added my last quart of oil and finished my lunch. I took Hwy. 50 west at the intersection with Hwy. 22, and crossed the state line just east of Lawrenceburg, IN.

I flew west on Hwy. 50, at 40 M.P.H., turned south on Michigan Road at Versailles. I passed the airport and landed in front of Barber's Boarding House shortly before lockout time, about 14 1/2 hours, 20 gallons of gas, and 3 quarts of oil.

I went in, Mrs. Barber was sitting in the parlor; I guess Inez and Harvey were out somewhere "smooching". She said she received my letter, and began to worry because I was late. She smiled and gave me a big hug, I ain't used to that hugging stuff. I took all my belongings up to the room. Mrs. Barber had it all fixed up for me, even saved the sheets of cardboard under the mattress. On my last trip in she insisted that I join her in the kitchen, have a cup of coffee and a piece of pie. It sure hit the spot. She wanted to know about the last few months I was gone, and then told me most of the students were gone. I gave her some of Dad's cookies, and some candy. I was beat, so I thanked her for the coffee and pie, went upstairs, left everything where I dropped it, undressed and fell asleep as I crawled under the blanket.

February 9, 1933 – Thursday

The weather was cloudy, rain, cold and the wind was pretty rough. There was no flying today.

I slept in this morning and thought I missed breakfast, but Inez cooked up some ham and eggs, rolls with butter and jelly and coffee. I'll bet this service won't last long. Although I was still "asleep" we talked for a brief spell and she told me Art Prosek, Dan Booth, George Hayden, Clair Kirsch, Walter Etchison, Howard Price and Dick Boepple graduated and have gone home. That was bad news. She said Bud Lawrence, Howard Brennan, Louis Benoit and Don Voison were still there. That was good news, but that meant there are only four students left. They should be getting a lot of flying time.

I drove out to the airport and chitchatted with Harvey and the students who were still there. They seemed glad to see me. Joe told me to have a cup of coffee, grab a chair and "shoot the breeze". I told him I received a list of airports, airlines, and plane factories from the Department of Commerce. And I spent all my time writing letters; I gave him a copy, so many that I finally talked my brother Paul into buying me a cheap typewriter. With all the writing and typing I've done so far, I haven't made a dent in the list. I told him that I stopped at the Steubenville, Columbus and Dayton airports on the way down here. The three I visited were not a bit encouraging and said unless things changed drastically; they won't be doing any hiring for quite some time.

I asked Joe what was the latest "scoop" at the airport. He turned serious and said it was not good, they were getting quite a few inquiries, but no new students. However, the correspondence course was doing quite well. He changed the subject and said I could come out to the airport in the morning and take the "Prep" up for thirty minutes to brush up on the maneuvers. He said "Usually in renewing a license there is no testing involved, but the Inspector can test a pilot at any time, so be prepared, just in case. I'll set up your flying schedule and as there is nothing for you to do around here you are free to leave after parking the ship. By the way, would you be interested in temporary work on the day you are not scheduled to fly?" This came as a surprise but I was quick to reply with a "you betcha". "I'll get in touch with Paul Wallace and ask him he can use a "handy man" for a few days. He has, no doubt, seen you around here. I know you can handle that kind of work; you have practically rebuilt your Cheve from the ground up. I know it won't pay much and it will be strictly temporary. I'll tell you what he says sometime tomorrow." I thanked him up one side and down the other. I gave Joe and the secretary some of Dad's cookies.

I went out in the shop where Harv and the students were "goldbricking" and swapped a few lies. I gave him some cookies and candy. Harve told me that Art arranged a deal with the school to buy the Curtis Wright, and flew it home. He still has a room at Richert's Rooming House, but naturally eats all his meals at Barber's and you can bet he gets special treatment.

I drove over to the Sachleben house, told them I arrived in town late last night, and would be in town for at least two weeks. We "chit-chatted" about the town and church for a time and then "Pop" said the future of the school didn't look very promising. Joe S. added, everthing has changed since you and the other "old timers" had gone home. I gave them some cookies, said I would see them in church Sunday.

I drove back to the Boarding House, and went in the kitchen where Mrs. Barber and Inez were preparing dinner. I told them Joe was going to get me temporary work down at the Cheve garage on the days I won't be flying. But on the day I fly I will have the rest of the day off. Mrs. Barber said all the rooms needed a good cleaning and if I wanted the job, it would pay my room, board and laundry. So we agreed, that I would clean the rooms from the time I finished flying to suppertime, except Sundays. I am to start tomorrow.

After dinner I walked down Main St., stopped in Vail's Furniture Store, and said "Hi" to Mr. Harry and Dale Wilson the bookkeeper. Then I walked down to the school office at 216 E. Second St., and told them I was back. A few minutes later I walked west on Second St. and stopped in Naill's Lumber Yard, and talked with Mr. Naill. A short time later I walked across West St. and visited the old gang at Maddox Garage. I left the garage and walked one half block east past the Bucknell-Wallace Cheve garage. I got back just in time for supper. I wanted everyone to know I was back in town.

We, the students and I, spent the evening roller-skating at the Riviera Gardens. There is safety in numbers, the west end "odd balls" are still hostile cause the girls like us. Had a great time. Wrote Mom a letter and hit the "sack".

February 10, 1933 – Friday

The weather was cloudy, cold, and the wind was real rough. I flew for thirty minutes and two flights today.

I ate breakfast with the whole crew for the first time this go-round. As usual, they all climbed in the old International truck, I followed them out to the airport in my Cheve. "Jack", our pup, about 130 pounds, greeted me like a long lost brother and darn near knocked me "hell west and crooked". I patted each ship a couple times and said "Hi" to my old "friends", one was missing, the Curtis Wright, Art flew home. I noticed the school hasn't sold any of the ships. I helped Harv and the crew get both "Preps" ready to fly. Joe came out and told me to take the "110" up for thirty minutes, two flights, and practice the maneuvers.

When Joe went back in the office, I asked Harv if Joe was going to "check time" me, it's been five months, you know. He just said, "You heard what the man said". I gave the ship a thorough pre-flight check, although I knew Harv kept the ships in tip-top shape. I cranked the prop, climbed in the cockpit; Harv pulled the chocks away and waved me down the runway. What a great feeling as the ship left the runway, this must be what heaven is like. I zipped through the figure eights; using Stephan's round barn and the hanger, and completed the left and right vertical power turns without a second thought. I made a nice 360-degree turn and landed in a normal three point landing, ignoring the line. I took off again, the engine cackling like a contented rooster as I went up, made the ship stall and turned it into a three turn spin to the right, and left facing the Ohio River. My time was about up, I quickly made a climbing power turn that put me in position to make a 180 degree turn and landed in a normal 3 point landing. I parked the ship, sat in the cockpit and thought that this sure would be a great way to spend the rest of my life. I was surprised how smooth everything went, I expected to be a little rusty. Harv came over to find out why I was still sitting in the ship, he just said, "I see you have learned your lessons well. Now get out of that crate so someone else can use it." I think he meant that I did ok.

I went in the office and Joe signed my pilot logbook. He gave me the flight schedule for next week, and said it could be changed at any time. He told me that he is going to get in touch with Joe Shumate and make arrangements to renew my license, and also make an appointment with the doctor in Seymour. I am scheduled to fly on Wednesday morning.

I went out in the workshop, told the crew I'd see them later, and drove down to Barber's, and said I was ready to go to work.

Inez mixed some stuff in a bucket, gave me a lot of rags, a scrub brush, a broom and we went upstairs in the back bedroom. As I knew absolutely nothing about this job, she showed me how it should be done, and went down to help her mother prepare dinner. I moved everything to one side of the room, and proceeded to wash all the woodwork, wrapped a large rag around the broom and swept the walls and ceiling wallpaper. I scrubbed the floor and washed the windows using a solution of ammonia and water. I was sure glad to hear Inez call to say it was dinnertime. This type of work must give you an appetite cause I ate like a hog.

After Harv and the students left, I went back up with a bucket of clean stuff. I was so full I had a hard time moving everything to the opposite side of the room. This is hard work and there ought to be an easier way, like flying, to make a buck. I rassled everything back in place just as Inez called and said it was suppertime. I was almost too tired to eat but managed to hold my own.

I soon forgot I was tired when I decided to get spruced up and drive out to see Cleone. She wasn't allowed to leave so we sat in the Cheve in front of the house, "chit-chatted" and pitched a little "woo" until it was time to go back in. There is nothing like a little "woo" to perk up a feller's spirits; I flew back to the house. The Cheve has been running great. Howard B is the night watchman tonight.

I have to be at the garage by 7:00 a.m. and I don't know what to expect, so without further ado I hit the sack.

February 11, 1933 – Saturday

The weather was partly cloudy, cold, and the wind was pretty rough. I didn't get to fly today.

I got up real early; I didn't want to be late the first day. I ate a big breakfast, it was great, and Inez said I cleaned the room just fine.

I drove down Second St., and parked the Cheve past the garage and saw the men in coveralls, one was colored, go in the garage area. At 7:00 am sharp I saw Paul Wallace and a tall silver haired man enter the store and office area. I jumped out of the car and hurried in after them, Mr. Wallace recognized me and we went in his office. He was all business and didn't seem interested in idle conversation. He said I would be paid $1.50 daily and the hours would be 7:00 am. to 5:00 pm, dinner from 12:00 - 1:00 pm. I am to report to Graham Harding, the mechanic, and he will tell me what to do. He introduced me to Teeb Taff, the tall silver haired man, as a car salesman, and said I would find Graham Harding out in the shop. I went out and saw a tall well built man leaning over the radiator working on a Cheve engine. I walked over and introduced myself and he said he had been expecting me. Mr. Wallace told him I was one of the "fly-boys" out at the airport that was going to work as a "grease-monkey for a few days. When I called him Mr. Harding he said, "Just call me Graham, I don't own a dog, Shorty". From then on I was "Shorty" at the garage.

He introduced me to Sid Schoolcraft, the auto body mechanic, and to Fred, the colored man, who washed, cleaned, and waxed the cars.

Graham H. drove a Cheve over to a hoist, showed me how to operate it, and raised the car high enough to walk under it. He showed me all the grease fittings, how to drain the engine oil, and how to check the transmission and differential oil levels. At the same time inspect the equipment that might need repaired or replaced. I lowered the hoist, filled the crankcase with oil, checked the radiator, hoses, belts, put water in the battery, and oiled the generator and starter. I washed the windshield, checked the air pressure in the tires, and parked the car in its stall. This is what they call an "oil change and grease job". Graham pointed to another Cheve and I went through the same "rigmarole", but this time he watched me perform all the "maneuvers", I mean all the tasks as before, and said, "Shorty, you catch on quick". I didn't tell him I knew every nut and bolt on my Cheve by name. I had one more car to service before dinnertime and now I know why they call a mechanic's helper a "grease monkey".

I left my Cheve parked, to save gas, and walked back to the boarding house and ate dinner. I told the students what an important job I had, hah! I walked back to the garage, and when the others returned Graham surprised me. He gave me a large shoulder pouch full of "flyers" advertising Bucknell-Wallace Cheve and Olds Dealership. I was to go up and down Main St. and put one on the windshield of every parked car, which I did. Cars were parked on Main St. and all the side streets cause this was Saturday. I got rid of the "flyers" pretty quick and went back to the garage for more. I got rid of the second batch shortly before quitting time. The circular had a picture of an Oldsmobile Sedan $785.00 FOB, and it looked great.

I walked to the rear of the garage and watch Sid paint a white 1/8" stripe, freehanded, about waist high, around the car. I couldn't believe it; to my untrained eye it looked perfect. I asked him if he needed any help, when he turned and answered with a curt no I smelled "bathtub gin". I walked over to Fred who was waxing a new Oldsmobile a doctor had ordered. He must have heard me ask Sid because before I could say "hi" he politely said he didn't want any help. It was time to put the tools and the equipment in the tool cabinets, wash up and go in the office and receive our wages. I thanked Mr. Wallace for the $1.50 and said that I would be back bright and early Monday morning. I think that was the first cash I've earned since I left home. I like Graham; he is friendly and has a good sense of humor.

I ate supper with Harv and the students, when they quit bragging about their exploits in the air I told them I slaved while they were 'goofing off". I got spruced up, walked downtown, and met Joe and Carl Sachleben. They were on their way to Metzger's Clothing Store so I went with them.

The clothes, from ties to trousers, and everything in between, seemed to be of good quality. I don't know whether the prices are high or low but Joe said they are, all things considered, the most reasonable in town. We cruised, on foot, up and down Main St., and I noticed a lot of cars still had the circulars on the windshield. We stopped and talked to a lot of friends including Ed Becht, and saw quite a few pretty girls along the way. Finally, I told Joe and Carl I would meet them at church in the morning. It suddenly struck me, on the way home, that I have never seen Joe V except at the airport and office. I didn't waste any time or ceremony on hitting the sack.

February 12, 1933 – Sunday

The weather was partly sunny, cold, and the wind was a stiff breeze. I didn't get to fly today.

I rolled out of the cot still kinda stiff from cleaning the room, and working nine hours at the garage yesterday. I struggled; grunting and groaning down the stairs, and didn't get any sympathy from those "oddballs". The hot breakfast woke me up and eased my aches and pains a bit. I washed, shaved, and put on my nicely pressed Sunday suit and walked to church. The Sunday school class was just the same as the day I left except all the eligible girls were spoken for, and no sign of Cleone. Everyone was nice, real friendly, and I think the lessons, especially the sermons, are starting to bear fruit, or should I say "rub off on me".

I walked back to the house, enjoyed a swell Sunday dinner, and told Mrs. Barber I might eat supper at the Sachleben's and stay all night. I drove out to the airport and loafed with the crew until I had a chance to tell Joe V. about my day at Wallace's garage. I watched him fly several passengers then drove over to the Sachleben's. I went out with Carl and Joe and watched as they fed the chickens, picked up the eggs, slopped the hogs, and milked the cows. They put the milk in the separator and hand cranked it until the cream separated from the milk. They will make butter out of the cream. That separator is some machine; when it gets up to speed, it really turns extremely fast and makes a high-pitched sound. The milk looks kinda thin, but the cream looks rich and thick. I now know that farming is very confining, the livestock has to be fed and the cows milked, no matter what.

Ma invited me to stay for supper and I accepted. She told Joe to go down in the cellar and bring up three jars of vegetables so I went with him. Joe took a kerosene lantern; we went outside, raised the door and went down the steps into the cellar. It was a great big room, high ceiling, dirt floor, creek rock walls, and four hallways going in different directions. The lantern didn't light up a very large area so I don't know where they went. Joe handed me two jars and asked me to stay all night, but I said I had to be at the Cheve garage at 7:00 a.m. We had a typical farm supper under a gasoline lantern, it is much brighter than a kerosene lantern, but you have to pump it up once in a while using the built-in pump. Again Joe asked me to stay overnight and said that I would have plenty of time to get to work, as they get up before daybreak to do the chores. Ma said Fred goes downtown to work at that time. So I stayed.

We sat at the kitchen table and talked about my folks, and what it was like living in a big city. Carl brought out the cards and we played for a spell, and had the usual liquid refreshments when Fred came in. It was late, so Joe carried the kerosene lantern and we went up in his room. The room was larger than any at the boarding house and so was the four-poster bed. He set the lantern on a small table, and we started to undress when he noticed a lump under the quilt. Joe pulled the quilt back and there was a gigantic snake curled up. He tried to grab it but the snake slid off the bed and disappeared in the shadows. He never did find it and I wasn't about to take any part in the search. It was at least four feet long and Joe said it was harmless. They have them in the barns to get rid of the rats and mice. I am no farm boy and I was ready to go back downtown, but I didn't want to be a "fraidy cat". I slid in bed but I didn't sleep a wink.

February 13, 1933 – Monday

The weather was sunny, very cool, and the wind was very brisk. I didn't get to fly today.

The Sachleben's got up before daylight to do the daily chores, except Fred and I. But I did get up when Joe did and kept my eyes peeled for that snake. I thanked "Ma" and "Pa" for everything and said I couldn't have breakfast as I had to go back to Barber's and prepare to go to work. I got there just in time to eat breakfast, put on my work clothes, and walk down to the Cheve garage. It was still locked up, Graham arrived first, unlocked the garage door, we both went in and Sid and Fred wasn't far behind. Graham wanted me to help him on a major engine overhaul and that suited me just fine. We used a hoist to lift the engine out of the car and put it on an engine cradle. We removed the various parts from the engine and I cleaned them with gasoline. Finally there was nothing left but the bare block and I steam cleaned it thoroughly. Graham honed the scratched cylinder walls smooth and I steam cleaned the engine compartment. That made a mess and it took me a considerable time to clean it up. I placed all the parts on the bench in proper sequence, and then it was dinnertime.

I walked back to Barber's, about four and a half blocks, ate dinner with the gang and took quite a bit of razzing. They wanted to know if I had my wings clipped, or if I was a Kiwi (a flightless bird). I didn't get "riled up" cause I had a license and they didn't; besides I was earning money for the first time in a long time. They went back to the airport, and it made me jealous. I walked back to the garage and helped Graham work on the engine until several cars came in for a grease job and oil change.

I put the first Cheve on the hoist, Sid S. came over, put a big wad of tobacco in his mouth and gave me a piece. I thanked him and said I smoke but don't chew. He kept coaxing me to chew a piece and Fred stood there grinning. Graham told me not to pay any attention to him and to work on the car. To my sorrow I finally gave in and put a small piece in my mouth. I raised the hoist and chewed up a storm, spitting with every chew. I looked up to drain the oil out of the crankcase, and before I could think, the tobacco slipped down my throat. For a minute or two I kept on working, then my stomach suddenly rebelled and I made a mad dash for the bathroom. I was so sick I thought I would never recover, but everything eventually comes to an end. I washed my mouth and face with cold water, came out, sat on the running board to catch my breath, and gave my stomach a chance to settle down. Those two clowns were still grinning when I went back to work. That was a powerful lesson, no more chewing tobacco, ever. I had to hurry to get the rest of the cars serviced by quitting time, but I made it. I went in the office and Mr. Wallace gave me the $1.50.

Graham H. invited me to go squirrel hunting before daybreak tomorrow and picked me up after supper. We spent the evening on the town, then went out to his dad's farm and slept a few hours.

February 14, 1933 – Tuesday

The weather was clear, chilly and the wind was mild.

Graham woke me up too soon to suit me; I put on my clothes and he gave me a gun and some shells. Still asleep, I told him I didn't know how to load or shoot it, and had never hunted squirrels. He loaded my gun, set the safety, and warned me not to shoot him or myself. I decided to let him do all the shooting. At the crack of dawn we walked into the woods and I could barely see him in front of me. I walked along still asleep as though I was on a sidewalk, and tripped over several branches. He stopped and said that if there were any squirrels around in this county, I was doing a great job of warning them. I tiptoed behind him and asked him to tell me the proper way to hunt them. He stopped again and with a forced smile, asked me to give him my gun and shells. He unloaded both guns and said that you can't go through the woods thrashing around like a mad bull, or talking a mile a minute, and expect the squirrels to sit quietly on a limb so you can take a shot at them. I would rather be sleeping in a nice warm bed. The sun came up over the horizon; we went back to the house and ate a swell breakfast.

He dropped me off at Barber's; I changed clothes, said, "Hi-di-do", to the gang, and walked down to the garage. Those two "jokers" were waiting and asked me how many squirrels I shot. To stop the razzing, Graham said he needed my help, and admitted he told them that I was a different breed of squirrel hunter. I helped him put all the parts back on the block, but he wouldn't let me do anything on my own. I handed him the parts and held them in place so he could bolt them securely to the block. When the engine was prepared to be installed in the compartment, we took the clutch assembly off the transmission. We inspected the pressure plate, bearing and found the pressure plate practically worn out. Graham removed the old lining and riveted new asbestos lining on both sides of the plate. We put the rebuilt clutch assembly on the transmission. We raised the engine above the engine compartment, but had a lot of trouble lowering it in place and lining it up so it would slide on the transmission shaft. After a lot of cussing and maneuvering, it slid in place and Graham bolted it to the frame.

We went home and had dinner. I returned to the garage and helped Graham install the radiator, battery, hoses, belts, etc. A Cheve came in for a grease job and oil change, and it didn't take me long to service it, but before I drove it off the hoist another Cheve came in. I spent the rest of the day servicing cars. Mr. Wallace came out in the garage at quitting time and told us that he had received the updated service bulletin from Cincinnati, OH today. Therefore, there will be a service school of instructions at 7:00 pm. tonight. It didn't concern me but Graham said I could come and learn the latest service methods of repairing and maintaining the new Cheve and Olds.

Mr. Wallace paid me $1.50. Graham and I walked out of the office, and as he locked the garage door, advised me to eat a light supper, they always have a lot of good food, he said, liquid and otherwise.

I did eat a light supper and Inez asked me if I was feeling poorly, or whether there was something wrong with the food. I told them I was going to an instructional meeting at 7:00 pm. at the garage, and they are going to have a mess of refreshments including the liquid type. Inez said to watch out for the liquid type. I shot the breeze with the gang, they said everything was just the same at the airport, but I still wanted to be out there with them. I waited with the rest of the mechanics for Mr. Wallace to show up and unlock the office. At precisely 7:00 pm. he arrived and unlocked the door and we staggered in. I noticed a man I had never seen before and Graham introduced me to Charles Robbins. He is a schoolteacher and part-time car salesman - full-time in the summer. We went upstairs, into a large room above the office, there were chairs along the east and west side. There was a large screen hanging on the north wall, a slide projector on a small table, and a long table. I don't think it had been cleaned for many a moon, and I was glad I didn't dress up. Mr. Wallace stood behind the projector table and talked about all the regulations and pressures the Chevrolet Corp. was forcing on all dealerships. He gave each of us a copy of the updated service bulletin, told Graham to operate the projector, and to explain each slide.

Graham was pretty darn good at explaining everything in good old plain English. Mr. Wallace then showed slides that pinpointed the improvements that were made in the new Cheve. He turned the lights back on as he concluded the training session.

Sid and Fred went into the other room and brought back enough snacks and liquid refreshments to feed us for a week. They placed everything on the table and said, "Come and get it". I followed Graham to the table, but was uncomfortable, cause I was just a temporary worker, but they urged me to dive right in. That was all it took, I filled my plate to the rim and wished it had "sideboards". I set it by my chair and went back for a glass that was already filled with soda. Graham sat beside me and we talked with our mouths full; eating and talking at the same time was no problem. But when I took a drink from my glass, my eyes popped out a good inch, and I coughed, spit and sputtered, it wasn't pure soda. They all died laughing while Graham patted my back. The drinks didn't seem to bother anyone else, I guess they are used to it. Between bites, I sipped my drinks very cautiously the rest of the evening. I had a suspicion they spiked the soda with "boot-leg" whiskey. It was amazing how fast the food and drinks disappeared, and everyone was a little boozy. With my stomach full, and in a mellow mood, I was ready to go home and hit the sack. The evening slowed to a crawl and I thought they surely would turn the lights out and lock up the joint.

Suddenly Sid said we needed some entertainment. Everyone, except Graham and I, thought it was a good idea. He said he would be back very, very shortly, and he wasn't gone very long. I heard him talking in the other room, he came through the doorway leading a woman in her birthday suit, naked as a "jay-bird", and walking like one. It ruined my entire evening! He said she was an exotic dancer, whatever that is, and would perform various dances for us, provided we chipped in to pay her. I didn't, I only got paid $1.50 today. It was terrible! In the first place she was not pretty by any stretch of the imagination. Second, she would look one hundred percent better with a few clothes on, her shape was nothing to crow about, and that is putting it mildly. Third, she must be about forty years old, and although I am a twenty year old, warm blooded, All-American boy, I sure am not interested in watching a forty-year-old woman make a fool of herself for money. And last, she just shuffled around the room waving her arms hither and yonder; she did have shoes on. I wonder where he found her? I felt sorry for her. I don't know anything about women, nobody does, but I did take a special art course in high school. We had three nude models, and a lot of clinical pictures of the human body, so I am not exactly wet behind the ears.

I told Graham I had to leave because I had to be at the airport early in the morning to do some flying. He left with me, took me home, and said this meeting was not to be broadcast. I am getting more education than I paid for since I took this temporary job, chewing tobacco and now this. I didn't waste any time diving in the sack.

February 15, 1933 – Wednesday

The weather was clear, very chilly and the wind was very breezy. I flew twice today, one hour and ten minutes and two flights; and fifteen minutes and one flight.

I woke up early this morning and felt like something the cat dragged in. I ate a nice hot breakfast and told Inez I would be back as soon as I finished flying. I followed the International truck out to the airport. During a bull session I told the fellers what I had been doing down at the garage, including the tobacco chewing experience. They really enjoyed that. I said they conducted a class on the newest method of repairing and maintaining the new Cheve last night. I described the food and drinks in great detail and was dying to tell them about the entertainment, but I kept my promise to Graham. The students didn't have anything exciting or different to talk about, just the usual wash and wax, fool around, horseshoes and school. They were anxious to finish school, get their license, and go home. I felt sorry for them. We received valuable experience repairing and rebuilding wrecked planes and saw the results of our labor when the ships took off and flew. These students get their experience on dummy airplane sections and engine. Harv asked me to join them tomorrow evening at school provided I let him do the teaching. I like the guy.

Joe drove up in a cloud of dust, slammed' the car door shut, and went in the office at a gallop. A few minutes later he opened the door and told me to take the "110" up for fifteen minutes, park the ship and come in the office. Bud Lawrence helped me fill the "Prep" with gas, and during the pre-flight test I noticed it was beginning to show its age. I pushed the throttle open and bounced down the runway, and the ship flew as per usual, it was well trimmed. The fifteen minutes flew by quickly, and I made a normal 360-degree turn and land maneuver. I parked the ship and went in the office. Joe told me to grab a cup of coffee and sit; and he would be finished in a few minutes. To pass the time I flirted discreetly with the pretty secretary, and she sure is pretty. He finished whatever he was doing and told me I was to be at Seymour, Friday, February 17th at 1:30 p.m. to take my physical exam. He also said he had been trying to get in touch with the Inspector to renew my license, and would keep trying. We went out to the office and he told Harvey to warm up the "Hisso" Waco Biplane and get it ready to fly, "Friday and I are going to test fly it". Was I ever surprised!

As we waited, he said it was time to take the ship up and blow the carbon out. I had never flown the ship before so he is going to check time me. "We have to stay close to the airport" he said, "because the ship isn't licensed, but I want you to have the experience". "We can't enter it in your log book, that is why I sent you up in the "110".

Harv warmed up the engine and it made the "Prep" sound puny. Joe V. told me to feel the controls as he did the flying, it's dual controlled, he climbed into the front cockpit and I climbed in the rear. It is an open cockpit biplane and the wind whistled past my face as we roared down the runway. It left the runway and climbed much faster than the "Prep", it has over twice the HP and you sure can tell the difference. Joe whipped through the "8's" vertical power turns, spins, stall and wingover. Then put the ship in a steep dive, pushed the throttle wide open, and pulled the stick back into a zoom. Instead of cutting the throttle, he held the stick back until the ship was upside down. When the nose went over the top, he cut the throttle, eased up on the stick, and recovered on the last half of the loop. I could feel Joe move the rudder during the entire loop as he kept the nose straight. It was simply great, great, great.

He pushed the throttle open, went up again, leveled oft, and motioned with his hands what I took to mean, that he wanted me to loop the ship. I took my hand off the stick, made a loop with my hands and pointed to myself; he nodded yes and put both hands up in the air. I grabbed the stick and throttle. I was surprised, but I pushed the stick forward and put the ship in a steep dive. The controls got stiff, I pushed the throttle wide open, smoothly pulled the stick back into a zoom and used the rudder to keep the nose straight. I held the stick back. With the throttle wide open, the nose left the earth and all I could see was the sky. When the ship was upside down and the nose went over the top, I cut the throttle, eased up on the stick, and used the rudder to keep the nose straight on the downward half of the loop. As the ship recovered, I pushed the throttle to cruising speed, and to my surprise, Joe clasped both hands together above his head. I hope as a sign of "well done". Both loops must have been ok because we remained in our seats during the entire maneuver. I closed my eyes and thought I died and went to heaven. He pointed down and mouthed a 360-degree turn and land maneuver, and kept his hands where I could see them. I spiraled down and as I approached the runway I felt a slight pressure on the stick and rudder. I know then, he was ready to correct any mistakes I might make, but the landing was better than I expected. I taxied back to the apron; Joe turned the ship around instead of parking it, and let the engine idle.

I thought he was going to go back up so I started to unbuckle my seat belt, when he turned around, pointed to his watch and said I had enough time to make two 180 degree turn and land maneuvers. He climbed out of the cockpit and waved me down the runway. I had to remind myself that all ships take off and land the same, although some are more powerful than others. I pushed the throttle forward; the ship rapidly picked up speed and wanted to leave the runway at the half way mark. The engine sounded like a brass orchestra, loud, smooth and with perfect rhythm. I wanted to try the vertical power turn maneuver with this 235 H.P. engine. I'll bet I could point the wings vertically straight down toward the ground and not loose or gain altitude. Not a chance, with Joe standing on the apron watching me. I completed the two 180 degree turn and land maneuvers with no problem, taxied back to the apron and parked the ship. Joe signed my logbook and said we might get a chance to fly a different plane tomorrow. I told Harvey he must have rigged the ship properly because it flew great. He just thumbed his nose at me as I climbed in the Cheve and drove down to Barber's.

Inez had the cleaning stuff prepared for me. We went up to the second room and she helped me move everything to one side. I cleaned and washed almost one half of the room before Inez called and said it was dinnertime. I washed and dashed down the stairs, and met the gang coming in, "Heaven help the hindmost". I listened as they complained about the problems they had with the major overhaul of the "122" engine.

Harv said they are considering replacing the fabric on the ship. I piped up, "don't look at me", I had plenty of that experience.

They left; I went upstairs and finished the one side of the room. I pushed and shoved everything to the side I had just cleaned and cleaned the rest of the room. I thought about that eight-cylinder crate all afternoon, and what a great time I had. I finished cleaning, except putting everything back in place, when I heard the gang come in. I talked them into helping me finish the job, and then we went down and ate supper together.

I planned to stay in stay in and rest tonight but Dave Gentrup stopped by and invited me to double date. We will take the girls to a free show, and refreshments at Cragmont, the state hospital. He picked me up at 7:00 pm and introduced me to the girls, Annie, apparently my date. Dave seems to have a new girlfriend as often as the sun comes up over the horizon, maybe a harem of them. We drove up to the hospital grounds and arrived a few minutes before the show started. The room was full, the patients and their attendants filled most of the seats. The show was ok, especially since it was free. Everyone went over to the dining room where the tables were stacked with the best of everything. Although I ate a big supper, I ate and drank my share. Alcohol was not allowed. I think my legs are hollow cause I don't put on any weight and one thing for sure, I don't have stomach trouble as long as I stay away from tobacco. I talked with several patients and attendants, just like the last time, it was a very interesting experience. The seriously ill are not allowed to mingle with visitors, and sometimes it is hard to tell who is the patient and who is the attendant. This is old stuff to Dave, he and his parents live on the grounds, so he entertained the girls. It was close to lock out time, he took Annie home, took me to the boarding house, and what he did after that was anyone's guess. I had a nice time, but he has to be a little more particular in choosing my dates from now on.

Bud L. is the night watchman tonight. I got a letter from Paul and Earl today, each contained a nice dollar bill that I will add to my collection. I went to sleep thinking about that "Hisso-Waco".

February 16, 1933 – Thursday

The weather was clear, very chilly, and the wind was slightly rough. I flew for 1 hour and 35 minutes, and 4 flights today.

I woke up early and was second in line for the bathroom. It is not the way it used to be when the whole gang was here. We had the usual hot breakfast and plenty of it.

The crew climbed in and on the International truck, I followed them out to the airport. Harvey followed me in his Cheve, and Joe showed up a short time later. Harv and the students unlocked the hanger, lined up the ships, and went in the office for today's schedule. He came out and said Joe wanted to see me pronto. I went in, had a hot cup of strong coffee that would probably take the varnish off the desk. We chitchatted briefly, and then he said it was time I climbed in a ship that I had never flown and take off solo. "The Eagle Rock, Hisso Biplane, hasn't been started nor flown for a time, and it is not licensed. It is like the ship you flew yesterday, of course all planes are somewhat different, but I think you can handle it without check time, don't you?" I said, "It is never too late to learn," and wondered if there was something besides coffee in his cup. This was a different Joe than the one I knew. He told me to watch for airplanes in the air, and if I saw one, I was to land as soon as possible, or he and I would be in trouble flying an unlicensed plane.

Harvey had the ship inspected, fueled up, warmed up, and ready to fly, when I got out on the runway. We looked at each other, and said at the same time, "what the heck is going on here?" He said I was to fly for 1 hour and 35 minutes, and 4 flights, make three 360-degree turn and land maneuvers first. If there were no problems, I was to go back up and perform maneuvers until my time was up. I climbed in the rear cockpit, buckled up, waved at him, and he pulled the chocks away. This ship felt quite different than the Waco, even the controls felt different, but it had the same engine and HP. I pushed the throttle open, we took off at the same place we did yesterday in the Waco. The engine didn't seem quite as smooth, but it did have plenty of pep. This ship, also a biplane, is quite different than the "Prep" as far as visibility is concerned. The "Preps" wing is above you and there is nothing below to obstruct your vision. On the other hand, the biplane has a wing above and a wing below. I went up about 2,000 ft. with the throttle wide open, circled the field, idled the engine, and as I spiraled down the engine began to run real smooth. Probably just needed the carbon blown out, or something. The Eagle Rock appeared to lose altitude faster than the Waco as I approached the runway. I gunned the engine briefly so I would not undershoot the runway and end up in the high lines. The three-point landing was tolerable.

I taxied back to the apron, turned the ship toward the west, saw Harv, Joe and the students standing in the doorway, Howard B waved "by-by". I pushed the throttle open, took off, and repeated the 360-degree turn and land maneuver, but this time I didn't need to gun the engine, and the landing was normal. Joe was the only one watching me as I took off and repeated the maneuver for the third time. The landing was smooth. This is heaven on earth. Wow! I took off, circled the field, Joe was nowhere in sight, and my time was about to run out. I climbed up to a very safe altitude above the airport, the controls don't respond as quickly as the Waco. Without thinking of the consequence, I put the ship in a steep dive, pushed the throttle open, pulled the stick back, kept the nose straight and we looped! It was not the prettiest but I did stay in my seat. I recovered, spiraled down and completed the fourth 360 degree turn and land maneuver. I taxied back to the hanger real slow so I could see whether there was any unusual activity.

I didn't see anyone outside, and I thought, if I was lucky, no one watched me fly. What a pipe dream. I parked the ship, walked through the hanger, the workshop, and into the office. Joe looked me straight in the eye as I opened the door, and he didn't smile, nor offer me a cup of coffee. He didn't give me a chance to open my mouth, but said, his voice rising, "I knew, without a doubt, you were going to do it," and slammed his hand on the desk. "Surely you didn't think I wouldn't hear that crate roaring up there and not knowing what you were doing? Did you stop to think how you would pay for that ship if it broke up, or how you could get down without it?" I thought he would tell me to pack up and go home, I just stood there and never said a word, and my heart fell down in my shoes. Finally, he said, "go out in the workshop and help them get it in shape for school. I want you to come out and attend the class tonight; and I hope the plane is ok." I was afraid I might say the wrong thing so I merely said, "yes sir', and walked out into the workshop. This was the old Joe. Harv and the students had it prepared for school and were out in the hanger goofing off. Harv asked me how the ship flew, and I told him the only thing I noticed, it seemed to lose altitude faster than the Waco.

No one commented on my visit in the office, although they must have heard Joe, he was quite perturbed.

Inez had all the cleaning equipment ready and told me to clean the hall upstairs. It was a lot easier than the two rooms, nothing to move around. I was getting a lot of experience I don't need. I finished the hall in time to wash up and take the cleaning material down to Inez. The fellers looked at me cross-wise during dinner as they discussed Joe's surly mood. Harv said I was lucky to be down here, a safe distance from Joe's wrath. I never opened my peeper. I knew they were trying to find out what happened in the office. I sure hope he cools off by the time I get back from Seymour tomorrow. I didn't realize he was that hostile, and I hope he don't show up at school. Harv said they expect me to be at the hanger at 7:00 p.m. sharp.

Inez fixed me another batch of cleaning solution and told me to clean my cubbyhole. I put everything out in the hall and cleaned it from top to bottom, side-to-side, and end-to-end. It was the only time I was glad it was so small. I cleaned up the mess, and put everything back in place, it smelled nice and clean. I washed, changed clothes, and was down in the kitchen aggravating Inez when the fellers came in. I even helped carry some of the food in the dining room, and it gave me a chance to grab the first chair.

I followed the truck out to the hanger and said a silent prayer of thanks cause Joe's car wasn't there, at least not yet. I sat in the back and had to remind myself that I was no longer a student.

The subject was "Meteorology", the only subject I don't know that is part of the Transport Pilot test. I am sure Harvey selected this subject for my benefit, and his, the students have not reached this level. He gave each one of us a copy of the lesson, as a reference. I listened intently all evening and tried to absorb and retain everything he said. He looked at me occasionally, either for a nod of agreement or because I was listening to every word. I think I surprised him.

The evening passed quickly, and as he finished the lesson, and answered a few questions from the students, I made a paper plane, the type every kid made and flew. He asked me if I had a question, and I said, "Seriously Harvey, why don't they build an airplane like this"? And I threw the paper plane in the air and it sailed around the room a long time before it glided to the floor. The students laughed and he seemed surprised, but I said, "Did you notice how long it stayed in the air?" He finally said, "How the heck would I know." "But I do recall that you turned in a sketch of an imaginary experimental plane as part of a test. It was like the one you just threw, and it is in your file." With that, we locked the hanger, the airport, and headed for the boarding house. Louis B. spent the night as the watchman.

Meteorology

The atmosphere is a gaseous envelope about the earth, being 150 to 200 miles thick and composed of nitrogen and oxygen. It has weight and is acted on by gravity. The lower layers of the atmosphere have the greater density, due to the weight of the air above; one half of the air is below 3 1/2 miles altitude. Atmospheric pressure is due to the weight of the column of air above the point where the measurement is taken, extending up to the limits of the atmosphere. Atmospheric pressure, since air is a fluid is exerted in all directions. The standard barometric pressure at sea level is 29.92 inches.

Weather maps, areas of high and low barometric pressure are constantly moving across the United States from west to east. A rapid movement and constant succession of highs and lows implies frequent changes in weather, while a slow movement portends continuation of the present weather.

Lines on a map connecting points having the same barometric pressure are called isobars.

Lines on a map connecting points having the same temperatures are called isotherms.

There are ten types of clouds, cirrus, cirro stratus, cirro cumulus, altostratus, cumulo nimbus, alto cumulus, cumulus, strato cumulus and nimbus.

A low-pressure area on a weather map. A region of low barometric pressure moves eastward at an average speed of 32 mph. The heavy arrows show the direction of movement of the low; the small arrows show the wind directions. The concentric closed curves are isobars; the broken lines are isotherms for every 10 degrees F.

The weather, (clouds, rain, fog, etc.) depends on the shape of the isobars. These shapes are seven classes, low or cyclone, high or anticyclone, secondary, v shape, wedge, col. or neck, straight.

In the path of the low, - at first the wind will be from the southeast, gradually veering to the south; when the storm center arrives, it sharply turns suddenly to the northwest.

North of the path of the low, - at first the wind is from the east, then the wind backs to the northeast and north.

South of the path of the low, - the wind is at first from the southeast or south, gradually the wind veers through the southwest to west and northwest.

The weather in a high barometric pressure is generally clear; the center of the high is perfectly calm with a cloudless sky.

The 'Secondaries – is a small center of low pressure and is attached to a primary low. Its feature is heavy rains, thunder storms and high winds.

V depressions - is an area of low-pressure thrust between two highs; rain, line squalls with high 'winds.

Wedge - is an area of high pressure thrust between two lows, the bad weather of the departing low clears up quickly. The weather is fine with cloudless skies.

Col or Neck - is an area between two high pressure areas, there being practically no pressure in the exact center of the Col, there is a calm there, while in the region surrounding this spot, the winds are blowing in various directions across the isobars. Violent thunderstorms are frequent. In general the weather is dull and overcast.

Straight - isobars are considered to be those running in a general east and west direction. The winds are from the west or southwest and are strong and gusty. As it travels eastward, showers, squalls and driving rain, is probable with winds and gusts.

The ration of the actual amount of water vapor present in the air to the amount which is the same volume of air would hold if it were saturated is known as relative humidity.

Questions

1. What is an isobar - an isotherm?
2. What is meant by a low-pressure area on a weather map?
3. What weather is likely to be encountered in a low area?
4. What is the general direction of travel, of low or high-pressure areas?
5. What are the types of clouds?
6. What is the direction of the wind as regards the center of a high-pressure area? Would this be the same in South America?
7. What is a weather map?
8. What is meant by relative humidity?
9. What principle gases form the atmosphere?
10. If you were flying due south and a low-pressure area was on your direct course, would you fly to the east or west of it, and why?

I got letters from Pearl and Bert, and accumulated two more bucks. I still have Mom's ten dollars. What a day! Tomorrow also is a big day.

February 17, 1933 – Friday

The weather was cloudy, cold and the wind was rough. I didn't get to fly today.

Ate breakfast with Harvey and the students. He reminded me that I had a doctor's appointment, and they left for the field. I told Mrs. Barber that I would have to leave for Seymour right after dinner. I thought I would not go out to the airport until I returned from Seymour with the doctor's report, to give Joe V. more time to cool off. But, by the same token, it would give me more time to worry, and I had to face the music sooner or later. I decided to grab the bull by the tail sooner; horns can hurt you, and go out to the airport. I helped Harv and the students line up the ships and get ready to fly. At the same time I kept my eyes pealed for Joe V.'s arrival and when he did I made myself as inconspicuous as I could. But he walked right up to me and said, "Beyer", I thought "Uh, oh", "You better not be late for the doctor's appointment, no exam no renewal". "Stop by the office when you return with the doctor's report." He never smiled, on the other hand, he never yelled or told me "to pack it in" as he and Harv went in the office. Don V. said, "You heard what the boss said", as they went about the business of goofing off.

I gave the Cheve a pre-drive check from stem to stern, and cleaned it inside and out. It runs great. I drove down to Maddox garage and bought four gallons of gas, 50 cents, enough to take me to Seymour and back. I ate dinner in a hurry and left before the rest had finished.

I was the only one in the doctor's waiting room, and I didn't have to wait very long in the doctor's office. He gave me the works, never missed a thing. He filled out an eight-page report and put it in a sealed envelope. He said I was in great shape, but my hearing and vision better not get any worse or I will need glasses; and that he was stretching it a bit this time. That shook me up and took the "wind out of my sails". I stopped at the hanger on the way home, gave Joe V. the report and told him I passed the test. He told me to be at the airport in the morning with Harvey and the students. He didn't act mad or glad, just the same old Joe V., at least I'm still here. Don V. is the night watchman.

Had a great supper with Harv and the students. There are not many outsiders at supper time. Once in a while the "towners" come in at dinner time, like Clarence Hoffman, Matt and Loyd Risk, Paul Gentrup and several others. I told Inez I would be back in the morning after I finished flying. I listened as the students recounted today's flying exploits, and to hear them talking, you would think they were the world's best. I'll admit, at that stage, I felt the same and no doubt recounted a few myself. We are a fine bunch of fellers, and that includes Harvey and Joe.

They all had dates, except me, and took off hither and yon. All alone, I got all spiffed up and drove out to Cleone's, she wasn't home so I drove down to Ed Becht's. Bob Shuck was there and we fooled around with the Hawaiian guitars. I'll never be any good at it, I don't practice. I went back to the boarding house and wrote Herbie Schwarz and Fred Moss a letter.

I got a package of Bugle tobacco from Ralph. I turned out the light and fell asleep. My sweet dream was interrupted by those jokers stomping up the stairs, they must have beat the "lock-out". I went back to sleep but the dream was gone.

Clyde H. Beyer

February 18, 1933 – Saturday

The weather was clear, cold and the wind was rough. I flew for fifty minutes, and two flights today.

I ate two eggs, sunny side down, a big piece of ham, two slices of toast, a small sweet roll and a cup of hot coffee. We went out to the airport, led by the International, followed by Harvey, and I brought up the rear. I helped with the usual routine of unlocking the gate and hanger, and getting the ships ready to fly. Don V. had the stoves roaring before he climbed in the truck and went down for breakfast. I watched Joe get out of his car and walk toward the office, to see what kind of humor he was in, but I couldn't tell. Harv followed him into the office and soon came out. He told Bud L. and I to help him get the American Eagle biplane, OX 8 cylinder engine, ready to fly. It is an ungainly looking plane, not one of my favorites, but should have plenty of power. It is similar to the Eagle Rock.

Bud L. got in the cockpit and Harvey cranked the prop. It started, but spit, sputtered, coughed up black smoke, sounded as though every other cylinder misfired, and caused the ship to shake. He yelled at Bud L. to push the throttle halfway open. I thought he was going to tell him to turn off the ignition switch, just then the engine started to fire on all 8 cylinders. It settled into a steady, smooth, light roar and Bud L. idled the engine. Harv told me I was to take it up for 50 minutes, and 2 flights. When I reached 1000 ft., I was to perform a 180-degree turn and land maneuver so he could observe how well the ship took off, performed and landed. I was to spend the rest of the time performing the routine maneuvers, Joe orders. "And my advice is, old friend and buddy, you better do just that", he said, "A word to the wise is sufficient." I climbed in, adjusted my goggles, the engine chuckled smoothly, I pushed the throttle open, we bounced down the runway, and took off with plenty of room to spare.

I completed the 180-degree turn and made a respectable landing considering the condition of the runway. Harv came over to the cockpit and asked me how it went, gave me the "0" sign and waved me off. I was surprised; the ship flew down right nice. It responded instantly to the slightest movement of the controls. Appearances can be deceiving I guess. I didn't have the urge to do anything dumb and when my time was up; I spiraled down and made a nice 3-point landing. I told Harvey the ship performed perfectly, but the runway sure needed a lot of work. Not to brag, but when I worked here it never was this bad. He told me that I did right well up there, and incidentally Joe V. wants you in the office. I went in and he asked me how the "Eagle" performed. I said the rigging and trimming seemed to be just right, in level flight the ship flies a steady course without the constant moving of the controls.

Then he said, "I called Joe Shumate at Louisville concerning the renewal of your license. He told me he was going to be away for some time, but could renew your license Sunday, February 19, provided you flew to Louisville, and met him at Bowman Field. I told him you would not have quite enough flying time by Sunday. He told me to send you down and he would make the necessary arrangements to take care of that, providing the doctor's report was favorable." That was a stroke of luck, taking that physical exam yesterday. I told him I was really sorry I messed up Thursday, and don't know what made me do such a thing. I thanked him for everything he had done for me, that I would never forget him, and said, "I think you are a great guy". I think he was embarrassed. He said, "Friday, that water went under the Madison bridge yesterday, tomorrow is a new day. Tell Harvey and the students to have the "110" in perfect condition to fly to Louisville tomorrow morning. I'll be here to see you off in the morning. To tell the truth, I was more concerned about you than that crate." We worked on the "Prep" til dinner time.

We were eating dinner when Inez came in and told me that Mr. Wallace called and said he didn't want me to report on Monday. Matt Risk, Loyd Risk, and Paul Gentrup sat opposite me. Matt asked me if I worked for Paul. I said I was a "grease monkey" at the garage during the days I didn't fly. Loyd asked me if I had worked on auto bodies, I had to say no but I was a good worker and always willing to learn. Paul said they had several cars that needed a lot of hand sanding. Matt said they would pay me the same rate as Paul, if I wanted to go to work at 7:00 a.m. Monday. I told him I would be there bright and early.

After dinner Inez gave me a mess of cleaning stuff and told me to clean the front bedroom upstairs. I had learned all the tricks of the craft by this time, and knew all the shortcuts, but it is still hard work. I finished in plenty of time to take a "Saturday night" bath, and shaved before the fellers came in from the airport.

During supper the usual lies were flying freely when Don V. suddenly hit his water glass a time or two with his fork. He announced with a grin that would stop a spin, that he had a heavy date. What a surprise! Everyone has a date, and I didn't get that "heavy" business. The lies and bragging resumed while I concentrated on cleaning my plate.

I didn't waste any time, I told them not to take any lead nickles, took off and flew down Main St. I went in Inglis Drug Store, sat down at the soda fountain, although I was full as a tick. I acted surprised when Nellie

H. came over and I ordered a cherry coke. I sipped the coke slowly as an excuse to chitchat, and to keep it from running out my ears. It was early in the evening and there were few customers, it gave us a chance to talk with little interruption. I felt great and thought I was making progress until she happened to mention that she had a boyfriend. My mind went blank, but I thought I heard her say his name was Frank Bird. We joked back and forth until the soda fountain became crowded with customers. I quickly said

I enjoyed talking with her and she gave me a great big smile. She is not only very pretty, but also has a wonderful personality. I gave up my seat to a customer and walked out of the store. They didn't make much on my 5-cent coke.

It seems as though I am always a step behind, or a minute too late with the girls, but as long as I have the planes to love I'm not too shook up. My time will come. I wandered down Main street toward Metzger's and met Joe and Carl talking to Harold Stephan, their neighbor. Joe told him I was the guy that always flies low over his house. Just then I saw Don V. and his "date" walking on the other side of the street. I don't usually talk about a person's physical appearance cause I'm no beauty. But I sure wouldn't want her to sit on my lap. I guess that old saying could be true, "beauty is in the eye of the beholder". I saw Millie and Jessie Ricketts waltz by with their boyfriends. I should say "men friends", they looked kinda old to me, and they made sure I saw them, with a big "how-di-do". I met Fred S.'s girlfriend, Lizzie, for the first time, and she seems to be a very nice woman. If you wander along the streets long enough on Saturday night, you will probably see everyone you know, but I never saw Cleone. I never saw the students, they are probably somewhere out of town. My guess would be, Harvey is sitting in the parlor with Inez, as he is not one to burn gas needlessly.

I finally gave up and walked slowly back to the boarding house. Sure enough, as I walked past the parlor and toward the stairs, I saw Harvey sitting in a chair, Inez was sitting on the couch, and they were talking like a couple of old folks. For crying out loud, he can't be over a year older than me. What a courtship, I'll bet she would like to wring his neck.

Tomorrow is going to be a great day. No one had to rock me to sleep.

February 19, 1933 – Sunday

The weather was partly sunny, very chilly and very gusty from the northwest. I flew two flights today, one for 1 hour and 25 minutes to Louisville, and one for 35 minutes to Madison.

I ate breakfast with the gang, put on my finest clothes, not counting my Sunday best, took my logbook, license, road map, goggles and helmet. Also, all the money I had, just in case, pipe, tobacco, (no smoking in or near the ship) and drove out to the airport.

They had the "110" parked on the line, filled with gas, warmed up and ready to go. Joe was in the office, one of the few days he was on time, and in a decent mood. He told me Harvey said the ship was in excellent condition, and he hoped I was likewise. He said, "Friday, I want you to fly the same route we flew May 16th of last year. Use the compass, but according to the windsock, you will be flying into strong crosswinds, which may cause you some problems. Use the Ohio River and the road map to keep you on course if necessary. Give this letter and the doctor's report to Joe Shumate when you land at Bowman field, he said he will be there until noon." We walked out to the ship, he gave me $2.00 to buy 5 gallons of gas for the trip home. He adjusted the compass and said; "I know you will bring the ship and yourself back in good shape." I said, "You can count on it, this is strictly a business trip". I climbed in the cockpit, put the letters, maps, pipe and tobacco on the seat beside me, waved, pushed the throttle open and headed south west toward Louisville.

The cross wind was pretty darn strong and gusty, and I had to compensate for it. At 1,500 feet I pulled the throttle back to slightly above cruising speed to overcome the strong wind that constantly pushed me southward. Although the engine ran perfectly, I automatically looked for places to land in case of an emergency, and for other aircraft. Several times, after checking the roadmap and the Ohio River, I had to make a slight change in direction. I had drifted too far south, so I headed a little west-northwest until I was back on course. I saw Louisville, noticed that I was still south of where I should have been, and I made another correction. I scanned the sky there were no planes in the area; and then I saw Bowman field. I circled the field and saw a plane on the runway. I didn't know whether it had just landed or was taking off, so I stayed up at 1,500 feet and circled the field. Evidently it had just landed, because it taxied off the runway and onto the hanger apron. I went down, and with no other plane in sight, made a nice 3-point landing on a concrete runway.

I taxied over to the hanger, parked the ship, walked in the Louisville Flying Service lobby, and saw Joe Shumate reading a newspaper. I introduced myself, apologized for being late, gave him the Dr.'s report, Joe V.'s letter, my license and logbook. He handed me the paper and told me to find a comfortable chair, while he checked the items I gave him. I sat down, lit my pipe, barely glanced at the paper, and waited for the Inspector to return. He finally came back, gave me my logbook, license and a letter addressed to Joe, but he kept the doctor's report and Joe's letter. He said Joe promised him that I would get the required flying time. Then he said that he was in the plane sitting on the runway, to see what I would do when I arrived, and thought I did a fine job. His attitude gave me the courage to ask him to go up with me and point out parts of Louisville. I was surprised when he said ok. I took off, went up to 2,000 feet, he pointed to Churchill Downs racetrack; it is big and elaborate, the downtown area, and several other places I never heard of. Louisville is smaller than I expected, he pointed over the Ohio River to Jeffersonville, Indiana. He said it was time to go back to Bowman field.

I parked the ship, thanked him for everything and asked if there was any chance for a guy like me to get a job in the Department. The answer was brief, "No", he said, "Only college graduates need apply, and even they are not being hired". He said he covers a lot of territory, and there are simply no flying jobs of any kind out there. I told him I have been writing and visiting airports on the list I received from Washington. "The best advise I could give you is get a job and buy flying time or buy a used plane. But never buy one that is not licensed or certified", he said.

We went in the Louisville Flying Service office, he introduced me, told the manager I graduated at the Dixie Flying Service, Madison, IN and he just renewed my L.C. license. I told the manager I was looking for a job in the aviation industry and hoped he had an opening. He laughed and said practically the same words the Inspector told me. I looked over at Mr. Shumate and he gave me a, "see what I mean", look. I thanked the manager, said I needed 5 gallons of gas, and would be on my way. As I went out the door, the Inspector said, "Good luck; and about your first landing, it was a test. If you had attempted to land, with the plane on the runway, or made any other maneuver other than what you did, we would have a problem." I thanked him again and said, "I think you are a mighty fine Inspector and hope to see you again". A young "grease monkey" put 5 gallons of gas in the tank and I paid him $1.50. He initialed and signed a receipt that stated "Louisville Flying Service, No. 815, 2-19-33, 5 gal. gas @ .30 = $1.50.

I cranked the prop; the engine sang a reassuring song of "fives", scanned the sky, all clear and took off cross wind. Good old Joe Shumate stood there and watched me head home at 1,500 feet. I was hungry, didn't have but one chance to smoke and emotionally worn out; job interviews always made me up tight. So I flew up the Ohio River with a strong tail wind, cut across the bends and curves in the river, and soon approached Cragmont and the bridge. Then I was over the airport; there were no planes in the air or on the runway. I circled the field, looked down at the rough runway, idled the engine, and made an ordinary landing. It took me 50 minutes to fly to Bowman field, due to the strong head wind, and 35 minutes to fly to Madison due to the strong tail wind. What a great feeling! I made all the decisions, was in control and responsible for how, where, and when I flew going, coming, and in between, as though I owned the plane.

I taxied back, parked the ship, Joe's ship was on the apron, and I didn't see any sightseers. I went in the hanger, Joe and the students were in the workshop loafing and shooting the breeze. I told them about the trip, no problems, and I was glad to be back. I showed them my renewed license, and gave Joe the Inspector's letter, the gas receipt, and the 50 cents. He gave me the gas receipt and the 50 cents that I was to use to buy my lunch in Louisville. He signed my logbook and told me to hang around, he went in the office to read the Inspector's letter. He came out and said, "Shumate was pleased to renew your license, and said you did a fine job flying around Louisville. He also mentioned the trick he played that you handled very professionally." I was surprised the Inspector mentioned the sightseeing trip and expected Joe to make a big deal of it. But he merely told me that I would fly twice on Tuesday and once on Wednesday for a total of 4 hours and 15 minutes. That will complete the flying time as requested by the Inspector.

I drove downtown, it was too late for dinner and too early for supper, so I went into a "greasy spoon" on Jefferson St. and ordered a 25-cent plate lunch. I acquired a reputation, over time, that I could and would eat anything that came down the pike, but I couldn't eat that stuff. I went back to the boarding house, told Mrs. Barber and Inez about my trip, and to call me at suppertime. I wrote Mom a letter and fell asleep. The students made enough racket to wake the dead, and I ain't dead.

The supper was great and I ate enough for both the dinner and supper. We went in the parlor, swapped a few lies, and played a few rounds of checkers. I should have suggested that we bet on the games as I won all but one game; Howard B. beat me once. On the other hand, they are as poor as I am. I went up and wrote Earl Seitz a letter. I told Mom I probably would be home this weekend. I am starting on a new job tomorrow so I hit the sack.

February 20, 1933 – Monday

The weather was cloudy, cold and the wind was rough. I didn't fly today.

I got up early, ate a big breakfast, and then walked a block and a half to Matt Risk's Auto Body Shop. It is located on the northwest corner of Jefferson and Third streets, and faces Jefferson Street. The Risk's home is a two-story house attached to the rear of the single story garage and faces Third Street.

I arrived early, Matt and Loyd R. was there, Paul Gentrup soon showed up, and Frank brought up the rear. After introducing me to Frank, Loyd gave me a pan of gasoline and several pads of waterproof sandpaper. He showed me how to sand the straightened and repaired sections of the body smooth. The edges of the paint adjoining the repaired area had to be sanded to a featheredge using a series of different grades of sandpaper. The sandpaper is to be immersed in the gas, as often as is necessary to keep it wet and clean of paint. I started using medium grit, and he showed me how to run my hand across the sanded edges to feel the places that needed more sanding. I sanded that area until I thought it was smooth, then went to another area and repeated the process. I sanded the last area on that body and Loyd came by and checked the sanded areas. He said they were ok for that grit and to sand them with a finer grit. I finished sanding all the areas and heaved a sigh of relief, but Loyd examined the areas and said they were ready to be sanded with the finest grit. I went through the whole rigmarole again and this time he said it was ready to be painted with a primer coat.

There were 3 cars in the garage and they were parked close together. He showed me where to sand the areas on the second car; and the gas began to irritate my hands. Matt's wife "Susie" soaked my hands in warm water, gave me thick salve to rub on them, and a pair of gas proof gloves to wear. She seems like a very nice woman, real friendly, and she bawled

Loyd out because he didn't give me the gloves before I started sanding. I couldn't help but notice her hair was dyed red, and was not becoming to her. She is a very nice looking woman. The noise, hammering, grinding, etc. made talking difficult as well as nerve wracking. I sanded the second car until noon, washed up, and walked back to the boarding house.

Neither Harv nor the students had anything exciting to talk about, no "ground loops", landing gear collapses, wrecks, or deaths, just the expected bragging. I sure didn't have anything to brag about, just a pair of sore hands. I walked back to the garage, Matt and Loyd and Paul were sitting on the running boards of the cars and I joined them. Matt asked me what did I do at Paul's garage and I told them I was Graham helper. He said auto body workers made a lot more money than mechanics. I never said anything but thought this isn't for me. Frank and Paul repair and get the bodies in shape to be painted and Loyd does the painting and finish work. Matt is the "boss man", "Susie" takes care of the office, and most of their business is with the insurance companies.

I finished the second and third cars before quitting time. Matt said I did a fine job, and showed me an area Loyd had painted with a primer coat that will be sanded with the finest grit. Then a coat of paint, matching the original paint, is applied and sanded. Usually at least 3 coats, and sometimes more, are applied and sanded. The last coat is sanded, then the final sanding is done with pumice powder, and finally the auto body is polished with wax. When they paint, all other work stops to prevent dust in the air. I washed up at quitting time and was thankful that all the noise had stopped. I went in the house, that's where the office is, and thanked "Susie" for the $1.50. She invited me to eat supper with them, but I politely declined and said I was pleased to meet them.

My hands burned, my wrists, elbows and shoulders hurt as I walked back to the boarding house. Nine hours of that kind of work is enough for me, but it did give me an appetite. I ate a good size supper and when I sat down in the parlor I had to loosen my belt. I listened as the students rehashed their exploits of the day. Apparently everyone got to fly, Bud L. and Howard B. spent the rest of the day repairing the runways. Louis B. flew through a telephone line and Joe had a "conniption fit'. Don V. practiced spins and had a hard time stopping the last one, Harv said he almost had something to write home about, it was that close. Louis B. and Don V. said they spent the rest of the day taking the training engine apart. Harv didn't list his activities, and I didn't add anything to the conversation, sanding bodies isn't exciting news. Directly all four students said they had to go "see a man about a horse" (dates) and took off. About that time Inez came in, and I know three is a crowd so I went up to my humble abode, rubbed the salve "Susie" gave me on my hands, and went to sleep wondering what tomorrow would bring.

February 21, 1933 – Tuesday

The weather was sunny, clear, and very chilly and the wind was pretty rough. I flew twelve flights, and 3 hours and 45 minutes today.

I was the last one to sit down at the breakfast table this morning. My hands still hurt but I managed to get the bacon and eggs, toast and grits in my mouth. The hot breakfast and coffee woke me up; I cranked up the Cheve and flew out to the airport. I shot the breeze with the fellers as they opened the hanger and lined up the ships. They have been having classes during the day cause there is nothing else to do. Bud L. said they have washed and waxed the ships so often that it is a wonder there is any paint left. Joe showed up with a young feller he introduced as the newest student, William J. Biehl, RR 1, St. Paul, IN is his handle.

Joe told me to get the "110" ready to fly, and told the new student to help me. As I pre-flight tested the ship, I explained what and why I was doing it. He watched me intently, and asked a blue million questions, as we filled the tank with gas and added about a quart of oil. Harv came over and told me that Joe said I was to take the student up for 25 minutes. He asked me if I was blackmailing Joe cause he was to get the Curtis-Robin ready to fly, and bets that I will be the one to fly it. To aggravate him, I politely asked, with a smile, if he would kindly crank the prop after Bill and I got in the cockpit. He just shook his head in disgust and walked in front of the plane as Bill and I climbed aboard. We fastened our seatbelts, and we went through the entire starting routine for the benefit of the student. I showed Bill, probably to impress him a little, my license and told him it had to be in your possession while flying.

I pushed the throttle wide open, the engine sounded great, I idled it and told Bill he must not touch the controls. Then I told him the story about the fire extinguisher, I pushed the throttle open, and we took off and performed all the maneuvers. I explained each maneuver in detail and said he would be required to execute them to receive a Private Pilot's license. I tried to imitate Joe but was not very successful and I soon gave it up. I flew over Madison to give him a view of the town from the air, then circled back and landed smoothly. He was as enthusiastic as I was the first time I took an airplane ride. He thanked me and told Harvey I was perfect, but Harv just rolled his eyes, shrugged and walked away. I think he is beginning to like me a little.

I noticed a Curtis-Robin monoplane sitting on the apron and the OX 5 radial air-cooled engine idling as we walked toward the office. Joe met us at the door and asked Bill if he liked the ride, without giving him a chance to answer told him to join Harv and the students. He closed the door and said he would fill my Cheve with gas and give me $1.50 if I would take Bill to Seymour in the morning for his physical exam. I didn't know what connection the idling ship might have had with my answer, but I didn't take a chance and said, "1 sure would". In addition to the gas, that would be the easiest $1.50 I've earned so far. He told me to take the Curtis-Robin up for one hour and perform every maneuver, including a spin, but not a single one that is not required! I was to watch for any abnormal reaction to the controls cause the plane hasn't flown for quite a spell. Am I being used as the amateur test pilot? I was to be careful.

I told Harvey I would be back I an hour, climbed in the cockpit, gunned the engine, and took off. The ship was slow in responding to the controls as I climbed straight ahead. I leveled off and had to constantly move the stick to keep the ship level. I adjusted the stabilizer, the adjuster has a limited range, and it seemed to help a little. I circled the field, idled the engine, spiraled down and made a wheel landing with a slight bump. I taxied back to the hanger and turned off the ignition switch. Harv came out and I told him that I had trouble keeping the ship flying level, that I adjusted the stabilizer and it seemed to help somewhat. He put the cockpit adjuster back in neutral, went back to the tail and adjusted and readjusted the stabilizer several times until he was satisfied. He said, "That should do it, you might have to fine tune it with the cockpit adjuster, I'll crank the prop for you." 'Thank you sir," I said and poked him in the ribs; I think he is beginning to thaw out a little.

He cranked the prop, I took off, leveled off and the adjustment of the stabilizer worked just fine. Although the Curtis-Robin and the "Prep" are monoplanes, they have different characteristics, and perform quite differently. All the maneuvers were normal, but halfway through the left vertical power turn the engine began to cough, spit and sputter. I leveled off but the engine continued to misfire, I idled the engine, spiraled down, landed safely. I tied the stick back and left the engine idle so Harv could hear it misfire. We decided to check the spark plugs and the number three plug's porcelain was cracked. We installed a new one and the engine sounded great. I took off for the third time and I completed the left and right vertical power turns. I went up and executed several very sloppy left and right spins cause I didn't compensate for the slow response to the controls. My hour was up, I spiraled down, and this time I managed to achieve a nice 3-point landing.

The OX 5 makes more noise than the Velie M5 engine that powers the "Prep". I parked the ship, started toward the office but Harv sent me back to the "Curtis" to cover the engine with a tarp. He doesn't think it will be flying in the near future. I went in the office before going downtown and told Joe we had a few problems with the "Curtis' which we corrected. But there is still a problem; the ship is slow in responding to the controls. He asked me if I had the time to check out the Travelair this afternoon. I said, "Yur durn tootin".

Harvey asked me, in jest, if I was going to fly the Fleet this afternoon. Acting serious, I said, "No, but I am going to fly the Travelair", and he almost choked on a piece of chicken. He said he was only joking, and I said I wasn't. I told Inez I had to fly this afternoon and can't do any cleaning. I don't think Harv believed me until he heard me tell Inez.

Joe drove up in a cloud of dust and Harv followed him into the office. He came out and said, "I give up, let's get the Travelair biplane ready to roll". Louis B. and Don V. helped us, Harv told me to climb in the cockpit and Louis B. would crank the prop. We went through the usual routine, but the "OX 5" would not start. Then Don B. cranked a spell, no response, Harv cranked until he was blue in the face and said, "One more time". I crossed my fingers cause I was dying to take this crate up. One cylinder fired, he cranked again and several cylinders fired belching black smoke, but after several revolutions the engine stopped. I held my breath, I was sure he would tell me to crank it, but Don V. grabbed the prop, gave it a mighty heave and the engine started with several cylinders firing. This time I kept it running by moving the throttle and choke back and forth. The black smoke changed to grey as the rest of the cylinders began to fire. The smoke cleared up and the 5 cylinders fired smoothly as the engine warmed up. I gunned the engine, it sounded great, I idled the engine as Harv came over to the cockpit. I said, "All that cranking sure wore me out", and he said "I don't see any sweat on your face, the next time you crank it".

History repeats itself, but I couldn't resist doing it cause all three sweated quite a bit. (And now, "Friday", be prepared to do your own cranking.) I wouldn't dare say or do anything like that to Joe I don't think he would appreciate it.

Harv thought we should tune up the engine before taking the ship up. I was afraid I would never get another chance to fly it, and asked him, as a favor, to forget we had trouble starting the engine. He said, "You are on your own", and pulled the chocks away. I guess he remembered the favors that I did for him. I took off, the ship felt real comfortable, and instantly responded to the controls. It was very easy to recover from the stalls, vertical power turns and spins. At 1,500 feet I idled the engine, completed the 360-degree turn and made a smooth landing. I taxied up to the apron to turn around and saw Harv and Bud L. working on the "Fleet" biplane, my dream plane. I took off and at 1000 feet I looked down in time to see a "Prep" take off. As soon as the "Prep" left the area, I completed the 180-degree turn and made a nice landing. They were still working on the Fleet as I prepared to take off.   The "Prep" was a safe distance north of the airport, so I took off and at 500 feet circled left, flew south of the airport, circled left, approached the runway from the east, idled the engine, came in high over the high lines and made a dandy side-slip landing.

The "Prep" was still up north; I took off for the last time and stayed south of the airport. I made several wing-overs, and had just enough time to make a picture perfect 3 point landing. The Travelair was the most comfortable and relaxing plane I've flown so far. It was the same feeling that you have driving your old car. It responds to the controls smoothly and practically flies itself in level flight. I love this ship. I parked the ship and covered the engine and cockpits with tarps; and watched the "Prep" land.

I walked over to the Fleet and asked Harv what he was up to fooling around with my sweetie. He said in a jocular mood, "I'm trying to get this baby ready to fly so you can go up and play around". I said, "you are kidding"? "Nope", he said, "you are to take it up for one hour and twenty minutes, no more, no less". It was my turn to be surprised, and I started to help put gas in the tank but he reminded me that I was no longer a student. I reminded him that I helped rebuild this ship from the ground up, he laughed and said those were the days. We pushed the ship on the runway, I climbed in the rear cockpit, we went through the starting routine, following his offer to let me crank her, that I declined, and after a few coughs the Kinner K5 radial, air-cooled engine started. It soon warmed up and settled into a nice smooth cadence. I gunned the engine, it barked sweetly, I pulled the throttle back to idle, Harv came over and said that if I put one tiny scratch or buzzard "do-do" on this plane he would wring my neck. I'm sure he is big enough to do it.

He jerked the chocks away and yelled, "Take it away". We flew down the runway, leveled off at 2,000 feet relaxed and proceeded to enjoy the scenery, the beautiful landscape below, the types and sizes of the farm houses, barns, and the shape of the land. It makes me wonder, who, what created this beautiful universe? I know! The plane flew like a dream, we apparently did a good job, and I can't understand why it hasn't been licensed. All the maneuvers were a breeze, and I executed the 360-degree turn and made a normal landing. I took off, the "Preps" were parked so

I had the sky to myself, I completed the required 180-degree turn and made another nice landing. My time was almost up, I didn't waste any time taking off, and at 500 feet approached the runway, came in high over the high lines and made a great side-slip landing. The biplane sideslips require a different timing and execution. One more required landing, I took off, completed a 180-degree turn, ignored the high lines and made a fine three point landing. I parked the ship, told Harv and the students that the ship sure is perfect. He pretended to examine it from stem to stern and admitted it looked ok.

I went in the office and told Joe the ships flew great. He then told me that I only needed 30 minutes flying time to fulfill Joe Shumate's requirements and he wants me to get it tomorrow. "Fill your car with gas before you go downtown, Biehl and you can leave for Seymour in the morning after breakfast." Harv helped me fill the Cheve with gas until it almost overflowed. It still runs great but needs to be polished.

I joined William, Bud, Howard, Louis, and Don, in the parlor after supper. We talked about the ships I flew today, which were the easiest to fly and which one was my favorite. I said the Fleet flew perfect for me, I helped rebuild it, so it is my favorite. The Travelair was a close second, I felt relaxed and comfortable flying it. he "Hisso-Waco" was a close third, it had the most power. But I liked them all, and would be glad to own any one of them; however, I am partial to the biplanes. I told William we would be leaving for Seymour right after breakfast.

I went up to my "cubby-hole" and rehashed this great day. I never flew 3 hours 45 minutes, and 4 different airplanes in one day. Ever since the day I started to tear the fabric off the Fleet, I've had a burning desire to fly her, and she is all that I thought she would be, she sure is a sweetie-pie. No reward or honor could compare to the feelings I have about today's events. If I knew how to cry, I think I would. I could not go to sleep, I flew each ship again and again and I finally did go to sleep.

February 22, 1933 – Wednesday

The weather was cloudy, cold and the wind was rough. I flew for 30 minutes and 2 flights today.

William and I took off for Seymour following an early breakfast, he had an 8 am appointment, and an hour later we were in the doctor's waiting room. I knew the tests were going to take all morning, so I went for a walk around town to kill time. It is situated quite differently than Madison, as it is quite level country. William was still in the examining room when I returned. I glanced through quite a few magazines, my eyes got tired and I dozed off. Wm. B. closed the office door and woke me up, grinning like the cat that just swallowed the goldfish. He didn't have to tell me he passed the test. We took off for Madison; he said a few of the tests bothered him, particularly the depth perception test. I told him it is considered the most important. The Cheve never complained as we sped down the highway at 40 mph. We stopped at the boarding house although it was well past dinnertime. Mrs. Barber fixed Bill B. and me a plate of leftovers, it tasted great, and we were starved.

We drove out to the airport, went in the office, Bill gave Joe the doctor's report. He told Bill to go in the workshop where Harvey was teaching class on aviation. He turned and told me to have a cup of (varnish removing) coffee, and sit. I placed my chair so that I was able to occasionally glance at and admire his pretty secretary. I don't know whether she is spoken for, but I do know she has never dated any of the students, and she must be fresh out of high school. I knew it was too late in the day for me, but if you can't or don't admire a pretty girl, or anything else for that matter, you must be over 50 and ready for the bone pile. Joe said he wanted to complete the promise he made, therefore I was to fly the final 30 minutes today. He told me to choose the plane, except "Peg", (that's off limits) that I wanted to fly as a going away present. Naturally I chose the Fleet and he wasn't surprised.

He turned serious and said, "Friday, I've tried to treat you fairly and I hope you will agree. I can honestly say I wish we had a job for you; first, because I know you would be a fine employee, you are a hard worker and trustworthy". I noticed the cute secretary smiled slightly. "You are a fine pilot, and have a natural interaction with people, especially with the students." Again, she smiled with a brief nod, Joe couldn't see her, and my mind kept wandering as he talked. She sure is a distraction. "And second, because it would mean the school was growing and prospering, you and Harvey would make a great pair. I'll admit it is a lot different around here since you and your "Grounded Club" members completed the course. For better or worse it is not as challenging".

"Before you get the idea that I think you are perfect, I must say you are too impetuous and too inclined to fly off the handle, I'll bet you used the boxing gloves more often than any other student." "Cutie" really nodded her head at that. I told him I was more than satisfied with the way he treated me, and I thanked him sincerely for everything he had done for me. I also thanked him for being so patient with my many transgressions, I said my ability as a pilot was a result of his instructions and I was grateful. "Joe", I said, "I will never forget you and I hope to see you again someday." "You too", I said to the secretary. He said, "Ditto". I said this was beginning to sound like a "Mutual Admiration Society". The secretary laughed out loud and that broke the ice, it began to get emotional. He gave me the $1.50.

Joe told Harv to help me fire up the Fleet for a 30-minute flight. I asked him why the "Hisso"-Waco, "Hisso"-Eagle Rock, American Eagle, Curtis-Robin, Travelair, and Fleet were not licensed. That was one of the few times I saw him get "riled up". He told me that every time Joe Shumate came, which was seldom, he always complained about some tiny defect. We would fix the "defects" and the next time he came, he would manage to find another defect. I'm sure Harvey took the rejections personally, because he is the A & E licensed mechanic, and is responsible for the planes. I told him I know the ships are air worthy and safe cause I helped rebuild and repair them. Maybe the Inspector is doing this because he thinks you are too young for the responsibility. "Hogwash", he said. He cranked the prop and I took off.

No more maneuvers, I cruised over Madison, Hanover College, Cragmont State Hospital and the hills across the river in Kentucky. I looked for buzzards over Clifty Falls State Park and flew over all the farms that were so familiar. This was really, really great; my dreams have come true. Now, I know the "Supreme Aviator" in the heavens scheduled me to be an Eagle (Buzzard?) but the signals were crossed and I was born a human. My time was up and I forced myself to idle the engine, spiral down and as a last gesture, I sideslipped in with a 3-point landing. I taxied out to the end of the runway, turned around and sat there with the engine idling. I looked toward the east and saw the hanger I helped build and all the planes I helped repair and rebuild. I taxied back to the hanger before I became too sentimental.

I parked the beautiful ship, shut down her engine, unfastened the seat belt and gazed toward the west. I saw a field I helped dynamite the tree stumps out, cut clumps of tree saplings down and burned piles of brush. I saw a field I hauled and spread truckload after truckload of creek gravel and cinders to create East West, and North South runways. I saw a field a sea of mud, the ships able to roll merely inches at a time and unable to take off. I saw a field where I worked nine hours picking tomatoes for one dollar and dinner. I competed against two brothers, Carl and Joe S., we each chose a row to pick, and they left me far behind. At quitting time I was unable to straighten up; I didn't have enough sense to quit. "Ma S." put towels, soaked in hot water on my back, then rubbed terrible smelling salve on, finally I was able to straighten up enough to drive home.

But today I see a beautiful airport of grass and beautiful level runways of creek gravel and cinders. This is my airport, I belong here; now, since it is up and running I have to leave. I don't know what the heck is the matter with me, I never felt this sad. Just as I began to feel sorry for myself Howard came over and brought me down to earth. That was scary. He said it was suppertime and they wanted to lock up the airport. We covered the engine and cockpits with tarps.

During supper I told everyone I would be going home early Friday morning. Don is the night watchman. Harv, Bill, Louis and I went in the parlor and shot the breeze. Bud and Howard had dates. Harv wanted to know precisely how each ship I flew performed and we discussed pro and con the good and bad features of each. He said he was going to retake the L.C. flight test soon. Inez came in and said Mrs. Barber wanted to talk to me. I went in the kitchen with Inez and her mother said I didn't owe any room or board. She asked me to stay another week or so and paint the woodwork and could pay me $3.00 a week, room and board. I said that I was tempted, and would like to, and it was a fair offer, but I wanted to go home and start on my career. I went in the parlor and the fellows were playing checkers. I played several games but was preoccupied and decided to go up in my "castle".

I got a letter from Earl S. today, and according to his letter, he has more ideas for making money than you can shake a stick at. He sure is a go-getter and a swell guy.

According to the local paper the federal government repealed the Prohibition Act today, now it is up to the states to do likewise. There was interesting news yesterday; the County Sheriff and his crew poured eleven kegs of bootleg whiskey down the sewer. I think they were ten-gallon wooden kegs. They destroyed the still, all the equipment, and the supplies and put the gentleman in the slammer.

I went to Social League at the church last week, they meet once a month, and had a swell time. Ma S. brings a bowl of food as my contribution; it is a pitch-in supper. Following supper, they conduct a business meeting and the members pay ten cents a month as dues that is used to pay for the meat and bread. After the meeting, the committee manages various simple contests, and the prizes are naturally small, like candy, chewing gum, etc., but it makes it interesting, like playing poker for matches. The whole family comes to these affairs.

I met two nice and interesting members, Andrew "Andy" Diedrich and George Grayson. They both work at the Post Office and look as different as are their jobs. "Andy" is big and looks like a typical farmer, he is a rural letter carrier, and delivers the mail out in the country by horse and buggy. He lives on West Second St. and keeps two horses, at his own expense, in a shed behind his house. His wife Florence, I think that's her name, reminds me of a typical farm wife and is very friendly. George is a trim, medium built, city letter carrier. He lives in the east end of town and delivers the mail on foot with a large leather pouch hanging from his shoulder. "Andy" never met a stranger and likes to talk. Whereas George is reserved, precise and is not in favor of wasting words, unless he is discussing his job. His wife Mayme is like that, stern, rarely smiles, and very outspoken, she intimidates me. I could listen all evening to the stories they tell about their jobs, and they do spend most of the evening doing just that. "Andy's" job sounds real interesting but I wouldn't know a thing about horses. George just walks up one side of the street and down the other putting mail in individual mailboxes. Working for the Post Office in Madison is considered a very good job. I hit the sack.

February 23, 1933 – Thursday

The weather was clear, very chilly and the wind was very brisk. I didn't fly today.

I waited until the students finished messing up the bathroom before I ventured in and did my share of messing. By the time I sat down at the table, the students and the few "locals" who eat breakfast here, had eaten most everything in sight. I still think Mrs. Barber favors me. As I scraped the bowls and platters clean, she brought me hot ham, two eggs, two pieces of buttered toast, a small bowl of grits, and a hot cup of coffee. That almost made me decide to take her offer and stay a week or so, but not quite.

I drove out to the airport, Joe was already there, and about to climb in the "Prep" beside Louis I asked him if it was ok to wash and tune up my Cheve. He said it was ok provided he didn't have to help me. He told me not to leave, and said we would have my Pilot Log book notarized after he check timed Louis, Don and gave Bill his first

flying lesson. I told Harvey that Joe seemed unusually "chipper" this morning and wondered if it was because I was leaving in the morning. "Could be", he said with a slight grin, "but I don't think so, didn't you know he kinda likes you"? "This would be an exciting place if you worked here" and "pigs could fly", I said.

I washed the Cheve, tuned up the engine and pre-road tested it. I have been neglecting her for quite a spell, but she seems to be in great shape. Occasionally I watched the "Prep" flying and noticed that when Joe and Don landed, they stayed out at the end of the runway a long time before returning to the hanger. I've been through that experience a few times and know what took place. Don had my sympathy. Harvey told Louis to be ready to hop in the "Prep" when they taxied up to the hanger. Don turned

386

the ship around and climbed out, Louis hopped in and they took off. If Louis was flying the ship, it was a real nice take off. I watched the "Prep" go through all the maneuvers and thought he was doing a nice job. They spiraled down in a 360-degree turn and made a fine landing. I guess Joe thought so because they turned around and taxied back without delay. They taxied over to the gas pump and turned off the ignition switch. Louis filled the tank with gas, Joe told Bill to be ready to fly and went in the office. He soon came out and told me to meet him at his dad's furniture store after dinner, about 1:00 p.m. and he would have the Log Book notarized. Then to everyone's surprise he asked Harv to fill my Cheve with gas and winked slightly at me (maybe he just blinked) as they climbed in the "Prep". They took off and it was obvious who was doing the flying. The instructor tells you to fly as he teaches, not as he flies, and there is a big difference.

Harv filled the tank to the rim, replaced the cap and said, "If I didn't know you so well I would say you were the instructor's pet", and ducked.

I went in the office and informed "Cutie-Pie" I was going home early in the morning, and that it was a privilege to have made her acquaintance. She said, "likewise", and that smile lit up the whole room. This was the last time I would see this hanger, I looked carefully around the office and remembered the numerous times I hauled coal to fill the stove. I looked intently around the workshop and remembered the times "Jack" and I slept there as a night watchman. I saw all the wings, fuselages, tail sections and engines I helped rebuild. I walked around the inside of the hanger and remembered the million times I washed the windows and the countless times I washed and waxed every plane.

I walked around the outside of the hanger and saw all the kegs of nails we used to install the siding and roofing, then painting it. I remembered climbing up on the roof and replacing the windsock from time to time. On the humorous side, I saw "Pop" jumping off the roof looking like a giant spider, with his long arms and legs waving every which way. Finally, I gazed up at the front of the hanger and saw, with pride, the words I painted, "DIXIE FLYING SERVICE, INC., MADISON AIRPORT". It covered the entire area above the sliding doors. Although I never painted letters that huge before, it looked great. I went to each ship and shook its prop, and thanked each one for bringing me safely down. I gave the Fleet a special hug. I filed these scenes in my brain so they can be retrieved at any time, and never forgotten. I jumped in the Cheve and hightailed it down to the boarding house.

I met Joe in Vail's Furniture Store; they were in the office talking as I walked in. I gave Joe my Log Book; he introduced me to Dale Wilson, the bookkeeper, who notarized it. I told Mr. Harry that I was glad I met him and he wished me the best. Joe and I walked out and as we approached his car, he said they were expecting a new student, O.E. Shirley, to arrive in a day or two. He shook my hand and said he will be glad to help me in any way possible. Then he said, "If you ever find the money, I will be glad to rent you a ship for $7.00 an hour". He climbed in the car and said, "Friday, I wish you the very best". I said, "That goes double", as I walked down to Naill's Lumber Yard, and informed him I was leaving town, and that I glad I had the opportunity to meet him. My next stop was the office at 216 E. Second St., where I informed Robert R. Foster and the office girls that it had been a pleasure to have known them. I back tracked down to Penn Maddox Garage and said "so long" to all my old friends, and "don't take any lead nickels, you guys are the best".

I don't know what happened to Clyde Weber, I heard a rumor that he joined the Army, I haven't seen him in a long time. That pretty girl wasn't in the drugstore; therefore I went back to the boarding house and arranged all my belongings in order to load up the Cheve quickly in the morning. I walked over to Matt Risk's Body Shop and informed them, including "Susie", that I appreciated their help. I drove down to the Cheve garage and informed Graham and the rest that I was leaving town in the morning, and thanked them for everything. I saw Dave Gentrup, Ed Becht and Bob Shuck walking up Main St. I parked the car and told them the same story. Dave G. and I recalled briefly the double dates we had, he still doesn't have a license. I drove back to Barber's just in time for supper and it was great as usual. I was lucky to have a "room" in this house. It was a home away from home, and the women even acted like a substitute mother, kinda irritating sometimes. They sure are fine people, and no doubt I should have listened to them more often. And the food was nearly as good as Mom's. Since leaving home I've eaten in a number of "Greasy Spoons", and it took a lot of imagination to call the stuff they served, food.

I stopped at Cleone's on the way to the Sachleben's. She wasn't home, therefore I informed her family that I was leaving town. All the Sachleben's were home and I had a great time talking about church, girls, D.F.S., and them. I reiterated that I had adopted them, and would never, ever, forget them. I invited every one of them to visit me and Joe S. said he would, for sure. It was late and I didn't want to get locked out; I thanked them for everything. "Pop" had Joe get me two "hands of long green". I fired up the Cheve, Joe S. handed me the tobacco, "so long brother". I shook his hand and said, "so long little brother" (he is 1 1/2 years youngerbut a lot taller). I got back to Barber's in time to meet Mrs. Barber locking the door. I was exhausted and hit the sack just in time to fall asleep.

February 24, 1933 – Friday

The weather was clear, cold and the wind was slightly rough.

I got up early and out of the bathroom before the students rushed in. I carried all my belongings down to the parlor and met Harvey as he came in. I thanked him for sharing his knowledge and experience with me; that everything I knew about planes is due to his teachings. I told him that he was a great guy and I considered him a good friend. I said, "Don't be surprised if we meet again someday". He said, "I hope so, it was an experience working with you". We went in the dining room and ate breakfast. They said, "So long" and took off for the airport. I put my stuff in the Cheve, double-checked my "humble abode" and I had forgotten my "Sunday best" suit under the mattress. I went down the stairs for the last time; Mrs. Barber was there and gave me an enormous lunch. That got to me; I had a hard time thanking her. She just gave me one of those motherly hugs and I hurried out to the car.

I took one last look at the boarding house, cranked up the Cheve and took off. I decided to drive, with the throttle wide open, past the airport and looked neither to the right nor left. But, as I approached within 1/4 mile of the hanger I couldn't resist, and changed my mind. I blew the horn constantly, slowed down to cruising speed, and finally slowed to idling speed when I reached the entrance to the airport. There they were, all of them, "Cutie-Pie", Joe,

Harv, Bud, Howard, Louis, Don, William, and "Jack" lined up like a bunch of crows on a fence, and waving up a storm. I was so intent on waving that I almost drove in the ditch but recovered in time, and flew up Michigan Road to Versailles. I turned right on S.R. 50 and drove east to Cincinnati 0.

I finally found the airport and talked to a Mr. Billman. I showed him my L.C. license and informed him that I wanted to apply for a job in the aviation industry. I explained that I had experience in repairing and rebuilding airplanes and engines. The response was the same as the others, I struck out again. I stopped at the Norwood, 0 gas station and had the tank filled with gas, and added a quart of oil. I continued east on Hwy. 22 through Wilmington, Washington C.H., Circleville, and stopped at the Lancaster airport. I asked Mr. Bookwalter if he was accepting applications for employees at the airport. His answer was a short and not so sweet "no". I thanked him, continued east on Hwy. 22 and stopped at the Zanesville airport. The interview with Mr. Carpenter was also a short and unsweet "no". I began to get the message. Although the school's brochure mentioned all the jobs available in the aviation industry, it must be some time in the future.

These rejections did not help my gloomy spirit I had since I left Madison. I stopped at the gas station, had the tank filled with gas, and added a quart of oil. I stopped at the east end of town, and although I was not in the mood, I ate part of my lunch. I drove east on Hwy. 40 and stopped at the Wheeling, WVa airport. I met a Mr. Cartwright and decided to try a different approach. I told him that I was passing through on my way to Pittsburgh, PA from Madison, IN where I graduated at the Dixie Flying Service - Dayton School of Aviation. We repaired and rebuilt airplanes and engines and I just renewed my L.C. license. He was curious about the school and we "chit-chatted" for a short spell. Then I asked him, if by chance, he could use an honest man willing to work hard at any job; me for instance. He said they didn't need anyone at the present time, but asked me for my name and address. He was the first one to ask me for my address and I thanked him.

I had a feeling that I had been through all this before. After I graduated from high school I heard the same phrase over and over, "We are not hiring anyone at the present time". One interview, in particular, still sticks in my craw. I sat in the waiting room, in a large office building, all afternoon and at quitting time the guy finally came in and said they were not hiring anyone at the present time. But I eventually did find a job. I hope the present situation turns out the same way. I stopped at a gas station had the tank filled with gas and added a quart of oil.

I drove north on Hy. 2 and stopped in Wellsburg, WV to visit my sister, Bert for a few minutes. I told her my school days were over and showed her my L.C. pilot's license. I said I stopped at the Stuebenville, Ohio and Wheeling, WV airports and inquired about a job. Bert lives about half way between the two airports, and said she would let me know when they began hiring. I felt slightly better after talking with her but could not shake a feeling of lost friendships. I continued north through Weirton, WV, merged with Hwy. 22 and drove northeast to Pittsburgh. When I reached the city limits, I decided it was time to get rid of the forlorn feeling I've had all day. Therefore, I turned my thoughts toward seeing Mom, Dad, Earl, Ralph and my full size bed. There might even be a letter or two for me, and that really perked up my spirits. It didn't take me long to reach home and park the Cheve.

I opened the door, said, "hi", and asked Ralph to help me unload the car. The next thing I said to Mom was in the form of a question, "did I get any letters?" There were only three letters, one from the Lincoln, NE airport, one from the Akron, OH airport, and one from the Harrisburg, PA airport. All three wrote practically the same message. "They appreciated my desire to become an employee of their organization, but did not anticipate employing anyone at the present time. They will keep my letter on file, and when an opening occurs, they will get in touch with me." I suddenly realized I was hungry, Mom gave me a cup of hot coffee and I ate the rest of my lunch. I said that I stopped at several airports on the road to Madison and the road home with the same results.

It was late, Dad had to work, I was tired, and so I went up to my room. My pride and joy ran great, never complained, 15 gallons of gas, 3 quarts of oil, $3.45.

I opened the book on this chapter of my life at 7:30 a.m., September 30, 1931 as we left Pittsburgh, PA and drove toward Madison, IN. I was eager in anticipation and expectation of attaining my dream of being a licensed airplane pilot, with a job.

Now I am closing the book on this educational, challenging, successful, wonderful, thrilling, exciting, comical chapter of my life with a Limited Commercial Pilot's License, and with a tinge of sadness, 11:30 pm, February 24, 1933. I am sure glad I took all those pictures!

Tomorrow I will open the book on a chapter in my life, in quest of a job in the aviation industry.

I hit the full size sack.

List of Students
Dixie Flying Service &
Dayton School of Aviation

Clyde H. Beyer, 4121 Main St., Pittsburgh, PA 10-2,1932
William J. Biehl, R.R. 1, St Paul, Indiana
Marlin Christianson, Box 22, Clarks Grove, Minnesota
John Hoover, 548 Montgomery Ave., Chambersburg, PA.
Kenneth B. Russell, 49 Emery St., Stanford, Maine
Arthur Prosek, Box 2, Hurley, Wisconsin
Sam Poletunow, Box 121, Bairdford, PA
E. Linwood Wright Jr., Churchland, Virginia
John Yellitz, Mountain Springs, PA
Pete Nemetsky, 315 Orchard St. Zeigler, Illinois
Lawrence Burklow, 920 W. Center St., Park Ridge, Illinois
David Swett, Box 212, White River Junction, Vermont
Frank Peternell, Box 185, Meadowland, PA
Chester T. McKinley, Ravalli, Montana
Rodney Pert, Bluehill, Maine
Andy G. Nufer, Jr., R.R. 3, Winamac, Indiana
Bob J. Parker Jr., Box 362, Galax, Virginia
Howard Brennan, 307 Bowman St. Wilkes Barre, PA
Bud Lawrence, 338 N. Defiance St. St. Mary's, Ohio
Daniel Booth, Tower, Indiana
George Hayden, Bangor, Michigan
Clair L. Kirsch, 117 Stone St. Gibsonburg, Ohio
Walter Etchison, Cana, NC
Howard Price, Darlington, Maryland
Max C.F. Sittner, Hudson, Kansas
Richard Boepple
Fred Works, 724 W. Lee St., Owensboro, NC
George Dykstra, 304 W. 15th St, Holland, Michigan
Al Thockle, Iowa
Louis Benoit, Fall Rivers, Mass.
Donald Voison, Colby, Kansas
O.E. Shirley (last student) 1933

16 states represented. Bud Lawrence became an aeronautical engineer. Walter Etchison became a flight engineer for Pan Am in North Atlantic Sector. Harvey Kattelmann – Douglas Aircraft Co., Santa Monica, CA. Aviation Technical School, Civil Aviation Authority, Federal Aviation Administration

Postscript #1

Two years later, March 1935, after intensive, relentless, exhaustive, and fruitless search for a job in the aviation industry, I accidentally wound up in Madison, Indiana; and stayed.

My Dad, as President of the Retail Baker's Association, had to attend their convention in Louisville, Kentucky, and asked me to drive him there. He didn't have to ask me twice as I looked forward to driving the Marmon. A day before leaving home he suggested that we drive the Cheve for a change.

We left home early and the Cheve seemed to be ok. We drove through Madison without stopping; then Dad said I could go back to Madison while he was at the convention, visit my friends, and maybe even find a job.

I was to pick him up at the hotel on Saturday at 4:00pm, but if I happened to find a job, I was not to drive to Louisville; he would take the Greyhound bus home if I didn't show up. We finally reached our destination at the Brown Hotel and he reminded me, "4 PM sharp".

I drove back to Madison and stayed at the Barber's Rooming and Boarding House. Everyone was surprised and seemed glad to see me. I spent the week having a great time, and went back to the Brown Hotel in Louisville, Saturday at 3 PM. It was wild; it was Kentucky Derby day. The lobby was crowded, and most of the people in the lobby had already had too much to drink. As I was early, I managed to find a chair where I could see the checkout counter. At 4 PM and no sign of Dad, I went up to the counter and asked the clerk if Martin G. Beyer had checked out. He said he would check as soon as possible; everyone wanted service, checking in and out, and jostling each other. Fifteen minutes later I went back to the clerk and he sent a Bellhop up to Dad's room. He came back and said the room was empty. By this time I was worried, angry, and demanded action. I asked for a phone to call the police. The clerk quickly checked the register and said that my dad had checked out at 2:00pm. I hurried across Broadway to the Greyhound station and asked the ticket agent if the bus to Pittsburgh had left. He informed me that it had left at 3:00pm but didn't know who boarded the bus in Louisville. Now I was really worried, anything could happen in that disorderly throng.

I jumped into the Cheve and drove back to Madison and told Mrs. Barber that I needed a place to stay, and what had happened. She said that I should send a telegram, so I did. "Did you get home ok? Send me money to get home." I got a telegram in reply and it said, "Got home, hope you are ok." That's all! No Money. I can take a hint when I see one. This was one of the two best advices my father ever gave me. The other was, "everything always turns out for the best in the long run". How true.

Neither kings, princes, potentates, nor anyone else could hold a candle to my father.

Postscript #2

One evening, October 1984, the doorbell rang; I opened the door, and was startled to see Harvey Kattlemann standing there. No one had seen or heard from him in 48 years.[33] We invited him in and after a short conversation; he asked me how he could find Inez Barber. I told him her name was now Inez Holcroft, but that her husband had died. He said that his wife had passed away recently. We looked in the telephone book and found her address, and he soon left. They were married on Valentine's Day 1985, and moved to California. They came back to Madison every year for a couple of weeks and never failed to stop by. They lived happily ever after. His tactics toward dating evidently had changed drastically from our days at DFS cause he quickly swept Inez off her feet.

One evening, a few years later, the doorbell rang, and I was surprised to see Graham Harding standing at the door. We had become very good friends while working at the Cheve garage, and at that time he wanted me to be a partner in buying the Cheve dealership. I didn't have two dimes to rub together and couldn't borrow any so that opportunity passed me by. He couldn't swing the deal by himself. I lost track of him for many years. We talked way into the night, he had moved to Indianapolis after his dealership was sold, then to California, and finally to Texas where he was working as an automatic transmission specialist. I never saw him again.

Joe Sachleben is the only one of the Sachleben family still living. He lives in Columbus, Indiana. We are very close, brothers you might say. We used to visit each other, but now we visit via phone and cards. None in that family was interested in learning to fly.

I guess that leaves me as the last living survivor of the Dixie Flying Service.

---

[33] Harvey barnstormed at the airport until 1934. In 1937 he was employed by Douglas Aircraft in Santa Monica, California. He taught at the Aviation Technical School, and worked for the Civil Aviation Authority and the Federal Aviation Authority.

Mildred Ricketts married Henry Mitchell, and Jessie Ricketts married Harry Burkhardt. Nellie Herrod married Frank Bird and they all lived happily ever after.

I lost track of Rosalee Darin and Cleone Grimes many years ago.

Me?As a little boy, before I knew airplanes existed, my first passion was railroad trains, bulldozers, cranes, road graders, and all types of earth moving equipment toys. For a number of years I earned a living repairing diesel-electric locomotives, diesel locomotive cranes, cranes, bulldozers, road graders, and other heavy equipment. In the process I learned to operate all of them. How often so you get paid to enjoy the real thing. So my one dream did come true.

I met the love of my life in church, naturally, a banjo eyed blond with a broken arm. After numerous pleadings, Sarah Jane Melton finally consented to marry me with one stipulation – we would stay in Madison. During the normal course of events, Diane, Martin, and Sallie came along, and we lived happily ever after.

The End

Joe D. Vail Widely Experienced Transport Pilot Will Train You

JOE D. VAIL, President

"Joe," as he is affectionately known to thousands, is a widely experienced Transport Pilot with many successful hours of passenger flying as well as student instruction to his credit. It is the close supervision of this skilled Transport Pilot which you need in order to advance rapidly. Joe's instruction, help and advise will come to you right from the practical experiences he gets from his active duties at the airport. When anything new is developed or when any urgent problem arises, Joe will be right here on the job giving you his personal experience which you could not possibly get in any other way than through his supervision, help and advice. When problems come up which must be answered and answered right, you want to know that your information has been gained from a practical expert who is meeting these conditions every day and whose decision you can depend upon. This is the type of instruction you will receive under Joe Vail.

TWO WEEKS GROUND TRAINING ON ACTUAL PRACTICAL JOBS

When you get to the field everything will be arranged for you. A good place for you to room and board will be ready and as soon as you are settled Joe will take you right out to the airport and you will begin your actual practical

training. These two weeks of what we call laboratory training, will thoroughly familiarize you with every part and piece of the airplane. You need this practical training in order to qualify you to hold down a good job in Aviation. You will receive instruction in the assembly and alignment of all parts of the plane, principles of the internal combustion engine, engine requirements, carburetors, cooling systems, and constructional details of aircraft engines, radial engines. Power and displacement, power and compression, engine overhaul. Emergency repairs and top-overhaul away from base of operation. Rigging, Wing Structure, wires, bracing and fuselage as well as wing repair and doping. All will be given you.

THE OPPORTUNITY OF A LIFETIME

Look over the illustrations on the opposite side of this page and you can readily see that this type of training and service is what you need to round out your Home Training Course. We do not know of another school in the country that gives two weeks practical training in connection with their Home Training Course. When you finish your two weeks here at our airport we feel sure that you will say it was the most profitable two weeks you have ever spent in your life.

2

We Use The "Learn By Doing" Method
It's the Only Way to Get the Best Results

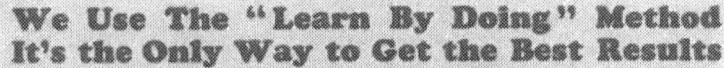

In addition to our school training, our company specializes in airplane repairing, flying instruction and airplane distribution. Students at our airport are allowed to work on the airplanes, engines, etc., under the supervision of our licensed mechanic. It is this lecture work combined with the actual practice of doing these things that helps you become thoroughly qualified in every respect when you have completed your training with us.

The Above Picture Shows Instructions In Servicing Airplanes On the Field.

[ABOVE] You will be surprised how quickly you will learn this work under capable instructors working on this fine up-to-date equipment. There is nothing like practical work to make you thoroughly capable of handling any kind of a proper thing that comes up after you are out on the job.

Wing Building and Repair

[ABOVE] You will be amazed how quickly you will comprehend this kind of work as never as you have had a few hours practical experience in wing construction and repair work.

Class Room Lecture Work

[LEFT] The picture shows the class room lecture work where you will be given a review of your theory tuned with the practical experience which you are learning while at the airport.

Engine Service Maintenance and Repairs

[RIGHT] As the picture shows you will be given lecture work on the theory and operation of engines as well as an opportunity to actually work on airplane engines. There are several different types of engines at the airport so that you can readily see you will get a thorough training in engine work.

Practical Instructions On Fuselage

[BELOW] You will be surprised how quickly you grasp this kind of work once you get on a torn down job and start building it up. This is the practical kind of work which is going to fit you to fill a top-notr job in aviation.

FASCINATING HOURS FOR YOU

We prepare men to be daring leaders in this great new industry. Your future begins with our famous Home Study Method, right in your own home, where you study and learn how to do any kind of Aviation ground work. Then when you have completed the Home Training Course we bring you right to the field where you get the actual practical experience which qualifies you to make your future in the fascinating field of Aviation. Every job in Aviation is an interesting one and no matter what part you choose, you will find it most interesting.

REAL INSTRUCTION
UP-TO-DATE EQUIPMENT

When you have finished your two weeks laboratory training at our airport you will know more about airplanes than you probably would learn in a month from the average instructor because Joe is one of the finest fellows you will ever meet and his whole heart is set on teaching you what you need to know. Joe has had a lot of experience in barnstorming, and passenger carrying and this gives him an insight into human nature which few people have and makes him a most agreeable companion as well as a thorough instructor. He will be here every day to help you, guide you, and encourage you.

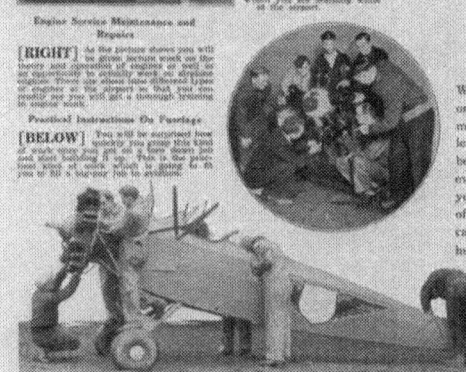

You Get Actual Service Work Under a Government Licensed Mechanic

Harvey Kattelmann is a government licensed mechanic and knows just what kind of mechanical training you

need to round out what you have already learned and he will work with you 100% to see that you get everything you need to know. You will have the opportunity to work with Harvey on engine repairs, fuselage, wings and all other parts of repairs in connection with the airplane.

This personal contact in servicing planes with this government licensed mechanic is the most practical type of training in existence today.

NOW IS THE TIME
AND HERE IS THE WAY

Few people realize the extent to which Aviation has already grown. The figures recently released by the Department of Commerce, Aeronautics Branch, show that in spite of the depression in 1931 the Aviation industry showed improvement and advancement. The total miles flown in the United States last year numbered 41,753,417 as compared with 31,992,634 during the previous year. Passengers increased from 374,935 in 1930 to 436,961. Mail payments showed an increase of a little more than $5,000,000 with a new record total during 1931 of $19,900,250.88 as compared with $14,762,655.71 in 1930.

Civil aircraft of the United States carried 2,339,862 passengers and flew 141,729,107 miles during 1931 in all types of operations. Col. Clarence M. Young, assistant secretary of commerce of aeronautics said.

PRACTICAL AVIATION TRAINING FOR A PRACTICAL JOB

The pictures show the ideal conditions under which you will receive your training. They show a few of the different kinds of work you will be prepared to do when you have completed our training. The kind of work, you, as an Aviation technician will be called upon to do as soon as you have completed our Home Training Course and the two weeks laboratory training.

4

You will be trained for the kind of work you like best

Showing one of the interesting steps in airplane construction — Wing Covering.

A new Waco is born. Putting on the finishing touches ready to take the air.

Adding the power plant. Adjusting the motor to the fuselage.

Wing Construction—very delicate work that calls for skilled hands and trained minds.

The AIR MAIL says,
"Give us More Trained Men at Once!"

Startling Air Mail Facts

It is estimated that the Air Mail Service will pay over $2,000,000 to its employees in 1932, and almost twice that amount in 1933. Over 6,000,000 pounds of mail will be carried in the same period. These staggering figures mean that more jobs will be ready for thousands of trained men!

When the United States Government makes the airplane a most important factor in the distribution of mail, you know that aviation is on the way to permanent success. Starting out with an experimental rate of 18 cents for every half ounce, and with but a few air lines connecting important cities, the government has cut the rate in half, and the number of air mail routes has been increased many times. Plans are now under way to reach every part of the country with the Air Mail. Already airplanes to Mexico, Central America and South America are operating. It is expected that the 1931 total of 2,000,000 pounds of mail will be doubled in 1932 and tripled in 1933. With the rapid advancement of the Air Mail has come an increased demand for men to pilot the ships and maintain them. Fliers, maintenance men, mail clerks, and other air trained men will be needed to keep pace with this important branch of aviation. Get the right training now—and you will be ready to start.

Air Passenger Lines
Increasing Daily
with More Opportunities for Men Who Know

In Europe passenger transport companies have been operating for 12 years with the same unfailing regularity as railroads in our own country. It is no longer a novelty for people to travel by air. In this country the same condition is rapidly approaching. Fifty-two passenger transport companies are now operating on schedule time. Last year over 2,000,000 passengers were carried by them. Close to 6,000,000 miles were flown! The big railroad companies, especially the huge Pennsylvania Railroad System, are operating air lines in connection with their trains. It is now possible to travel by air from one coast to the other in many hours less than in trains.

The Start of a Great System

With over a thousand new landing fields being constructed in cities everywhere, with night flying made practical and safe, with new route-guiding devices being invented every day—like light-signals, radio, etc.—the next few years will see the creation of a passenger transport system that will rival even the railroads in completeness. People will travel by air with as little concern as they now travel by rail. The increase in volume will reduce the fares and put air travel within the reach of everyone. This, my friend, is not a dream—one only needs to consider the single fact that huge airplanes, capable of carrying twenty to twenty-five passengers each, are now being built. Around the corner is a great industry that will revolutionize all passenger transportation. Now is the time for you to get your training so that you will be ready to step into a good Aviation job when business conditions have changed and trained men will be in big demand.

Interior view of Passenger Transport

[Courtesy Boeing Airplane Co.]

Typical Twelve-Passenger Transport used on Chicago-San Francisco Air Lines.

A typical Airline Schedule Sheet showing extensive use of Passenger Travel

R. R. Foster One of the Founders of D. S. A.

R. R. FOSTER, Vice President

Is the man who has arranged for you to get your two weeks ground training and your first flying lesson without extra charge.

The very minute you finish our Home Training ground course we will notify you when to report to our airport. These will be the most interesting and profitable two weeks you have ever spent and particularly will it be a thrill for you to get your first flying lesson. It has been a great pleasure for us to be able to work out these two weeks ground training and your first flying lesson without extra cost to you and we feel that you will appreciate the trouble and expense we have gone to in making it possible for you to get this additional training without costing you one cent more than we have ever charged for our Home Training Course.

WHAT THIS TRAINING MEANS TO YOU

In making your enrollment for our Home Training Course in Aviation you will not be going into something new and untried. We have been giving our Home Training Course over a period of years and since adding the two weeks laboratory and your first flying lesson we have had an opportunity to judge the merits of this training and find it the most satisfactory addition we have ever made to our already well rounded out course of training.

NO ADVANCED EDUCATION OR PREVIOUS EXPERIENCE NEEDED

It isn't necessary for you to have had a high school or college education. Neither is previous experience necessary. Joe Vail knows exactly the kind of training you need and will give you that training. He will give you the training you need to become, in a short time, a skilled air man.

10

ONE BIG HAPPY FAMILY

In this photograph you see some of the boys in the living room of their home. We say home because Mrs. Barber, who keeps the boys has turned her entire 13 room house over to them. In this photograph you see a group of them around the piano and believe me they spend mighty happy evenings under the supervision of Mrs. Barber right here at home.

THERE'S NO TIME LIKE THE DINNER HOUR

What a happy time it is when the boys gather around the abundantly set dinner table. It is a revelation how Mrs. Barber finds time to cook such excellent food for the boys. Every day she has standing orders for special delicacies from those who are completing their training and are about to go out on the job. You will find Mrs. Barber a marvelous hostess and will enjoy every minute of your stay with her.

Joe giving final instructions before taking off with student for first lesson.

MAIL YOUR ENROLLMENT AT ONCE AND GET YOUR TWO WEEKS GROUND TRAINING AND YOUR FIRST FLYING LESSON WITHOUT EXTRA COST.

It is a real joy and pleasure to step into the plane beside Joe and get your first flying lesson without costing you one cent extra. This thrill of receiving your first flying lesson is just one of the many added features which we combine with our Home Training Course to give you the most complete, thorough and practical training obtainable. What a thrill it is going to be for you to take the stick for your first flying lesson—not as a novice—not as a "cub" looking for excitement, but as one who knows a great deal about Aviation and who is familiar with every operation of the plane.

11

The Louisville Courier-Journal
April 26, 1935

Joe Vail dies in Plane Accident

Government inspectors were enroute here Friday to investigate circumstances surrounding a plane crash near Bowman Field Thursday night in which the pilot lost his life and his two companions were injured gravely.

The body of Joseph Vail, 27, the Pilot, was removed Thursday night to Madison, Ind. where all three occupants made their homes.

His companions, David O. Gentrup, 17, and John Carl Sachleben, 25, were taken to the Kentucky Baptist Hospital. Gentrup's skull and right arm were fractured. Sachleben suffered fractures of the skull, two cracked ribs and arm.

Motor Sputters

The trio took off at 6:10 o'clock Thursday night from Bowman Field in a Warner Cessna monoplane. The motor sputtered and missed fired several times and Vail went into a vertical bank in an effort to turn back to the field. The plane's speed was not sufficient to accomplish the dangerous turn and the ship nose-dived from an altitude of about 100 feet into a field adjoining the airport.

The field was on the farm of John L. Hettinger, Cannon Lane, who heard the motor sputter, saw the plane attempt to turn and stood by, horrified and helpless as the plunged to the ground, burying its nose two feet deep in the plowed field.

Several men who ran from the airport to the scene of the crash extricated Vail who had been pinned by the motor, and placed him in a police patrol. He was dead when he reached the hospital.

The Department of Commerce inspectors were expected here from Indianapolis during the day. A tractor pulled the wreckage of the plane to a hanger to await their inspection.

The three Madison men flew to Bowman Field at 1:30 0'clock Thursday afternoon and spent several hours inspecting a new Porterfield monoplane, which was demonstrated for them by I.M. Henry, 2341 Cherokee Parkway, a pilot for the Louisville Flying Service.

Those who rushed to the aid of the men after the crash included Henry, Lieut. John Russ, Sergts. Chester Layton, Hugh O'Daniel, 2041 Ridgeland Boulevard. Russ and Layton took one of the men in an Army Corps ambulance.

Vail, who had been flying six was in charge of the Dayton Flying School, Madison, and Gentrup had been one of his pupils. Vail leaves his parents, Mr. And Mrs. Harry Vail, and a sister, Mrs. C.B. Carlaw. His father is an undertaker and furniture dealer in Madison.

Sixth Death at Field

Death of Vail was the sixth fatality at or near Bowman Field since it was founded. The vertical bank which he attempted was the same maneuver which resulted in the first fatalities at the field. Sergt. Louis Opperman and Lieut. Robert O'Hanley were burned to death after their plane crashed while demonstrating a vertical bank at Bowman Field's first air circus held in 1923.

THE MADISON COURIER, MADISON, INDIANA
April 26, 1935

MADISON MEN IN PLANE CRASH AT LOUISVILLE

Joe Vail, 27, Madison airport pilot, was killed and David G. Gentrup, 18 and Carl Sachleben, 17, were critically injured at Louisville late night when their plane nose-dived to the ground near Bowman field. Vail died in an ambulance before he reached the hospital, His chest, right leg and the side of his face was crushed,

Gentrup remains unconscious at the Baptist hospital where the three were taken. He suffered a compound skull fracture and his right arm was broken, but physicians held hope for his recovery after a two-hour operation last night.

Sachleben's right arm was broken in five places and his skull was fractured across the forehead, but he remained conscious throughout the accident and during the treatment at the hospital. He was "doing fine", according to a report from the hospital this afternoon.

One of Gentrup's fractures ranged downward over the right eye to the sinus and the other was behind the right ear. Sachleben submitted to an operation and was able to talk to the surgeon while it was in progress.

The three left the Madison airport located on the Sachleben farm, three miles north of this city, yesterday at 2:45 p.m. At the Bowman field, Vail demonstrated a new Porterfield monoplane to J.P. Wilkerson, a contractor who built the Hanover Hill road. Wilkerson met Vail at Louisville and they examined and flew the new plane that Wilkerson planned to purchase.

After the demonstration, the three Madison fliers took off at about 6:15 p.m. to Madison in a plane Wilkerson owns, which had been in Vail's care at the local airport. Wilkerson was not in the plane as the three Madison men started home.

Vail was piloting on the takeoff and had gained about 350 feet altitude when the motor started sputtered and missing. Vail started to turn back to the landing field. He banked and in the wingover the plane took a nosedive when the motor failed to provide the necessary power for proper handling.

Diving to the ground the plane buried its nose in a potato field owned by J.L. Hettinger, 32, of Cannon Lane, who witnessed the accident. He said the plane tore a hole in the ground several feet wide.

Vail was pinned by the motor but was released by four or five men who rushed from the airport. An Army Corp ambulance in charge of Lieut. John Russ and Sergt. Chester Layten, took him to the hospital.

I.M. Henry. A pilot for the Louisville flying service, said the Wilkerson plane, a 1929 Warner Cessna monoplane, had been sold to the contractor by Orville Puckett, former Bowman field pilot.

Among the first to reach the wrecked plane were Henry; Major J.C. Bennett, manager of the Louisville Flying Service; the two Army Reserve Corps men; Hugh O'Daniel, a pilot, and Caesar Kaiser, who had been flying at the field

Major Bennett said he heard the motor missing shortly after the plane began its takeoff. He said the accident apparently occurred as the pilot attempted to turn back. The wrecked ship was pulled to a hanger by a tractor and held there pending an investigation by Department of Commerce men, assigned to investigate all air crashes.

Vail, a pilot of several years experience, formerly was an instructor at the Dayton School of Aviation and was known throughout the state as a thoroughly trained and experienced pilot.

Those who witnessed the tragedy said the crash was in no way Vail's fault, and Sachleben claimed a supercharger, which forces hot gasoline into the carburetor was out of order and they were unable to keep the motor functioning properly on cold gasoline.

Gentrup, a member of the North Madison High School graduating class who are to receive their diplomas next Tuesday evening, has had three or four years of flying experience. He made the trip to Louisville with Vail yesterday to take a medical examination for a limited commercial license as a pilot. He is a son of Mr. & Mrs. Grover Gentrup.

Sachleben, a companion of Vail's, was a passenger on the trip. He is the son of John Sachleben, who owns the field on which the airport is situated.

Gentrup is one of Vail's students and has about 75 hours of flying service to his credit.

Joe was at the controls and the other two passengers sat behind him. To save the plane and passengers he remembered instructions he learned years six years ago, and turned off the ignition to prevent the plane from bursting into flames. When retrieved from the wreckage, Vail's clothing was saturate with gasoline and oil.

The plane was crumpled mass of battered wreckage. Joe is said to have been very fond of the new plane he was demonstrating for Wilkerson yesterday, and was pleased with the prospects of operating it for Wilkerson instead of the one in which he went to his death.

Vail had approximately 1,400 hours of flying time to his credit. He held a transport pilot license, having attended the Parks air college at East St. Louis, Mo., for his early training and finished his course at the Shambaugh airport at Lafayette. It was while he was at Lafayette that he purchased his first plane, a Waco three-place biplane which he affectionately referred to as "Peg".

He was an officer and instructor for the Dixie Flying Service, which is affiliated with the Dayton School of Aviation. For four years he had been an instructor and had become well know in aviation circles throughout the state. He piloted a plane in the annual Indiana air tour a few years ago and had flown on numerous trips chartered from the local airport.

Surviving besides his parents, Mr. and Mrs. Harry Vail, is a sister, Mrs. C.B. Carlaw, who is in Mexico City. His father is a member of the firm of George C. Vail & Sons, furniture dealers in this city. Joe's full name is Joe Daniels Vail, a nephew of Miss Sarah Daniels.

His remains were brought from the Louisville hospital last night at 11:30 o'clock to the Vail funeral home where funeral services will be conducted tomorrow morning at 10:00 o'clock by Rev. Jesse M. Tidball. Friends are invited to the services.

Sachleben families went to Louisville last night and remained at the hospital today.

About the Author

Clyde H. Beyer was born and raised in Pittsburgh, PA. At the age of 19 years he quit his job as an assistant bookkeeper, in the depth of the "Great Depression" to pursue his lifelong dream of becoming a commercial pilot.